SCENARIO LOGIC AND PROBABILISTIC MANAGEMENT OF RISK IN BUSINESS AND ENGINEERING

Springer Optimization and Its Applications

VOLUME 20

Aims and Scope
Optimization has been expanding in all directions at an astonishing rate during the last few decades. New algorithmic and theoretical techniques have been developed, the diffusion into other disciplines has proceeded at a rapid pace, and our knowledge of all aspects of the field has grown even more profound. At the same time, one of the most striking trends in optimization is the constantly increasing emphasis on the interdisciplinary nature of the field. Optimization has been a basic tool in all areas of applied mathematics, engineering, medicine, economics and other sciences.

The *Springer Optimization and Its Applications* series publishes undergraduate and graduate textbooks, monographs and state-of-the-art expository works that focus on algorithms for solving optimization problems and also study applications involving such problems. Some of the topics covered include nonlinear optimization (convex and nonconvex), network flow problems, stochastic optimization, optimal control, discrete optimization, multiobjective programming, description of software packages, approximation techniques and heuristic approaches.

Scenario Logic and Probabilistic Management of Risk in Business and Engineering

Second Edition

By

EVGUENI D. SOLOJENTSEV
Russia Academy of Sciences, Institute of Mechanical Engineering,
Petersburg, Russia

Springer

Evgueni D. Solojentsev
Institute of Problems of Mechanical Engineering
of Russian Academy of Sciences
V. O. Bolshoy pr, 61
St-Petersburg, 199178, Russia
risk@sapr.ipme.ru

ISSN: 1931-6828
ISBN: 978-0-387-77945-4 e-ISBN: 978-0-387-77946-1
DOI 10.1007/978-0-387-77946-1

Library of Congress Control Number: 2008938965

Mathematics Subject Classification (2000): (Primary) 6502, 65F10, 65N22, 65N55; (Secondary) 65F50, 65N30, 65Y20, 65H1

The first edition of this book appeared in 2004, published by Kluwer Academic Publishers as Volume 93 in the Applied Optimization Series.

Printed on acid-free paper

springer.com

About the Author

Evgueni Dmitrievich Solojentsev

Evgueni Dmitrievich Solojentsev was born in 1939. He is head of the "Intelligent Integrated Automatized Design Systems Laboratory" of the Institute of Problems in Mechanical Engineering, Russian Academy of Sciences; Doctor of Technical Sciences, Professor of St. Petersburg State University of Aerospace Instrumentation; and an Honored worker of Science of the Russian Federation.

E. D. Solojentsev graduated from Kharkov Polytechnic Institute (1960), defended the candidate dissertation (1967, Central Research Diesel Engine Institute, St. Petersburg) and the doctoral dissertation (1983, Institute of Cybernetics of UAS, Kiev), and became Professor (1994) and Honored worker of Science of the RF (1999). From 1967 to 1985, he worked as Head of Department of Automated System CAD/CAM/CAT in industry (Gorkiy, Sumi). Since 1986, E. D. Solojentsev works at IPME RAS.

E. D. Solojentsev is the author of about 180 scientific papers including 6 books. He is the founder of scientific bases of construction of the automated debugging test systems. He developed the logic and probabilistic risk theory with groups of incompatible events for problems of classification, investment, effectiveness, and bribes. E.D. Solojentsev is the Chairman of the National Organizing Committee of the International Scientific School "Modeling and Analysis of Safety and Risk in Complex Systems" (St. Petersburg, IPME RAS, 2001–2007).

E. D. Solojentsev is an expert in the area of management of risk at stages of design, test, and operation in complex banking, economical, and engineering systems.

Contents

Foreword

The first Russian edition of this book was rather popular and sold many copies. The book was also published in English by Springer-Verlag. The success of the first edition substantiates publishing of a new extended edition.

In the new edition, the discovered mistakes are corrected, definitions and notations are clarified, some new topics are included, some sections are updated, the list of references is extended, and the subject index is renewed.

In the forewords to the books "Logic and probabilistic valuation of banking risks and frauds in business" (St. Petersburg, Politechnika, 1996) and "Logic and probabilistic models of risk in banks, business and quality" (St. Petersburg, Nauka, 1999) by this, E. D. Solojentsev, and V. V. Karasev, V. E. Solojentsev, I wrote that the authors opened new fields for application of rigorous analytical methods of estimation, analysis, and investigation of the risk in economics and engineering. In those forewords, I expressed the hope, which I am glad to express again, that the new logic and probabilistic methods of risk estimation will be successful.

In many respects, the occurrence of this new book is stimulated by E. D. Solojentsev's activity in organizing the International Scientific Schools meetings. "Modeling and Analysis of Safety and Risk in Complex Systems" (St. Petersburg: June 18–22, 2001; July 2–5, 2002; August 20–23, 2003; June 22–24, 2004; June 28 to July 1, 2005; July 4–8, 2006; September 4–8, 2007). Russian and foreign scientists and experts presented more than 500 papers devoted to the problems of safety and risk in economics and engineering.

For many years, the author worked in industry in the field of designing and testing complex engineering systems. Now he works at an academic institute, where he is engaged in risk problems in engineering, banking, and business. His achievement in the risk field were noticed by universities in Germany, Japan, and Switzerland, where he was invited for scientific collaboration.

The experience and the knowledge allows the author to propose the uniform logic and probabilistic (LP) approach to risk estimation and analysis both in engineering and economics and to lay a foundation for systematization and formation of the risk LP-theory and also to create the scientific principles of the scenario LP-management by risk.

The titles of the author's papers such as "the logic and probabilistic **estimation**," "the logic and probabilistic **models**," "the logic and probabilistic **approach** to the risk analysis," despite the clearness of the terms separately (they are well-known for many people, who are far from the risk analysis in engineering, economics, politics), require some explanation for their combination "logic and probabilistic".

Unfortunately, most of books in the field published in Russian, including "Mathematical encyclopedia dictionary" [Moscow Soviet encyclopedia, 1988, 846 p.], avoid definition of the probabilistic logic, as a logic of statements, accepting a set of degrees of plausibility, that is the values are contained in the interval between "true" and "false."

As the revolutionary break in the development of **the inductive logic**, George Boulle's paper "Mathematical analysis of the logic being experience of calculus of the deductive reasoning," published in 1847, should be mentioned. The **calculus** of statements is the essence of mathematical logic and a new step in the development of the formal logic.

One of the fathers of the mathematical theory of information, Clod Elwud Shannon, succeeded in closing the gap between the logic algebraic theory and its practical application. In his D.Sc. dissertation (1938), he developed principles of the logic model of the computer, *by connecting Boolean algebra with the functioning electrical circuits*. The success of his ideas concerning connections between the binary calculus, the Boolean algebra, and electrical circuits, Shannon explained as follows: "Simply it happened so, that nobody else was acquainted with both areas simultaneously."

The necessity of quantitative estimation of non-failure operation of complex technical structures at the beginning of the 1960s century stimulated **the so-called logic and probabilistic calculus (LPC)**, which is a part of mathematics treating rules of calculus and operating with statements of two-value logic. LPC is based on the logic algebra and rules of replacement of logic arguments in functions of the logic algebra (FAL) by probabilities of their being true and rules of replacement of the logic operations by the arithmetic ones.

In other words, with the of help of LPC, it became possible to connect the Boolean algebra with the probability theory not only for the elementary structures but also for the structures whose formalization results in FAL of iterated type (bridge, network, monotonous). This original "bridge of knowledge" includes some proven theorems, properties, and algorithms that constitute the mathematical basis of LPC.

Investigation of the safety problem has resulted in development of **the original logic and probabilistic theory of safety** (LPTS), which allows one to estimate quantitatively **the risk** of system (as a measure of its danger) and to **rank** the contribution of separate arguments to the system danger (in the case of an absence of truth probabilities of initiating events). The ranking arguments under their contribution to system reliability was proposed by me in 1976 in the monograph "Reliability of Engineering Systems. Principles and Analysis". (Mir Publishers, Moscow, 1976, 532 p.) with the help of introduction of concepts "**Boolean difference**," "**weight**," and "**importance**" of an argument.

The aim of the author, from my point of view, is the connection of the logic and probabilistic calculus used in the field of technical systems with questions of risk in economics and organizational systems.

Studying the works by the author, I realized that these economical and organizational systems essentially differ from technical ones, and the direct carrying of the knowledge and results of LPC from the area of engineering into the area of economics is not effective, and sometimes it is not even possible. It is likely that much time and effort will be needed so that the new approaches in logic and probabilistic calculus could make the same revolutionary break in the financial market as was made by George Boulle in development of the inductive logic in the middle of the 19th century and by H. Markowitz in the choice of the optimal security portfolio with the help of the analytical theory of probabilities in the middle the 20th century.

The author, presumably not wishing to simplify solutions of real problems of risk, has selected the algorithmic method as the basic method. In this connection, it is useful to quote the Academician Ya. Tsipkin: "Algorithmic approach to resolving extreme problems enables one to use modern computers and not to squeeze the problem conditions into the Procrustean bed of the analytical approach, which usually moves us far from those real problems that we really wanted to consider."

The existing publications on management LP-theory by risk are not complete, have small circulation, and are not known by a wide community of experts. The typical difficulty in mastering scenario LP-management by risk in economics and engineering can be explained by the fact that the risk LP-theory and such scientific disciplines as the LP-calculus, the methods of discrete mathematics, and combinatorics are not usually included into the educational programs of high schools. Therefore, publication of the given monograph devoted to the LP-management by risk seems to be useful.

Academician of Russian Academy
of Natural Sciences,
Professor I. A. Ryabinin

Introduction

Back to basics, to logic and arithmetics (sets), so that we might solve complex problems.

The author

What is new in the current edition

The first edition of this book caused a great interest. Numerous discussions caused the necessity to solve the following problems through a second edition of the book:

- To compare the probability logic by John von Neumann and Nils Nilsson with the logic and probabilistic calculus by I. A. Ryabinin.
- To explain the epigraph of the first edition of the book "Back to basics, to logic and arithmetic (to sets), so that we might solve complex problems."
- To develop the transfer from the statistical database (DB) to the knowledgebase (KB) in the risk LP-theory with GIE.
- To establish a place for discrete and continuous mathematics in the risk theory.
- To give the definition of the dependence of the events in the risk theory.
- To specify the concept of the structurally complex object.
- To formulate the criterion of a transparency of techniques for the estimation and the analysis of risk.
- To introduce risk models for the limited set of objects states.
- To introduce integrated risk models.
- To give proof of use of algorithmic iterative methods of optimization (identification) in risk problems.
- To define the risk through the attributes of the set of objects/states, the object/the state, the parameter of the object/the state, the gradation of the parameter.

- To investigate the discrete and continuous criteria of the identification of the risk model according to the statistical data.
- To define the rules for the calculation of the derivatives in the group of incompatible events (GIE).
- To take into account the dependence of the events in the risk model on the basis of the band *the copula.*
- To elaborate the non-success risk LP-models of the company management.
- To develop the risk LP-models for revealing.
- To outline the peculiarities of the use of modeling by Monte Carlo, the Bayes' formula, the Weyl's theorem, and the VaR methodology in the risk LP-theory.
- To develop the risk LP-theory on the basis of I. Ryabinin's LP-calculus, A. Mozhaev's structurally logic modeling, and the author's risk LP-theory with GIE.
- To develop risk LP-models of bribes and corruption.
- To state the formal risk LP-theory with GIE.
- To state the contents of the training course "Modeling, estimation, and analysis of the risk in economics."
- To describe software for solution of risk problems in economics and for fulfilling of laboratory works in the training course.
- To underline features, similarities, and distinctions of the risk LP-theory with GIE.
- To describe the author's contributions to the construction of the risk LP-theory.

Nowadays, many people deal with risk and safety in the state structures and legislative bodies, in politics and economics, in business and banks, in management and engineering. Practically every day, thousands of specialists take decisions on the basis of the estimation and analysis of the risk. However, contradictions and ambiguity in the understanding of the notion of risk result in considerable difficulties when teaching at the universities and in the mutual understanding of the risk problems by specialists. That is proved by the fact that there are more than 20 different definitions of the word "risk."

In the new edition of this book, the author did his best to make his contribution to the solution of those problems on the basis of his personal experience and taking into consideration the tendencies in publications by Western scientists in the field of risk. In the new edition, the mistakes of the former have been corrected, definitions are formulated more accurately, some new material has been added, and some chapters and sections have been combined.

Special attention was given to the role of the works of John von Neumann and Nils Nillson on probabilistic logic, of N. A. Kolmogorov on the logic and the discrete and continuous probability theory, and of those by V. V. Alexandrov on forecasting the risk in the developing process.

New chapters have been written: Chapter 7, "Transparency of methods for the risk assessment," Chapter 17, "Scenario logic and probabilistic models for revealing bribes," Chapter 20, "LP-models of the company management non-success risk," Chapter 22, "The formal non-success risk LP-theory with GIE," and Chapter 23, "Training course: Modeling, estimation, and analysis of risks in economics."

Chapters 10, 11 and 8, 16 of the previous edition have been combined. Chapter 17 of the previous edition is presented here by four chapters: Chapter 15, "LP-model of the credit risk of natural persons," Chapter 16, "LP-model of the credit risk of the juridical persons," Chapter 17, "Scenario logic and probabilistic models for revealing bribes," and Chapter 21, "LP-models and scenarios of risk of frauds in business and in the interaction of companies."

The list of literature is added, and the subject index is remade. The volume of the changes and of the new text makes a revision of more than 30%.

What is stated in the new edition

To the author's knowledge, the phenomenon of risk in complex technical, economic, and organizational systems has not yet been studied thoroughly and has not been solved properly for applications. In complex systems, non-success occurs rather often with human victims and economic losses. The management risk problem is current and challenging; it forces us to carry out new investigations and to seek new solutions for a quantitative estimation and for the analysis of risk.

Risk is a quantitative measure of such fundamental properties of systems and objects as safety, reliability, efficiency, quality, and accuracy.

Risk is also a quantitative measure of non-success of such processes and actions as classification, investment, designing, tests, operation, training, development, management, etc.

We shall consider the theory of the scenario logic and probabilistic management of the risk in the multicomponent systems with groups of incompatible events for the problems of the classification (the credit risks of natural and juridical persons), of the investment (the security portfolio risk), of the efficiency (the analysis of social processes, the risk of the losses of markets, of quality and accuracy, etc.), of the management of companies (the company management risk on functions, directions of an activity, the achievement of aims), and of bribes and corruption. The named problems comprise the description and the statement of the optimization of the tasks.

Generally **Risk** is characterized by the following quantitative parameters:

- The non-success probability;
- The admitted risk;
- The maximum admitted losses or the minimal admitted efficiency;
- The value of the losses or of the efficiency parameter;
- The number of different objects or conditions of the object in the system;
- The number of dangerous objects or conditions of the object.

As was marked by the founders of many fields of modern science — by John von Neumann and Norbert Wiener — the behavior of complex technical, economic, and social systems cannot be described by means of differential equations. However, the description can be made on the basis of the logic and the set theory, instead of the theories of chaos, accidents, bifurcations, etc. (See the book by Morgenstern and Neumann, "The game theory and economic behavior," Moscow, Nauka, 1970, sec. 1.2.5. and 4.8.3.)

The analysis of the theories of management and risk development and of the interaction between man and risk in complex systems proves the correctness of this point of view. In complex human-machine systems, the logic and probabilistic theory (LP-theory) reveals considerable achievements in estimating, analyzing, and forecasting the risk [1–4].

The LP-theory attractiveness is in its exclusive clearness and unambiguity in the quantitative estimations of risk; in the uniform approach to problems of risk in economics and engineering, in big opportunities when analyzing the influence of any element, including personnel, on the reliability and safety of the whole system. The risk LP-model may include the logic connections *OR*, *AND*, *NOT* between the system elements and cycles. Elements of the system under consideration may have several levels of conditions. The system risk dynamics can be taken into account by considering the change in time of the probabilities of conditions.

The estimation of the risk as the probability of the non-success demands special knowledge of the non-simple logical transformations and also the use of the special logical software.

In complex systems, the technology of the scenario risk LP-management is based on the formalization of the risk scenario, on the construction of the structural, logical, and probabilistic risk models, and also on the identification of the risk P-model on the statistic data on the estimation and analysis of the risk. Generally, it is impossible to control the risk without a quantitative analysis of the risk. The estimation and the analysis of the risk, and also the search of the optimum management, have been carried out algorithmically with calculations that are very time-consuming even with modern computers.

The risk LP-theory comprises: Ryabinin's LP-calculus, Mozhaev's methodology of the automatized structural and logical modeling, and the author's risk LP-theory with groups of incompatible events (GIE).

The LP-calculus is a special part of discrete mathematics, which should not be confused with the probabilistic logic or other sections of the mathematical logic. Therefore, it is useful to give a brief outline of the history of the publications on this subject. To the author's knowledge, the idea and development of the subject should be attributed to Russian authors. The contents and formation of the LP-calculus originates from the work by I. A. Ryabinin, "The Leningrad scientific school of the logic and probabilistic methods of investigations of the reliability and safety" (in the book: "Science of St. Petersburg and the naval power of Russia," v. 2, 2002, p. 798–812).

The LP-calculus was created in the early 1960s in connection with the necessity of a quantitative estimation of the reliability of complex structures (annular, network, bridge-like, and monotonous). Scientific literature of that time could suggest nothing suitable to deal with. The experts in reliability could only perform calculations for the consecutive, parallel, or treelike structures.

In 1987, Kyoto University published the book by I. A. Ryabinin and G. N. Cherkesov, "Logic and probabilistic methods of research of reliability structural-complex systems" (M.: Radio and Communication, 1981, 264 p.), translated into the Japanese language. In the book, the theoretical and logical parts of the LP-calculus were improved. In the new book "Reliability and safety of structural-complex systems" (SPb., Polytechnika, 2000, 248 p.), Prof. I. A. Ryabinin has generalized his forty-year experience in the research of reliability and safety by the LP-calculus. There is a review of this book in English (Andrew Adamatzky "Book reviews" — Reliability and Safety of Structure-complex Systems. — Kybernetes. Vol. 31, No. 1, 2002, p. 143–155).

The current publications on the risk LP-theory and on the risk management do not represent the state-of-art in the field of science, they have a small circulation, and the knowledge is confined to a small group of experts. The risk LP-theory and such scientific disciplines as the LP-calculus, the discrete mathematics, and the combinatorial theory are, as a rule, not included into the educational programs of the higher school. So the active mastering of the scenario risk LP-management in business, economics, and engineering presents a great difficulty. The publication of the current monograph, devoted to the scenario risk LP-management, seems to be quite timely.

The current book is of applied significance. The purpose of this book is to acquaint the economists, engineers, and managers with the basis of the scenario risk LP-management, with examples of scenarios and models of risk in different fields of economy and engineering.

The important feature of the suggested presentation is an attempt to embrace different fields: the logic and probability theory, the discrete mathematics and combinatorics, the set theory and Malzev's formal theory, Shannon's entropy and the non-linear optimization, the statistical modeling and the algorithmic calculations on modern computers, Bayes' formula and Markowitz's and VaR theories for risk of security of the portfolio.

It is the first time that the basic principles of the modern risk LP-theory (the LP-calculus, the LP-methods, and the risk LP-theory with GIE) are stated in one book, using the uniform methodology and terminology and with a practical orientation for use both in engineering and in economics. With the permission of Prof. I. A. Ryabinin, some mathematical results and examples from his book [2] have been reproduced here. With the permission of Prof. A. S. Mojaev, the technology of the automated construction and the analysis of LP-models have also been presented here [5, 6].

The methodology of construction of the non-success risk scenario in different fields for all stages of the life cycle of the systems has been introduced here. For this purpose some concepts, principles, some experience, scenarios and examples of risk management in business and engineering at the stages of designing, debugging, operational tests and trial have also been considered and systematized. It should be emphasized that the imperfection of the risk management of the operations, mentioned above, as well as the non-sufficient financing of the testing may lead in the future to failures and accidents. The development of non-success scenarios is a basis for the construction of the risk LP-models and for the quantitative analysis of the non-success risk.

The non-success risk LP-theory with GIE, which has an application in business and engineering, is introduced. The theory considers the risk for the systems with several discrete conditions of elements and for the system with a multidimensional distribution of its output, dependent on the initial random events with arbitrary distributions. For the credit risk estimation, the risk LP-model has shown a twofold higher accuracy than any other known method, and also a sevenfold robustness. When the choice of an optimum security portfolio is performed, the risk LP-model gives the same accuracy as the theories by Markowitz and VaR, but it allows us to solve a wider range of problems of the portfolio risk analysis and to use arbitrary distributions of the security yield (not only the normal law).

The description of the special logical software for the risk LP-modeling and analysis is given. The logic transformations and algorithmic computations are very complex and time-consuming even with modern computers, and they cannot be carried out manually. The software for the automation of the construction of the risk LP-models (package by Mozhaev), the identification of the non-success risk LP-models with GIE (package by Solojentsev), the orthogonalization of L-functions by the cortege algebra (package by Kulik), the optimization of the security portfolio risk (package by Solojentsev) are described. The software by A. Mozhaev and E. Solojentsev has obtained the national registration and certification.

Examples often teach much better than a pure theory. The examples have been given of the application of the risk LP-theory and the scenario risk LP-management in complex systems. Applications of the risk LP-models have been considered in different fields of business and engineering with a demonstration of their high accuracy, robustness, and transparency. The abilities for the risk analysis of one object and of a set of objects and the management of the risk are discussed.

We consider the following examples: credit risks of persons and organizations, bank credit activity analysis, frauds of managers and clerks, speculations with investments, the management of the state and the development of the company by the risk criterion, the struggle of building companies for profitable contracts, the financing of building projects by several banks with reservation, the risk of security portfolio, the management of efficiency, the risk of the non-success of the management of the company, the models for the revelation of bribes and of corruption, etc.

The presentation is organized as follows:

In Chapters 1–7 the methodological aspects of the scenario logic and probabilistic non-success risk management are considered, including: connections of the management, of risk and of human being, concepts of risk management stages of the design, test, and operation of complex systems, requirements for the transparency of the methods of estimation of credit risk.

Chapter 1 considers the problems of management and risk, of controlling the risk and insurance, monitoring and risk. It also considers sources of failures and accidents and the fields of applicability of the methods of nonlinear mechanics, the probabilities theory and LP-methods for the estimation, analysis, forecasting, and modeling accidents.

Chapter 2 discusses the intentional and unintentional actions of the personnel resulting in failures and accidents. It is proved that it is necessary to take into account the behavior of the personnel for the development of scenarios of non-success, of failures, incidents, and for the design of safety systems.

Chapter 3 considers the principles of risk management for the design of complex systems on the basis of generalization and unification of knowledge, of technologies, and of the practical experience of risk management in different fields of human activity.

Chapters 4 considers the technologies of risk management at the stages of debugging and operational tests. They are based on forecasting any possible troubles and on the development of the LP-scenarios for the occurrence and development of accidents and failures.

Chapter 5 considers the technology of risk management for the functioning complex system. The technology is based on the monitoring of wear and deterioration of the equipment and includes the construction of the LP-scenarios of the occurrence and development of incidents and of the appropriate risk LP-models.

Chapter 6 considers the basic concepts of the risk management on the dangerous plant.

Chapter 7 contains the formulation of the requirements to the quality of the methods of estimating the credit risks, namely, to the accuracy, robustness, and transparency. It gives the substantiation of the individuality of banks and of their models for the estimation of credit risks. The chapter considers the scheme of the bank management on the risk criteria and procedures of the technology of bank management. Conclusion has been driven on the expediency of using the LP-approach to estimation, analysis, and management of credit risks.

Chapters 8–14 contain the theoretical basis of the scenario non-success risk LP-management in business and engineering, including the LP-calculus, the methodology of the automated structural and logical modeling, and the

LP-theory with groups of incompatible events (GIE). Examples are given of the risk LP-models with logical connections *OR*, *AND*, *NOT*, cycles and GIE, which are hardly known to most mathematicians, economists, engineers, and managers.

The non-success risk LP-theory with GIE is considered in detail, namely, the non-success risk LP-models with GIE, the identification of the non-success risk LP-models with GIE on statistics, the LP-analysis of non-success risk, and also special logic software.

Chapter 8 contains the statement of the basis of the logic and probabilistic calculus and the information from the algebra of logic, the basic logical operations, and theorems.

Chapter 9 discusses questions of modeling and of the analysis of risk in engineering, the basic concepts of the LP-calculus for problems of risk and safety, the expression for the logical functions, formed on the minimal paths of the successful functioning and the minimal sections of failures; concepts of the weight and magnitude of the element in the LP-function. There are some examples to illustrate it.

Chapter 10 considers the basic concepts and the technology of the automated structural and logical modeling and introduces the scheme of functional integrity, fictitious and inverse tops. It also gives the effective algorithms of orthogonalization.

Chapter 11 contains the basic concepts of the non-success risk LP-theory with GIE, stated in literature for the first time. It proposes the structure and the tabular presentation of the data, and adopts the discrete arbitrary distributions of grade-events in GIE, gives the connections of probabilities in GIE, and shows in tables the orthogonalization of the L-function for different objects. Examples are given of the structural models as well as of the risk L-models and P-models. The chapter also considers the connections of different risk parameters, describes the features of different areas of the application of the risk LP-theory with GIE, and names the problems of the LP-theory with GIE. It derives the basic equations for GIE, introduces the risk measures and the price for the risk, it considers the connection of GIE and Bayes' formula, and describes the dynamic risk LP-models. It also considers the connections of database (DB), the knowledgebase (KB), and sets (SETS) in the risk LP-theory with GIE.

Chapter 12 gives the statement of the problem of the identification of the risk LP-models with GIE according to the statistical data, and states the basic regulations of the algorithm of the identification. It gives the description of the features of the identification by the random search method and the gradient method. It suggests the integer and continuous criteria of the identification of the risk LP-model, and results on the research of the identification of the risk LP-model. It proves the essential advantages of the risk LP-models on criteria of the accuracy, robustness, and transparency in comparison with the well-known methods of the estimation of the risk and the classification of objects.

Chapter 13 presents the techniques of the statistical, combinatorial, and logic and probabilistic analysis of the risk in systems with GIE. The essence and attributes of the transparency of the risk LP-model with GIE and the results of the estimation as well as the analysis of the risk are considered in detail.

Chapter 14 gives the description of the special logical software for the automated construction of the risk LP-models, for the identification of the risk LP-models with GIE, and for the orthogonalization of L-functions on the basis of the cortege algebra.

Chapters 15–21 contain the description of the non-success risk LP-models with GIE and their applications. They are described for the following purposes: to estimate and analyze the credit risks of natural and juridical persons, to reveal bribes and corruption, to estimate and analyze the risk of the portfolio security, for controlling the quality, efficiency, and the management of companies, for modeling frauds and the interactions of companies.

Chapter 15 gives the structure and the tabular presentation of data, scenario, credit risk logic, and probabilistic models for natural persons. It also presents the method of the analysis of risk of the credit and credit activity of the bank on crediting natural persons. Examples are given, demonstrating the transparency of the LP-theory and of the results obtained.

The LP-theory of estimation of the credit risk was compared with other methods and it has shown essential advantages in the criteria of accuracy, robustness, and transparency.

Chapter 16 gives the description of the methods of the estimation of the credit risk of juridical persons in the Western and Russian markets. It proposes the scenario and the credit risk LP-model of juridical persons. Essential advantages of the risk LP-theory have been shown on the accuracy, robustness, and the transparency of the results in comparison with other methods.

Chapter 17 describes the LP-models for the revelation of bribes on the basis of the statistical data. The following scenarios and the LP-models for the revelation of bribes are described: in an institution according to the results of its functioning, for the officials on the basis of the descriptions of their behavior, in institutions and for the officials on the basis of the analysis of the service parameters.

Examples of identifying and of the analysis of the bribe LP-models according to the statistical data are given here. Problems of bribes and corruption are of great computing complexity and are solved only by means of special logical software.

Chapter 18 describes the LP-modeling and analysis of the portfolio security risk on the basis of the discrete arbitrary distributions of the yields of the shares. The method of the portfolio selection on *VaR* and the suggested methods of selection and analysis of the portfolio on the basis of *LP–VaR* are considered. Examples of using the method of *LP–VaR* for the portfolio with independent yields of shares, with dependent yields of shares, with yields of shares dependent on the external factors are given. Comparison is made of the methods of modeling portfolio by the LP-theory. Examples are given of the optimization and of the analysis of the portfolio by the risk LP-theory and of the LP-estimation of efficiency of controlling the portfolio by *LP–VaR*. The technology has been considered of the risk estimation of the portfolio with the dependent shares yields on the basis of the *copulas*

Chapter 19 gives the risk LP-models of losses of the quality and efficiency in the systems with GIE. It states the general problems of controlling the quality in business and the particular problem of estimating the reliability of the production of the type of "bridge." It also considers the modeling and the analysis of the risk in problem of the efficiency, when the output random parameter of the efficiency depends on other random processes.

In Chapter 20, the risk LP-models of the company management non-success are presented. The statement of the problem is formulated. The risk LP-models of the management non-success on functions and directions of activity are described. The scheme of management by the company as the complex object is suggested. The non-success risk LP-model in the achievement of one aim and in groups of aims and the estimation of the quality of functioning company are described.

Chapter 21 considers the risk LP-models of fraud in business and in the interaction of companies. It gives the scenarios and risk LP-models for frauds of the manager and the wage laborer, for a speculation with an investment, for the struggle of building firms for the profitable contract, and for the financing of projects by some banks.

Chapters 22 and 23 contain the formal risk-theory with GIE on the base of the theory by academician Malsev and the description of the training course "Modeling, estimation, and analysis of risks in economics."

Chapter 22 states the formal risk LP-theory with GIE. It gives the description of the probabilistic space. It describes sets, relations, signatures, axiom, and rules of outputting for the formal risk LP-theory with GIE. It states the connection of the bases of data, the bases of knowledge, and sets. It also contains the methodology of the search of the functionally and topologically dangerous and weak places in complex infrastructures.

Chapter 23 states the training course on the risk theory, include features and advantages of the risk LP-theory, areas of application of the risk LP-modeling and analysis, themes of lectures, contents of laboratory works, the list of indexes, and the software for identification of risk models, estimation and analysis of risks, LP-estimation and analysis of security portfolio risk, and automated structurally logic modeling of risks.

The *Conclusion* underlines the features, similarities, and distinctions of the risk LP-theory with GIE. The answers to the presented question are given.

A review of the applications of the risk LP-models in engineering, economics, banks, and business is made. The basic developers and their contributions into creating the risk theory are named.

Some formulas and pictures are repeated in different chapters, as they are used in various subject areas of the risk. It allows one to study the risk in these areas independent of the contents of other chapters.

The author paid much attention to developing the Subject Index and the list symbols, to the names of chapters and sections, to computing attributes of risk for the object set, to every object, every sign of the object, every grade of the object sign as the basis of the risk LP-theory — the new scientific discipline.

In writing the book, the author proceeds from his own research in the field of the design and in testing complex technical systems [7] and investigating the application of the risk LP-theory in economics and business [3, 4]. Besides, he made use of some results of the scientific school of LP-theory created by I. Ryabinin. The author was one of the editors of the book "Theory and information technology of modeling safety of complex systems" (issue 1–5, 1994–1995) and of the First to Sixth International Scientific Schools meetings "Modeling and analysis of safety and risk in complex systems" (2001–2007). It is natural that the author tries to inform the reader of the most useful ideas, principles, and methods developed by his colleagues in the field of risk management.

The book "Scenario logic and probabilistic management of risk in business and engineering" contains some ideas that the author expects the readers to consider rather valuable. The definitions and terms printed in italics, the Subject Index, and the list of acronyms and the general notations will help to concentrate attention at the most important concepts. Conclusions at the ends of the chapters will allow the reader to repeat the basic ideas.

The book is intended for experts in the field of risk management in business, in technical, economic, and organizational systems at the stages of designing, testing, debugging, and operation. It will also be useful to students, to post graduate students, and to the teachers who work at the economical, financial, and technical universities.

The author wishes to express his thanks to Prof. I. A. Ryabinin for his active interest in publishing this book and for his valuable remarks. The author thanks Dr. O. V. Motygin for his critical reading of the manuscript, for his significant contribution to editing, and for his improving the translation from Russian. The author is also indebted to his former students Dr. V. Karasev, Dr. N. Stepanova, Dr. N. Lebedev, V. Solojentsev, A. Rukin, A. Rybakov, V. Alekseev, I. Mashkantsev, V. Shokolov, D. Strokov, and A. Shiryaev, who have made their contribution to the appearance of this book.

The author expresses special gratitude to Professors Eberhard Stickel (Germany), Hiromitsu Kumamoto (Japan), and Giovanni Barone-Adesi

(Switzerland) for the given opportunity to visit at their universities and for teamwork in the field of risk.

The author thanks the Scientific Council on Program of Fundamental Investigation of RAS for their financial support in 2003–2006, which made it possible to write and publish this book.

The author obtained much knowledge while writing this book, and he hopes that you will obtain considerable knowledge while reading it. Although the author has done his best to eliminate mistakes, experience shows that it is impossible to reach absolute perfection. Therefore the author will be thankful to readers for their constructive remarks, which he asks to be directed to the address: 191178, St. Petersburg, V.O., Bolshoy pr., 61, Institute of Problems of Engineering of the RAS; e-mail: risk@sapr.ipme.ru.

Acronyms and General Notations

SET	set of objects, states, parameters, grades
DB	database
KB	knowledgebase
DNF	disjunctive normal form
CNF	conjunctive normal form
$ODNF$	orthogonal disjunctive normal form
$PDNF$	perfect disjunctive normal form
GIE	group of incompatible events
L	logical (for example, L-model, L-function)
P	probabilistic (for example, P-model, P-function)
LP	logic-and-probabilistic (for example, risk LP-model)
VaR	Value-at-Risk (by Markowitz)
$LP - VaR$	Logic-and-probabilistic Value-at-Risk
$k = 1, \dots, K$	components of the system
$i = 1, \dots, N$	index of different objects (or object states)
$j = 1, \dots, n$	index of different signs or parameters influencing the object efficiency or its states or its parameter-events and logic variables
$r = 1, \dots, N_j$	index of different grades of signs
N_{max}	maximal number of different objects or object states
N	number of objects or states of the object in statistics
N_j	number of intervals in the discrete distribution of the parameter or grades in the parameter (in the sign Z_j)
N_y	number of intervals in the discrete distribution of the efficiency parameter or grades of the one
Y	value of efficiency parameter or logic variable of one
$Y_r, r = 1, \dots, N_y$	grades of the efficiency parameter or logical variables for grades of the efficiency parameter
Y_m	the mean value of the efficiency parameter on sets of objects or states
$Z_1, \dots, Z_j, \dots, Z_n$	random values of influence parameters or parameter-events (logical variables)

Z_{jr}	the mean values of influencing parameters on interval or grade-events (logical variables) for parameter-events
$P2_{jr}$	relative frequency of grades in a set of objects of systems
$P1_{jr}$	probabilities of grade-events in GIE for non-success of sign-event
P_{jr}	probabilities of grade-events in GIE for non-success of object
P_m	mean risk of the object on statistics
P_{av}	mean risk of the object on the risk LP-model
C_i	price for the risk
F	objective function of training the risk LP-model
N_{opt}	number of steps of optimization during training the risk LP-model
N_{mc}	number of attempt of optimization on one stage
E_{gb}	coefficient of recognition asymmetry of good and bad objects
E_g	error of recognition of good objects
E_b	error of recognition of bad objects
E_m	mean error of recognition of objects (the accuracy of risk LP-model)
Ks	robust coefficient of recognition of the risk LP-model
$\triangle P_{ij}$	contribution of the sign j in the object risk i
$\triangle P_{jm}$	contribution of the sign j in the mean risk of objects
$\triangle F_j$	contribution of the sign in the object function F
E_{jrg}	error of recognition by the grade-events Z_{jr} of good objects
E_{jrb}	error of recognition by the grade-events Z_{jr} of bad objects
E_{jrm}	error of recognition by the grade-events Z_{jr} of objects in average
$x_1, \ldots, x_j, \ldots, x_n$	relative parts or weights of parameters, influencing on efficiency parameter
Y_{ad}	admitted value of efficiency parameters
P_{ad}	admitted risk for objects or states of objects, or for the efficiency parameter
$Risk$	risk (probability) to have value efficiency parameter less then Y_{ad}
N_{ad}	a number of dangerous objects or states of objects in "tail" of distribution
H_{ad}	entropy of probabilities in "tail" of distribution
D_{jr}	contributions of grade-events Z_{jr} to Y_{ad} and N_{ad}
C_{jr}	contributions of grade-events Z_{jr} to $Risk$ and H_{ad}

1

Management and Risk

The behavior of economic, social and organizational systems can be described only on the basis of the logic and the set theory.

Norbert Wiener, John von Neumann

In the current chapter, the history of development interrelation between theories of management and risk is stated. Causes and consequences of large catastrophes and accidents are considered: the most dangerous industries are indicated, and risk values and possible damages are shown. A classification of sources of catastrophes and accidents is given. Two different approaches to risk management on the basis of active actions and insurance are considered, and the role and place of monitoring in risk management is discussed. General theses of the State Safety Program of Russia are presented. The role and place of the nonlinear mechanics methods, of the theory of probabilities, and of the logic and probabilistic risk theory in modeling and risk management of catastrophes, non-success, and accident are considered.

1.1 History of interrelation of management and risk

Management and risk existed at all times from the moment of the appearance of mankind. Management provided existence of each human being and the whole human community. First, the management was empirical, it was performed on the basis of intuition, experience, and common sense taking into account the risk. At later stages of human history the states appeared. Management was performed by the supreme governor of the country on the basis of the statute book and the aims of religion. The basis of such management remains both in society and engineering to this day. Later, for more efficient management in the practice of solution of particular problems, people began

E.D. Solojentsev, *Scenario Logic and Probabilistic Management of Risk in Business and Engineering*, DOI 10.1007/978-0-387-77946-1_1,

to use the elements of mathematical management theory and mathematical optimization theory.

Classic management theory appeared during the Industrial Revolution to manage single mechanisms, devices, and processes. It was based on the description of dynamics of objects in terms of differential equations. In management, the risk was taken into account indirectly by using criteria of stability, opportunity of the resonant phenomena, destruction, etc. The success of the classical theory of management is enormous; for example, management of starting and movement of a spacecraft. The founders of the classical theory of management are H. Chestnut, R. W. Mayer, F. R. Bellman, L. S. Pontryagin, J. Z. Tsypkin, etc.

Norbert Wiener's cybernetic management theory appeared after World War II. In this theory, mathematical model of "black box" object was built by its input and output parameters [8]. Such management was used to solve particular problems of optimal management. The risk of such management was considered as probability of failure in achievement of purposes due to the inadequacy of the model and the presence of hindrances.

John von Neumann's probabilistic Logic appeared at the end of the 1940s for evaluating reliability of computational process in a computer having given logical function of reliability and probabilities of device failures [9, 10].

John von Neumann is among the most outstanding scientists. He was a specialist in the field of mathematics and mathematical physics and he also was one of the founders of cybernetics. He considered cybernetics as a general theory of automatic machines and computers. Von Neumann took part in creation of the first computers. Some fundamental ideas in this field belong to him. A wide circle of interests and a variety of talents were peculiar to von Neumann as a mathematician. These interests and talents put him in the same line with the greatest specialists in the field of applied mathematics. He was well acquainted with the most urgent questions of technology, natural sciences, and abstract methods of pure mathematics. He kept in touch with scientists and engineers.

John von Neumann was engaged in the theory of projection of computers and built probabilistic logic, where components failures were considered as an essential and integral side of computer working. The theory of probability is called a combinatorial branch of science. The common method in mathematical logic consists in studying whether a specific procedure is done by computer in a finite number of steps. As the number of steps grows, the probability of machine failure increases, and probability of reliable result decreases. The theory studying quantitative aspect of calculating must rely on both discrete and continuous mathematics. Formal logic means "all or nothing," i.e., discrete.

Von Neumann called his theory of automatic machines a logical theory of automatic machines. He also called it a theory of automatic machines and information, emphasizing the important role of the theory of information. He marked out in the theory of management and information two parts: exact and probabilistic. The exact (or strong) part includes mathematical logic expanded to cover final automatic and Turing machines. The probabilistic part includes Shannon investigations [11] on the theory of information and probabilistic logic.

Von Neumann considered probabilistic logic as a generalization of formal logic. He considered probabilistic argumentation as a branch of general logic. But this point of view is neither trivial nor common. It is also not the principal understanding of probabilities. It competes with frequency interpretation, in terms of which logic is considered as single valued, and conclusions expressed by frequencies are made in relation to phenomena, the information about which is not complete.

Laplas mentioned opportunity of two points of view on a probability: frequency and logic. He also understood the difference between these points of view. This difference was clarified by economist Keins [12]: having analyzed the problem, he has shown that side by side with widespread frequency point of view of the probability, the logical point of view exists.

Shannon's theory of information [11] is like formal logic but it is closer to the ordinary mathematics than to logic, because contemporary formal logic has extremely nonanalytic and non-mathematical character. It occupies processes like "all or nothing," which are weakly connected with analysis, which is the most advanced and well-known part of mathematics, and associated with the theory of combinations, which is the most unknown part of mathematics.

Accuracy of a computer is determined by repeated calculating and detecting results discrepancy. Computer stops just after first error. The error must be localized and corrected by an engineer, but it is difficult for him especially if there are several errors. If it is a single damage, it is possible to divide the area of the error in two parts and to determine the one where it has occurred. This procedure can be repeated while the damage is determined. This method becomes complicated when there are several errors.

This can be described in words, and can be accomplished by neurons (computer software). It is not needed to provide nerve formations and computer devices with supernatural powers or unusual complication. The description of a computer and a brain has a higher logical type (complexity) than does the description of their functions. So there may be extremely long description (for example, the description of visual analogy). While studying the human nervous system, it was revealed that each element can be easily described individually, but you will be amazed by the whole volume that is to be described.

Von Neumann's theory of automatic machines must be mathematical and logical in significant degree. Frankly speaking, it is possible to divide mathematics to discrete and continuous. Logic belongs to the discrete part and has a combinatorial character. Von Neumann believed that mathematics in the theory of automatic machines must be rather continuous. It is relied on in the analysis. He suggested that analytical approach in mathematics has certain advantages over combinatorial.

Formal logic is cut from the most developed fields of mathematics and belongs to the field of the theory of combinations, which offers the worst difficulties. This observation is interesting because von Neumann made a great contribution in discrete mathematics. It is claimed in the work of Morgenschtern and von Neuman [13]. Mathematical methods that will be developed for social sciences will rely on theory of combinations and set theory rather than differential equations (Section 1.2.5 in this work).

In his theory of investigations of automatic machines, John von Neumann started from discrete and went to continuous. The example is his probabilistic logic. After appearance of the probabilistic logic, he suggested discrete-analog computing system that is associated with it. His first models of self-reproduction were discrete, but he hoped to get continuous models later [9].

Numerating is a simple method to get good accuracy from bad. Writing 30 binary ones by 30 devices with two positions (with a mistake about 10%), it is possible to produce a number with accuracy up to one billionth. The main advantage of numeric system is that another method to get the same accuracy doesn't exist.

Von Neumann's probabilistic logic was not evolved widely because the contemporary semiconductor computers became up to several levels more reliable than tube computers in the 1950s and 1960s, and urgency of providing reliability of calculating essentially decreased.

Kolmogorov's probability theory and logic. Taking into account a great scientific and philosophic importance of the problem of use of discrete and continuous mathematics, it is useful to introduce to the readers the ideas of the great Russian mathematician and logician Andrey Nikolaevich Kolmogorov [14, 15]. In 1938 he wrote in the redactors' paper [15] and before in his paper "About analytical methods in the theory of probability" (see pp. 5-41 of the work), "Today the theory of probability is going through a period of rapid developing and rearranging. Works of Einstein and Smoluhovsky about the theory of Brownian motion, American authors about "congestion problems" in technical statistics, Fisher about mathematical theory of natural selection and many other investigations, appearing in *specific sciences*, only with the theory of difficulty are moving into limits of *classic schemes* of probability. By request to systemize big materials, the new direction of investment has appeared. Here an attempt is made to design general methods for investigating the system with random changing process and arbitrary system, the state of which at any moment may belong to a set of possibility conditions."

Norms of *the classical theory of probabilities* are more satisfied with considering conditions of the studied system only at the time moments. They form a discrete sequence, that is the reduction of the random process of changing the system to *discrete* sequence of passing from one condition to another. In other words, using the language of the classical theory of probability, the sequence of separate "tests." *Discrete* random transfers from one condition to another, passing, however, in any moment with *continuous time* the distribution of probabilities, form naturally the first single type of the new, *continuous on time*, schemes of random processes. For example, the probabilistic scheme of radioactive decay of atoms; or investment on capacity of telegraph and telephone lines, etc.

More essential change is suffered in classical theory in case when the set of possible conditions is a continuous variety and the process of random change is *continuous* (without leaps). That scheme appeared first in physics in the works of Einstein and Smoluhovsky named above.

> "During studying *continuous* events of nature, and also *discrete* the schemes are idealization of real processes. In the probability theory, after creating and studying in detail the continuous schemes, it was obvious that the vast classical area of *limit theorems* of the theory of probability we can consider as *a connecting link* between *discrete and continuous* theories.
>
> According to this point of view, the substance of limit theorems of the theory of probability (theorems by Lyapunov) consists in learning how regularities of *discrete* random processes at increasing the number of the single steps and decreasing the size of each from them passes in regularities of *continuous* random processes".

All quotations, belong to A. N. Kolmogorov and text in italic to I. A. Ryabinin, showing the utility of reading "old" works [15, 16], which allows one to see *the narrowness* of every idealization after 66 years from the new point of view.

H. Markowitz's portfolio theory appeared in 1952 for investment the risk managing by means of selection of optimal securities portfolio [17]. The yield as average of distribution and risk as mean square deviation and measure of yield vagueness was considered for any security in portfolio. Such new conceptions as a diversification, indifference curves, achievable and effective multitudes of portfolios were introduced. The contribution of H. Markowitz was significant and he was awarded a Nobel Prize in economics. Further, the portfolio theory was developed as *VaR* theory in works of D. Tjubin, D. Marshall, W. Sharp, S. Ross [18].

V. M. Glushkov, V. I. Skurihin, etc., informational management appeared in the beginning of the 1970s due to the creation of powerful computers. Computer-aided management systems (CAMS) [19, 20] had a well structured database, information technology with the window interface, software for solving the certain type of optimization problems, expert systems for decision making, software for forming reports and printing illustrations. The systems allow one to give out any information at inquiry or to solve problems, to find areas of optimal admitted decisions, to choose the most effective solutions. Acceptance of the final unique decision belongs to an expert. Within the framework of CAMS, the problems of numerical risk estimation were not solved.

D. A. Pospelov's situational management appeared in the 1970s. It uses logical-linguistic models [21]. It was shown that the management of complex objects is impossible in principle without taking into account the qualitative semantic information that cannot be expressed quantitatively. For the first time in the theory and practice of management the *logic*, *sets*, and *logic connections* of objects and events were introduced. Various approaches were suggested for description of situations that were under preview, based on languages with advanced semantics; various methods of construction of knowledge models were presented, allowing one to reflect in the models' qualitative proportions and the rules that were inherent to the object; various procedures were given for finding solutions to problems of management, based on logical-linguistic models. The considered theoretical results find applications in problems of operatively dispatching control in seaports, airports, etc. Problems of risk in systems of situational management were not studied. Further, the concepts were developed in [22].

I. A. Ryabinin's logical-probabilistic theory appeared in the 1960s [1, 2, 23, 24] for the purpose of quantitative modeling and analyzing reliability of complex technical systems. Logical-probabilistic calculus (LPC) is a special section of mathematics. LPC allows ranging elements of complex system by importance. These methods were approved in real projects of the Navy. They became the intellectual core of systems of managing reliability and safety of complex system.

Klod Elvud Shannon was able to liquidate the gap between the algebraic theory of logic and its practical application. He connected Boolean algebra with electrical scheme working. By help of LPC, he was able to connect Boolean algebra with probabilistic theory not only for primary structures but for structures formalization that leads to logic algebra functions of repeated type (bridges). This original "bridge of knowledge" consists of several theorems, properties, and algorithms that is forming the mathematical base of LPC.

Complaining in the paper [25] about absence of definition of *probabilistic logic* even in "mathematical encyclopedia" (1988), it should not be surprising that *LPC phenomenon* is known only to specialists of technical profile concerned with problems of reliability, vitality, and safety [2]. Absence of mention about LPC in mathematical encyclopedias and reference books is the evidence that these problems are not studied by "pure" mathematicians. Practically all creators of the contemporary LPC conception, John Boolean too, did not have special mathematical education, but were high-quality engineers.

Neumann's probabilistic logic and Ryabinin's LP-calculus appeared in the 1950s and 1960s independently in different data domains. In principle it was possible to use one term suggesting LP-calculus as evolution of *probabilistic logic*. However, LP-calculus gave rise to the whole scientific school and became a base of LP-theory of failure risk, involving LP-calculus, structural-logical modeling, LP-theory of risk with GIE, and risk theory based on cortege algebra. The LP-theory is richer in ideas, results, and publications than the probabilistic logic, and it is impossible to get back to this term.

Nils Nilsson's probabilistic logic appeared in the 1980s for evaluating probability of the truth of some logical statements having given probabilities of other statements [26]. At first, that problem of identification was formulated on basis of solving problem of linear programming.

Probabilistic logic is a semantic generalization of logic in which values of true statements are the values of probability (between 0 and 1). Generalization can be applied for any logical system that has a final set of statements. The method unites logic with probabilistic theory so that probabilistic logic conclusion is weakening to usual logic conclusion when probabilities of all statements are equal to either 0 or 1. The value of true statement in probabilistic logic is considered as probability of statement in general logic.

To determine what is implied under *probability of statement*, area of samples is considered in which probabilities are determined. Statement S may be *true* or *false*. If the interest is only in one statement S, then one may imagine two sets of possible universes: a set W_1 containing universes in which S is *true*, and a set W_2 containing universes in which S is *false*. The real universe must be in one of these sets but it is unknown where. One can create a model of uncertainty about real universe having assumed that it is in W_1 with probability p_1 and in W_2 with probability $p_2 = 1 - p_1$.

Nils Nilsson's probabilistic logic did not get wide application because proposed methods of evaluating probabilities by set of logic statements could be realized only for primary cases with sets of assumptions.

A. S. Mojaev's automated structural-logic modeling appeared in the beginning of the 1990s [5, 6] as a development of logical-probabilistic methods. The methodology made it possible to use all logical connections (*AND*, *OR*, *NOT*) and introduces schemes of the functional integrity. The latter allows one to represent the scenario of successful or unsuccessful functioning of a technical or organizational system as a graph including fictitious nodes.

The software for numerical structural and logical analysis of stability and efficiency were developed. They were successfully used for the educational purposes and for solving various applied problems of the analysis and management on the bases of suggested methodology.

E. D. Solojentsev's LP-theory with groups of incompatible events was created in the middle of the 1990s and at the beginning of the century [3, 4, 27–30] on the basis of logical-probabilistic approach.

LP-theory of non-success risk with the GIE involves LP-models of failure risk with the GIE, identification of LP-models of risk by statistical data, LP-analysis of failure risk, and also applications of LP-theory of failure risk in such problems as credit risk, securities portfolio risk, risk of efficiency lack, company management failure risk, etc.

LP-theory of failure risk with GIE allows us to create a model and to analyze risk in the systems where the elements have multiple states, and to apply LP-models with GIE to quantitative modeling and analyzing risk not only in technical but also in economical and organizational systems. The states of the elements in systems are described both quantitatively and qualitatively. High accuracy, robustness, and transparency of LP-models of risk are relied on using Bayes' formula and well organized probabilistic polynomial of risk. Discrete distributions of probabilistic events are used in LP-models of risk with GIE that allows carrying out computing with many-dimensional distributions, and each of them may be different.

LP-model of risk with GIE allows implementing active scenario management of failure risk in complex systems at the stage of design, test, and operation on basis of risk analysis.

V. V. Vitlinsky's problems of risk in economy and production are stated in the work [31]. Organizing and planning risks of production, distribution, and service of produce, risks of financial activities of banks and custom house in unstable conditions, and transition to market economy are described here.

R. M. Yusupov's scientific information fundamentals were formed in the beginning of the century [32]. Conceptual bases of the information theory are considered, its essence, purposes, and principles are determined and formulated, problems of information theory and ways of their resolving are shown, the basic stages and directions of development of information theory are also determined, dual nature of mutual relationship between science and information theory is revealed, inevitably the problems of informational safety are considered, too.

V. V. Alexandrov's investigations on forecasting upcoming processes are the contribution in theory of risk of upcoming processes with changing structure on discrete basis [33–35]. Analysis and processing of empiric data demand non-linear models of investigations.

The new conception "memory effect" as constituent of non-linear paradigm allowing one to analyze time series taking into account "prehistory" of forecasted event on basis of method of normalized amplitude (R/S-analysis) was introduced by H. Herst. "Memory effect" is inherent in natural time series of floods, earthquakes, tsunamis. When modeling evolving processes the statistical data are represented by time series (TS).

The commonality for these processes is that law of irregular growth is satisfactory description of process just within certain limits of characteristics changing, whereupon parameters changing take place in spurts. Power law $y = a \cdot x^{\gamma}$ sufficiently well describes different upcoming processes merely on single regions of empirical data with a priori unknown intervals of contingency.

Technique of risk management on basis of active actions and passive insurance and its optimal combination is created in the works of N. K. Pechenin, etc. [36]. Insurance is a great invention of humanity and if it could have been applied always, then catastrophe losses would not exist.

Methodology of risk management on basis of monitoring is represented in works of O. V. Krasnov, etc. [37]. It was created on the basis of the LP-approach, which allows making decisions with opened eyes.

Nowadays, two different components in the *theory of risk management on the basis of active operations and passive insurance* are intensively developed and their optimal combination is sought. Here, the great achievements of development and use of monitoring risk management in business and engineering, which allows one to make decision with open eyes, should be mentioned.

N. V. Hovanov's theory of synthetic money for decreasing currency risks on basis of summation of random processes with optimal weights was proposed in the work [38].

Interdisciplinary and philosophical approach to risk phenomenon is described in the works of A. A. Muzalevsky [40]. It is given as examples more than twenty resumptive and special definitions of risk. The knowledge in fields of danger, safety, and risk were discussed, integrated, and critically analyzed. Risks of natural and man-caused catastrophes related to losses and other social-economic consequences were described.

Optimal investment to provide for safety of technical systems is suggested in the works of V. A. Melnikov [47]. Approximate analytical problem solution for complex systems is proposed. Application examples are given.

Methodology of coordinate switching to manage safety in complex infrastructures of the enemy's country is suggested by A. Yaroshenko. The technique and software for searching functionally and topologically weak locations in complex systems is developed [43].

National program "Safety of Russia." In Russia, works on strategy of risk management with application of new approaches from the area of fundamental sciences started in 1997. In the book "Risk Management" [39] by famous scientists, who are also the authors of the *State Program "Safety of Russia,"* special attention was paid to problems of strategy of risk management. The concept of the authors is the assumption that the mathematical theory of safety and risk can be constructed on the basis of the accumulated experience of the new science. This theory would take place between the level where political and strategic decisions such as laws are made and the level of development of concrete technical systems. As a methodical basis for creation of such theory, it was suggested to use nonlinear dynamics. We note that the latter point can be true only for accidents such as earthquake, floods, snow avalanche, etc., characterized by slow accumulation of energy or weights with their further very fast release.

In most cases, accident in human–machine systems occurs when some events happen simultaneously or risk of condition of system and its elements as result of "deterioration" exceeds the admitted value. Even example of a human being clearly shows that the person becomes tired, requires rest and food in order to prevent him/her or a technical system, which he/she controls, from accidents. Here another approach is necessary to model risk of failures and the accidents, which would be alternative to methods of the nonlinear mechanics. We shall name such an approach logical-probabilistic or scenario approach for management of risk of non-success.

1.2 Reasons and consequences of large accidents

Development of the environment created by technologic activity of mankind in the 20th century occurred at much higher rates than in previous centuries. It has resulted in two opposite consequences both in industrial countries and in the rest of the world [39]:

- outstanding results in electronic and nuclear industry, airspace, power and chemical engineering, in gene engineering, which advanced mankind to essentially new boundaries in all areas of activity, were achieved;
- unprecedented potential and actual threats to a human being, to objects created by people, to local and global environment, not only in war, but also in peace time, were created.

Thus, the center of attention moved from dangers to risks, from typhoons, flooding, earthquakes, and other natural phenomena, to man caused, ecological, social disasters, stipulated by decisions, accepted by people.

For the first time, the special attention of the public and scientists to large industrial failures was attracted after disasters in the 1970s and 1980s at the chemical enterprizes in Flixborough (England, 1974) and Seveso (Italy, 1976); then, as result, hundreds people were affected, there was essential, irreparable damage to environment, and huge resources (material, human, time, etc.) were spent for liquidation of their consequences. In the 1980s, the tragedy in Bhopal (India, 1984) and Chernobyl (Ukraine, 1986), perpetual virus attacks in the Internet, and large-scale acts of terrorism in the USA (September, 2001) continued the list. As a result of accidents, enormous damage to environment was caused, and the amount of lost people was measured by thousands [41, 42].

Strengthening of two types of dangers [43–46, 48, 49] is observed in natural and technogenic spheres. First, it is the well-recognized ecological dangers for nature, as the living environment, caused by persistent negative anthropogenic pressure on environment. Increase of these influences in combination with global natural processes of change of climate and environment can result in ecological disasters of global and national scale. Secondly, the rapid scientific and technical development in civil and defensive areas in many countries of the world has resulted in essential gap between exponentially growing threats in natural and technogenic spheres and ability of each country and the whole world community to withstand these threats.

The level of a person's safety, of safety of states and of all mankind, of the natural environment from all increasing dangers of natural and technogenic accidents does not rise yet despite the efforts undertaken everywhere in the world. It is notable that natural and technogenic accidents are able to create and strengthen threats in sociopolitical, economic, demographic, and strategic spheres.

The insufficient ensuring of safety results in annual losses, measured by billions Euros. Problems of safety and risk in ecology, engineering, finance and economics, terrorist and information danger have become actual problems on a state scale.

Today in Russia there are about 45 thousand dangerous industries, a great number of constructions, whose destruction can result in disasters not only of regional, but also of national scale.

Many countries, including Russia, are facing with necessity of liquidation in the shortest possible time of large-scale extreme situations (ES) having non-military character. If the extreme situation arises in an industrial area, or a large city, it inevitably causes significant destruction and losses; hundreds and thousand of human beings can be lost.

A great number of ES happen annually in the world. In 1994 in the Russian Federation, 1076 technogenic ES occurred. The most ES happened in industrialized territories. A number of technogenic ES essentially increased in Northwest (91%), Central (48%), and Baikal (41%) regions.

1.3 The most dangerous industry branches

According to the level of potential danger resulting in accidents in technogenic civil sphere, it is possible to give extra attention to objects of the nuclear, chemical, metallurgical, and mining industry, unique unusually large-scale engineering constructions (dams, oil storages), transport systems (space, water and underwater, ground), which carry dangerous cargoes and a large number of people, gas and oil pipelines. Many military objects such as space-rocket and aviation systems with nuclear and traditional charges, nuclear submarines, large warehouses of usual and chemical weapons should be mentioned, too.

For providing the technogenic safety on the boundary of the 20th and 21st centuries, it should be taken into account [39], that in global technogenic environment, both in civil and military sphere, there are about 10^3 objects of nuclear engineering for peace and military purpose, more than $5 \cdot 10^4$ nuclear ammunitions, about $8 \cdot 10^4$ tons of chemical armament of mass destruction, hundreds of thousands of tons of dangerous explosives and strongly acting poisonous substances, tens of thousands of objects with high reserves of potential and kinetic energy of gases and liquids [39].

In analysis of safety of technogenic sphere along with the above mentioned damages, it should be taken into account whether the corresponding potentially dangerous objects are made in series. The heaviest accidents are characteristic for unique objects, i.e., produced in the single copy or in small series. The number of nuclear power reactors of the same type is 1–10 with their general number 450–500 in operation, the number of the same space-rocket systems is from 3–5 to 50–80. Medium-series potentially dangerous objects are estimated by the hundreds and thousands, and large-series are made in tens and hundreds of thousand (cars, agricultural machines, etc). In connection with the above stated, the integrated economic risks, which are determined by multiplication of individual risks by the number of objects, are comparable for accidents of big objects and for accidents of many small objects.

Of high importance, the level of substantiation of safety of potentially dangerous objects is achieved in designing. With reference to failures of large-series complex technical systems, where dangerous damages arise in usual conditions of operation, the level of forecasting safety and reliability is 10–100%. Dangerous and catastrophic destructions of large- and medium-series complex technical systems in conditions of normal operation are predicted in much smaller measure — from 1 to 10%.

From information about probabilities and risks of technogenic failures and accidents on objects with extremely high potential danger, it follows that the difference in the levels of required and admitted risks, from one side, and the level of realized risks, from other side, reaches two and more orders. At the same time, it is known that increase of the level of security of objects from accidents by one order only requires huge efforts in scientific and technical sphere and the expenses being comparable with 10–20% of the project cost.

1.4 Sources of accidents depending on humans

Generally, as complex systems (CS), we shall understand the structural complex human–machine systems consisting of the equipment, computers, and actions of the personnel both having elements and output with several conditions.

The appearance of emergencies, failures, and accidents in such CS as nuclear power plants, starting rocket systems, oil- and gas processing and other chemical manufactures, pipelines and transport systems, is usually classified as rare casual events. However, in view of the consequences such as emission of radioactive and toxic substances, explosions with scattering parts of construction, extensive fronts of flame, and pollution to the environment, the biggest of the disasters can be compared with large-scale natural ones.

The reasons for failures and accidents in CS, depending on their developers, manufacturers, and consumers, are

- Insufficient quality of projects;
- Insufficient quality of development tests;
- Insufficient quality of operational tests;
- Insufficient quality of operation monitoring;
- Deterioration and aging of the equipment in operation;
- Decrease of quality of personnel due to influence of social factors;
- Mistakes and swindles of the personnel;

- Terrorist actions;
- Attacks of hackers.

Actions of these reasons both separately and in their combination results in failures and accidents with human losses (both personnel and the population of the region), with large material damage, with danger for the environment, and decrease of living standard of the population.

We note that both experts and the public pay insufficient attention to some of the mentioned reasons of failures and accidents, because of their appearance with delay; the latter explains the absence of interest of developers in spending extra money on project safety and the tendency of owners to hide the true reasons of failures, unsatisfactory quality of testing systems. As an example of such underestimated reasons we mention the following.

1.5 Risk management and insurance

We consider features of risk management using a historical example of approaches to estimation of danger of sea pirate attacks, the so-called *Bernoulli's* and *Columbus'* approaches. Two hundred fifty years ago, *Bernoulli* found a way to reduce the insurance tariff, at insurance of merchant. Using low tariff, he drew the clients, and due to the big number of clients, he could achieve sufficient accuracy in calculation of probability of loss of the goods or the vessel, and with the low insurance tariff he could get a good profit [36].

Two hundred fifty years earlier, *Columbus* started searching for a way to India. For his ships, as well as for the merchant ships of *Bernoulli's* time, the main threat was the pirates. The probability of attack of pirates was high, but was it necessary for *Columbus* to know the value of this probability? *Columbus* equipped the ships with rectangular sails of the maximal area. He lost maneuverability, but this essentially increased speed of the caravan. On the second day of expedition, a pirate sailing vessel approached *Columbus'* ships, however, some days later, it lagged behind hopelessly. It is necessary to notice that the pirate ships had greater maneuverability than the trading ones, and high speed. But their sails were universal, adapted to fight maneover, and had no such large area as the sails of Columbus' ships.

The given facts from history illustrate two approaches to the risk estimation. The first approach *(Bernoulli)* assumes that a process, in which failure risk is necessary to estimate, cannot be adapted or it is not controlled consciously. The second approach *(Columbus)* is applicable to processes where failure risk should be reduced ad infinitum by appropriate adjustment.

Bernoulli's approach does not demand an investment of money and efforts to transformation of process, which failure risk is estimated. It is the passive financial approach. Permanent updating occurs because a new process is generated instead of unsuccessful process. The approach is applicable to processes where the failure costs are lower than those of the process adjustment.

Columbus' approach, on the contrary, should be applied to processes where failure costs appreciably exceed the process adjustment costs. This approach is troublesome, but expenses for its realization grow linearly depending on complexity and danger of process, and costs from failure of complex and dangerous processes grow in geometrical progression. Thus, with some complexity and danger of process, the approach of Columbus appears to be economically reasonable.

Nuclear insurance pool successfully illustrates absurdness of *Bernoulli's* approach to the insurance of nuclear and radioactive dangerous objects: even for a hundred years it is impossible to generate the pool, sufficient for liquidation of consequences of failure of the Chernobyl type, as the enterprizes are not able to pay insurance tariffs.

The aspiration of the insurance company to be prepared for failure of the Chernobyl type is nothing but an attempt to resolve the *Columbus'* problem by *Bernoulli's* methods. *Bernoulli's* approach is applicable in its original form, if:

- insurance cases come frequently, values of insurance premiums are not significant, insurance tariffs do not constrain economically the activity of the insured enterprizes and cover costs of the insurance company, which can work effectively;
- insurance cases come rarely, values of insurance premiums are big enough, but insurance tariffs for the large number of the same objects of insurance cover costs of the insurance company, which can work effectively;
- insurance cases are coming with any period but the size of insurance premiums changes over a wide range and from time to time can put the insurance company on the face of the crash. In this situation, work of the insurance company in *Bernoulli's* approach assumes inevitable bankruptcy when the most serious insurance cases occur.

Application of *Columbus'* approach in the insurance of dangerous and expensive objects eliminates the possibility of the appearance of failures such as Chernobyl.

1.6 Monitoring and risk

Monitoring is the integral part of safety security in technical, economic, organizational, and social systems. An example of monitoring is given by world economics. A large number of daily and weekly economic newspapers inform us about costs or stock indexes of companies, about exchange rates, sales volumes, etc. There are numerous independent institutions and agencies that estimate and publish rankings of banks, countries, and branches, the reliability of capital investments.

Now using the Internet, it is possible to follow in real time (with a delay of minutes) the situation on all main financial and commodity exchanges of the world in New York, London, Chicago, Tokyo, etc., including sales volumes, a pent-up demand, exchange rates, indexes of stocks, the prices for grain, cotton, petroleum, gas, gold, copper, and other metals and the goods. The same detailed information can be obtained for any period in the past, in minutes, hours, days, months and years. Everything in business is made with open eyes. The openness of information is the reason why the world economy for the last 70 years has not been in such sharp crises, as in 1929.

Monitoring such kinds of sports as chess and tennis allows sport organizations to rank players according to their results and, thus, to solve the problem of formation of the lists of participants and optimal scheduling of tournaments. Monitoring in medicine based on patients' disease records, including their cardiograms and analysis data, allows physicians to organize effective and safe treatment. Monitoring the society via public-opinion polls on various subjects makes it possible to reveal the most urgent problems of the society to prevent social explosions and to plan effective programs of reforms.

For complex technical systems and constructions, intended for long-time operation, failures and accidents can be caused by degradation of properties of materials, by reaching limit levels of the accumulated damages, by formation and uncontrollable propagation of cracks, by cavitation wear, by breakdown of tightness of flanges, by reduction of resistance of isolation of cables due to ageing polymeric coverings, etc. For potentially dangerous objects and manufactures, the essential exhaustion of the design resource is characteristic. In crucial branches (power, petrol, and chemicals plant), potentially dangerous objects have exhaustion of designed resource at the level of 75–90% [39].

1.7 The state safety program of Russia

As a rule, failures and accidents are followed in a short time by a flash of activity of "government officials" on creation of the commissions for investigation and distribution of welfare payments. Charges of the Russian Ministry on Extreme Situations are going to take soon a quarter of the budget of the country because of increased number of failures and accidents.

Their "work" on overcoming consequences is visible "on the face." To ensure the work on decrease of a risk level of failures and accidents is much more difficult, as it needs new approaches, strategy, principles and methods, new culture and means. Results of these efforts will be visible only in some years or remain unnoticed if serious failures and accidents do not occur.

The analysis of results of examinations of large man-caused failures and accidents of the 20th century shows that the further development and realization of programs of scientific and technical development of the modern civilization and operation of CS is impossible without the system scientific approach to solve the problem of maintenance of safe functioning similar objects and development of the methodical apparatus for quantitative risk estimation.

Creation of fundamental scientific, lawful, and economic bases of providing safety is one of the purposes of the state scientific and technical policy and the state scientific and technical program on safety of natural and technogenic spheres, intended to increase safety in the industrial, energy, transport, building, oil–and–gas, mining, and defensive branches, in production of new materials and technologies.

The state program "Safety" defines and fixes transition to the analysis and management of risks, as the basic system of regulation and safety, instead of the existing approach of maintenance of absolute safety [39, 42–52]. The state strategy is intended to provide formation, acceptance, and use of scientifically approved methods and criteria of management of conditions of systems in the parameter space of admitted risks. The purposes the state strategy are as follows:

- controllable and normalized state, regional, branch, and object management of creation and functioning of the CS by new risk criteria;
- optimization of actions in extreme situations for minimization of their immediate and distant consequences.

Ways for reduction of risk and softening consequences of extreme situations follow from the general principles of safety in the natural and technogenic sphere: priority of safety, high level of state regulation, use of risk analysis methods, inevitability of personal responsibility, obligatory compensation of damage, availability of information, declarative way of activity, analysis of extreme situations.

The fundamental problem of modeling and analysis of safety of CS includes various tasks: creation of scenarios of failures and accidents and construction of mathematical risk models, and development of methods for providing safety of an operator, working personnel, and population in case of emergencies in the CS.

1.8 Methods of nonlinear mechanics and probability theory

Nonlinear mechanics methods. In the state program "Safety of Russia," hopes are laid for use of methods of nonlinear mechanics for forecasting and modeling accidents [39],[54]. For that, a number of possible approaches and models are considered: regimes with intensive development as analogues of the catastrophic phenomena, strong turbulence as a mechanism of origin of the accidents, the self-organized criticality as the universal mechanism of accidents, the theory about channels, etc. Because the formulation of these approaches in the program is rather declarative, we shall describe these approaches to estimate their applicability for modeling and forecasting accidents.

Regimes with intensive development as analogues of the catastrophic phenomena. In order to forecast the catastrophic phenomena in complex organized systems, it is necessary to answer a number of key questions:

- Whether the structure of a system permits phenomena of this kind?
- In what elements (areas) an accident can happen?
- When it will take place; whether it is possible to estimate time of development of accident?
- What part of structure of the system is determining for occurrence of accident?
- Of what kind and how heavy could damage be?
- How should the structure of the system be changed or how should the governing influences be regulated to prevent the accident?

It turns out that these questions stated in different terms are answered in the theory of the nonlinear parabolic differential equations [39]. The parabolic equations make a basis of the mathematical models describing evolution of various processes in physical, chemical, biological, social, economic, etc., systems. For example, the equations are used in the theory of nonlinear heat conductivity, diffusion of the charged particles in plasma, filtration of gases and liquids in porous environments, in chemical kinematics, and in problems of description of evolution of populations.

Strong turbulence as the mechanism of origin of accidents. It is suggested to develop an approach to investigate the problem of occurrence of rare catastrophic events in systems with complex behavior on the basis of synergetics principles. The essence of the approach is the observation that fast processes in systems are often determining, or at least, essential and seriously influencing behavior of slow processes. Therefore, of most interest are the charge, which spontaneously arises in the system and quickly develops in large scales. An example of such process is the development of a crack in a solid body. However this example is not interesting, because as a result of the process, the initial system disappears. Another example is appearance of large typhoons in the system "atmosphere–ocean." In this case "large-scale perturbation" does not result in destruction of the system, but leaves an essential trace.

Nevertheless, it is difficult to create models for the phenomena of this type. In view of the events, the conditions of their rise being rare, even if they are determined, are usually treated as the extremely rare combination of improbable events. In the models, it is very difficult to find the key factors and parameters of order. Among the models with exponential distribution, the model of "heap of sand" is most popular in the theory of self-organized criticality. As events the theory considers massive avalanching from the heap on which separate grains of sand fall. Such models are described using strong turbulence concept in Ginzburg–Landau equation.

The self-organized criticality as the universal mechanism of accidents. Here event is treated as catastrophic or dangerous, if it appears unexpectedly (i.e., it cannot be predicted or if it is extraordinary (i.e., it is distinguished from set of events related to it), or both. In either case it is possible to conclude that the system inducing this event is a complex system, because from simple systems it would be natural to expect a clarity and predictability from one side and uniform behavior from another side.

Though rigorous definition of concept of complexity does not exist, experience by development of synergetics and studying real systems intuitively determined as complex allows us to state some common ideas about properties of any complex system at different levels of the description.

At the mathematical level, complexity is intricacy related with nonlinearity of the description, as for linear systems we apply the principle of superposition, allowing us to consider independently various working factors, parts of system, etc., that guarantee its simplicity.

At the physical level, the description, as a rule, is possible only in statistical terms, such as the density of probability, correlation, the mean of distribution, dispersion, etc. It occurs either due to chaotic behavior, specific for many nonlinear systems, which limits possibilities of the determined description, or in view of very large number of elements, combining the system, what makes such description practically useless.

At the philosophical level, the following observation is essential: the more sophisticated and specific a mechanism of some phenomenon, the less often it should be realized. Besides, because virtually any thing in nature is somehow connected to complexity, the mechanisms, laying in its basis, should be simple and universal.

From the above stated follows that the investigation should be concentrated on the universal nonlinear mechanisms, resulting in complex behavior, demanding statistical description. Thus, in the study it is possible "to find a back door" — to generalize the data on the investigated complex systems and on the base of this material to try to give the description of the mechanisms laying in their basis. Below we shall consider manifestation of complexity, and the theory of the self-organized criticality. Besides, we shall give a review of some self-organized critical models.

It is typical (see, [39]) that despite loud promises and declarations of applicability of the above mentioned and other nonlinear methods of mechanics in physical, chemical, biological, social sphere, etc., no concrete example of modeling real accident is given. It is easy to explain because it is impossible to write down the differential equations for laws of conservation of energy, mass, and amounts of movement for complex systems; it can be made only for the simplest systems and elements.

As early as 50 years ago, outstanding scientists John von Neumann and Norbert Wiener wrote about impossibility to write down the differential equations describing behavior of complex systems. They stated that mathematical methods that would be developed for CS would be based on logic, combinatorial theory, and the set theory, but not on the differential equations.

Let us recall the rule "Occam's Razor" [55], which is applied when in conditions of uncertainty or incomplete information for description of a complex natural or social phenomenon it is necessary to choose one of two or several theories (methods). The meaning of this rule is that simpler explanations of the phenomenon have a high probability to appear correct than more complicated ones. In other words, it is reasonable to choose the theory that includes least of possible number of assumptions or involved argumentation. The sense of the metaphor, giving the name of the rule in question, is in cutting off the superfluous principles and reduction of model to possible minimal number of assumptions. "Occam's Razor" is an extremely useful but seldom used methodological tool.

We omit discussing other non-linear and logical and probabilistic methods that have proved to be successful. We only note that all the methods meet demands of "Occam's Razor" principle and recommendations of John von Neumann and Norbert Wiener.

Probability theory methods. For modeling the risk, it is proposed to use the informational-statistical approach to formation of risk models and their identification from limited information on the basis of analytical laws of distribution of random values [56–58]. There is a lot of different distributions, but they cannot rigorously and precisely predict rare events of the real world, and it is proposed to improve these distributions by considering parameters of distributions as random values. At this, methods of randomization of Poisson parameter, generating functions, principle of maximum uncertainty, and Lagrange's probabilistic distribution are used.

In the way, the following distributions are obtained:

(1) The Poisson distribution, where ν is a quasi determined value;

(2) The modified Poisson distribution, where ν is distributed by the normal law with known parameters;

(3) The modified Poisson distribution, where ν is distributed by the normal law and estimations of parameters m_ν, S_ν^2 are known;

(4) The modified Poisson distribution, where ν is uniformly distributed over a known interval;

(5) The Pascal distribution (negative binomial distribution), where the law of distribution ν is approximated by the gamma distribution with the form parameter m and the scale parameter λ;

(6) The non-central negative binomial distribution, where the law of distribution ν is approximated by the gamma distribution with the form parameter m and the scale parameter λ;

(7) The Poisson distribution of the degree k, where the law of distribution ν is approximated by the gamma distribution with the form parameter m and the scale parameter λ;

(8) The beta geometrical distribution of the degree k, where the law of distribution ν is approximated by the gamma distribution with the form parameter m and the scale parameter λ;

(9) The beta negative binomial distribution of the degree k, where the law of distribution ν is approximated by the gamma distribution with the form parameter m and the scale parameter λ;

(10) The modified Poisson distribution, where ν is distributed by the normal law and parameter estimations m_ν, S_ν^2 are known (volume of sample $K < 10$);

(11) The extreme distribution, where ν is distributed by the geometrical law and the mean estimation ν is known.

The given one-dimensional parametrical models of distribution do not solve problems of estimation and forecasting non-success risk or accidents. We can do uncountable quantity of curves through points and it is difficult to say which curve is better. Many works appear that are devoted to more sophisticated apparatus of one-dimensional analytical distributions and take into account "heavy tails" of distributions. The applications using "heavy tails" are also practically absent.

The main flaw of the classical theory of probabilities is by using virtually only one-dimensional distributions, that is the influence of many factors (their real number reaches hundreds) are not taken into account. Let us consider this defect in more detail.

Multidimensional distribution in the probability theory. In real systems, the risk depends on many factors. For example, the security portfolio includes tens of valuable papers of different yields and risks. Often different factors have different dimension, the laws of distributions of factors are different and not normal.

Now there is the mathematical theory only for multidimensional normal distributions, that is, each individual factor is distributed normally and its mean value and dispersion is known. The dispersion matrix of all factors is also known. The theory of calculation of the risk in real multidimensional systems, with influencing factors having different non-normal laws of distribution, is yet created on base of connections of the type *copula* [59–62].

The probability theory, as the applied science, is also named the "urn" theory, because the basic scientific results were obtained in experiments with urns and spheres of different colors. In those experiments, the probabilities or relative frequencies of random choice of different combinations of spheres of different colors were estimated. Thus, connection between the probability theory and the combinatorial theory was found.

Here two things are important. First, the analytical formulas for estimation of event probabilities appeared for convenient estimation but basically were not obligatory, as almost all results could be obtained without these formulas, having the table of records of previous random samples.

Second, the combinatorial analysis has not obtained sufficient development because of astronomical number of combinations. The logic and probabilistic calculus can help combinatorics. We have got modern computers and algorithmic methods for solving difficult problems. Certainly, it is necessary to prove the basic results of combinatorics by the "large" computations. However, for applications the astronomical computations are not obligatory if the appropriate numerical methods and algorithms for PC are developed.

Below in different sections of the book we shall show opportunities of the logic and probabilistic theory of modeling and analysis of non-success risk in complex systems. This theory confirms the rule “Occam’s Razor” and concepts by John von Neumann and Norbert Wiener on applicability of the differential equations for the description of risk in complex systems.

1.9 Power rating distributions of data of developing processes

One should estimate the risk for developing processes. The behavior of such systems is acyclic with the difficulty predicted change of functioning modes. Forecasting similar processes consists in the definition of the interval of the possible contingent event, for which the probability is easily calculated. The analysis and processing of the empirical data demand nonlinear research models [33–35].

For the restriction negotiation of classical approaches of the statistical and correlation analysis and the moment calculation, in present time we start to apply the rating distributions and the power functions at the processing of time series. We enter the concept of “effect of memory” as the component of the nonlinear paradigm, allowing us to analyze time series with taking into account “back history” of the predicted event on the basis of the method of the normalized amplitude excursion (R/S analysis) entered by H. Hurst. The natural time series of floods, earthquakes, and tsunamis have “effect of memory.”

The known example of the developing process is the death distribution of inhabitants (Fig. 1.1). Here on the abscissa axis is the age, and on the ordinate axis is the number of people from 1000, dying in the given age. The form of this curve practically coincides with the dependence character of the occurrence frequency of breakages in automobiles and other biological and artificial systems, formulated as laws of growth and development. Let’s consider these diagrams in details. Relatively high death rate at the life beginning (the left part of the diagram) corresponds with the world statistical data. Children’s death still remains concerningly high even in developed countries.

Further (in the age from 10 to 40 years), we observe the rather low death rate. Increase in the death rate after 40 years is caused by worsening of the physiological condition of the person, increase in “degrees of deterioration” of his organism. Here again we have the direct analogy with operational characteristics of the automobile.

At modeling and forecasting evolutionary processes, the statistical data is represented by time series (TS) of numerical values, for example, the dollar exchange rate, the productivity of the fields, diseases, etc. At modeling similar processes it is actually the forecasting problem of the further behavior

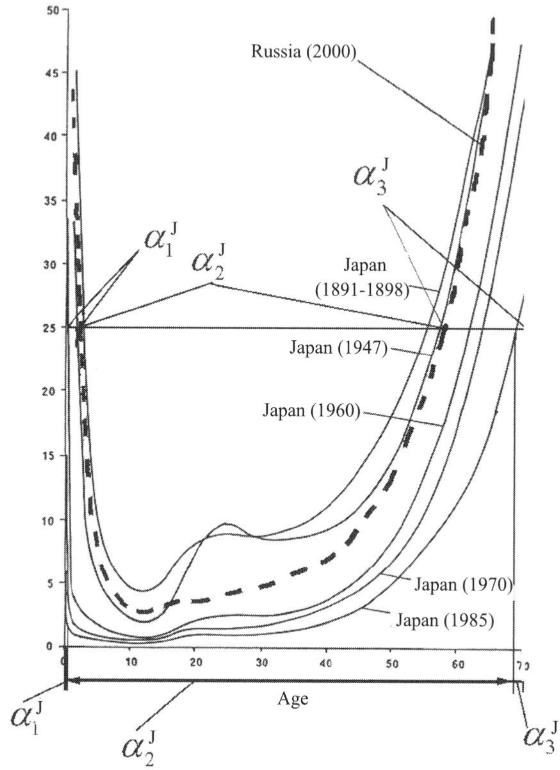

Fig. 1.1. Distribution of population mortality in Russia and Japan against age

considered TS. It is important to know the interval of trustworthy positive forecasting. The R/S analysis, the self-similarity and the power laws are used for finding-out presence or absence the interval hierarchical cyclic memory in considered TS. The processes of the developing type are often met in engineering,economy, and, in essence, are the power laws of the development, which changes the discrete values of parameters a and γ in the power law. The law of the irregular growth is satisfactory for the process description only up to the certain limits of the characteristic change, then the parameters stepwise change.

The power law $y = ax^{\gamma}$ well enough describes the various developing processes only on the separate parts of the empirical data with a priori unknown intervals of contingents. In developing systems, we show the non-uniformly scaled manifestations of the causally and investigatory connections, local and global variable, power laws and rating distributions that have the different physical, economic, and social nature.

There are the estimation methods of the parameter of the power rating distribution for the processing of polytypic empirical data with the revealing purpose of the event contingent and the estimation of its prognostic importance. The power law properties are investigated at processing heterogeneous data types. The algorithm is offered for constructions of approximating distributions of developing processes on empirical data. It is shown that for the various types of the developing processes, the parameter of the power law belongs to the interval $0 \le \gamma \le 2$. It is shown that the stable condition of developing systems is characterized by the deviation of the empirical estimation from the Fibonacci parameter. The forecasting technique of bankruptcies of companies is offered on the basis of the contingent cluster-analysis.

Properties of power rating distributions of the polytypic empirical data are investigated. The estimation methods of the power parameter at the construction of the approximating distributions of the empirical data also are developed with the purpose of revealing event contingent and estimation of its importance for the forecast. The example (Fig. 1.2) of the bank rating distribution by the number in groups is the result.

The revealing method of the event contingent is offered as the criterion of the cluster-analysis, which basis uses the construction of the approximating distributions on the empirical data.

The formalized approach of the revealing self-similar (fractal) structural empirical data is developed. It summarizes the Eliot's waves on the basis of the power approximation of experimental data as the deviation functions from the Fibonacci and Koh parameters.

The applied importance of the developed approach is illustrated on processing of experimental data of the following processes: the demographic development; the fluctuation of the water level in the Neva; the dependence of

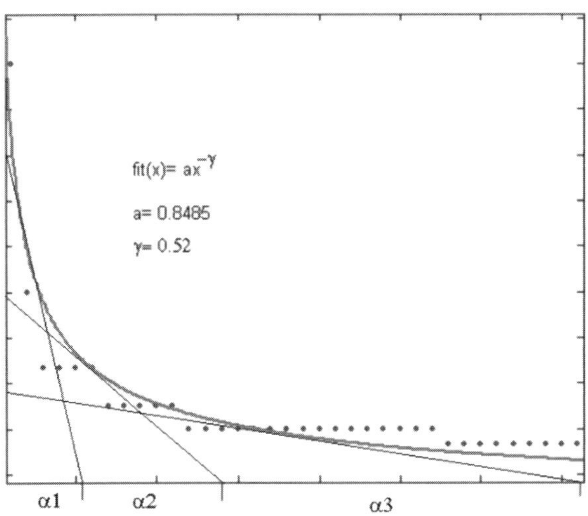

Fig. 1.2. Rating distribution of banks on the number in groups

the wage level from the education; the productivity of grains; the distribution of capitals on banks; the research of text semantics.

1.10 Concrete mathematics

Outstanding American mathematicians Roland Graham, Donald Knut, and Open Patashnik in 1970 have found out that the mathematics required for the proved interpretation of computer programs completely differed from classical abstract mathematics. The abstract mathematics began to degenerate and to lose the connection with the reality. The authors [63] write: "The pursuit for generalizations appeared so fascinating, that the whole generation of mathematicians has lost the ability to find the charm in particulars, to take pleasure in the decision of numerical problems or to appreciate the role of mathematical methods."

The authors understand the *concrete mathematics* is the mix of the CONtinuous and discrete mathematics. It is the ordered set of the tools allowing one to operate with discrete objects. The book [63] gives the introduction in the mathematics that forms the basis of *the information science and the analysis of algorithms*. This book opens the secret of *the phenomenon of American education* — how to transform uneducated schoolboys into fine mathematicians.

The concrete mathematics arose in the troubled and restless decade. In the trouble years, the seemingly valuable constantly were exposed to doubts: campuses turned in the centers of the hot discussions. Curriculums were challenged, and mathematics did not become an exception. Just at that time, John Hammersli wrote the polemic article "About decreasing the level of mathematical education at schools and universities..."

Others troubled mathematics were asking themselves the question: "Is it possible to save mathematics?" One of authors of the present book has planned to make the series of works under the name "The art of programming for the computer," but he has found out that there are not the important mathematical tools in his arsenal: mathematics that was required for the proved interpretation of computer programs completely differed from the studied one in the college in the kind of the main subject. Therefore he has entered the new course, containing material that he would prefer to hear in due time.

The name of the course "Concrete mathematics" meant its opposition "to abstract mathematics", as the concrete classical results were quickly washed away from the modern mathematical education by the new wave of abstract ideas. The abstract mathematics is the wonderful subject, and it has not anything bad: it is beautiful, general, and useful. However its adherents have run in the error, that all other mathematics occupies the lower position and is not worthy of notice. The rush for generalizations appeared so fascinating, that the whole generation of mathematicians has lost the ability to find the charm in particulars including to receive pleasure from the decision of numerical problems or to appreciate the role of mathematical methods. The abstract mathematics began to degenerate and lose the connection with reality. The mathematical education required the concrete counterbalance for the restoration of the stable equilibrium.

The concrete mathematics has arisen as the reaction to other tendencies in mathematics. As this course continued to take the appropriate place in the educational process, the contents of the course proved the value in new applications.

What is actually represented by the CONCRETE mathematics? This is the intelligent operating by mathematical formulas with use of the certain set of methods for the solution of problems. Studying material of this book will allow one to calculate the awful sums, to solve the confusing recurrence relations, to reveal the artful laws in the data, to seize the algebraic techniques in such degree that it will be easier to receive the exact results, rather than to be satisfied with the approached answers that are fair only in the limit.

Calculation of the sums, the recurrence relations, the elementary theory of numbers, the binomial factors, the generating operator functions, the discrete probability theory, and the asymptotic methods are the most important topics of this book. At the same time, the authors give preference to the technicality, instead of existence theorems or combinatory reasonings. Our purpose is training each reader on the discrete operations (such as computing function of "the largest integer" or the final sum), as far as the studying of the analysis are familiar with the continuous operations (such as the calculation of the certain integral).

1.11 Scenario of LP-management of non-success risk

In complex systems, the scenario of failures and accidents have logical and probabilistic nature. Therefore, we write a scenario of non-success or catastrophe. Further, we build the risk logic function and the risk probabilistic polynomial [2].

In the scenario, the elements of the complex system have the logical connections OR, AND, NOT, cycles and groups of incompatible events (GIE). Using the non-success risk LP-model, we can fulfill the quantitative analysis of the non-success risk at the given probabilities of the initiating events. The risk analysis allows us to manage the risk.

Probabilities of element failures can be changed in the course of time (the elements wear out, age, collapse, depreciate). The risk LP-models with dynamic interpretation are much more constructive and transparent than with differential equations, so they are true with high probability.

Each complex system has some value of risk of safe functioning (and each element of the system has value of risk of non-success or failure). If the risk becomes more than the admitted one, the system either cannot be supported by itself, or it is useless, or harmful. Then the system ceases existence (for example, a bank bankrupt). Or there occurs a serious structural reorganization of the CS when some elements disappear and some new ones are brought to the system. The latter changes logical connections and probabilities of failure, so a new CS appears.

Thus, it is possible to simulate visually all past and modern catastrophes on the basis of the LP-approach without using the mathematical apparatus of nonlinear mechanics [39] and the classical theory of probabilities. Many examples of real catastrophes can be interpreted by LP-risk models [42–46, 48, 49] with changed probabilities of elementary events and the objectively existing admitted risk for the CS. One should build only the scenario of the connection of the events and further write the catastrophe risk logic and probabilistic functions, which show with large transparency how it can occur and with what probabilities.

The non-success risk LP-theory with GIE [3, 4, 30] allows one to model and analyze the risk in systems, which elements have several states, and to apply LP-models with GIE for quantitative modeling and analysis of the risk in engineering systems and in economic and organization systems. Element states in systems are described as quantitatively as quality. Non-success risk LP-models with GIE have high estimates of quality and namely accuracy, robustness, and transparency.

The scientific basis of the risk LP-theory and technology of the scenario logic and probabilistic management by risk are the risk LP-theory with GIE. We use also the LP-calculus, the theory by Markowitz, and VaR for the risk of the security portfolio. Besides, we use also the logic, the discrete mathematics, the combinatorial theory, the nonlinear optimizations, the modeling of Monte Carlo, the Bayes formula, the Shennon entropy, the algorithmic calculations, and the special logical software.

In the current chapter, the history of development interrelation between theories of management and risk is stated. Causes and consequences of large catastrophes and accidents are considered: the most dangerous industries are indicated, and risk values and possible damages are shown. A classification of sources of catastrophes and accidents is given. Two different approaches to risk management on the basis of active actions and insurance are considered, and the role and place of monitoring in risk management is discussed. General theses of the State Safety Program of Russia are presented. The role and place of the nonlinear mechanics methods, of the theory of probabilities, and of the logic and probabilistic risk theory in modeling and risk management of catastrophes, non-success, and accident are considered.

2

The Human Being and Risks

The science and techniques are crystallization of wisdom, but frequently wisdom carries in itself seeds of madness.
Sato Susumu, Kumamoto Hirimitsu

Science and engineering are a crystallization of wisdom, but often wisdom brings seeds of insanity, and, therefore, rapid stormy development of a science and engineering results in a wide spreading of scientific and technical evil [64, 65]. Growing destruction of the environment everywhere over the world, accumulation of nuclear waste products, disasters, such as AIDS, failures and accidents in engineering, crises in economy, the political and information terrorism, etc., are symbolizing this fact. Modern industrial civilization or society of automation and information is characterized by fetishism of money and sciences, alienation of people, and growth of dementia.

A human being and his participation as a risk element stand in the center of typical disasters in the modern world.

2.1 Frauds in business

Let us consider some statistics on frauds in business in the USA [3, 65]. The Chamber of Commerce informed that losses due to wastes of hired workers are estimated $ at 20–40 billion annually. The volume of such stealing is much greater than those by house-breaking, hijacking, robberies, and the usual thefts in total. Federal services estimate the common annual damage by swindle with a sum from $60 up to $200 billion. The losses due to telephone swindle in 1991 in the market were estimated at $10 billion.

The Federal Trade Commission (FTC) and the American Association of Health Insurance (AAHI) estimate at 10% the number of fraudulent accounts on health services in the field of public health services. By the end of the 20th

E.D. Solojentsev, *Scenario Logic and Probabilistic Management of Risk in Business and Engineering*, DOI 10.1007/978-0-387-77946-1_2,

century, fraud in this area caused more than $160 billion losses. By another estimation, fraud is absorbing up to $75 billion of all expenses in the USA on public health services.

Scientific researches show that three of ten workers are looking for possibilities to steal something, three others of ten will steal as soon as they have opportunity, and only four of ten will stay fair in any circumstances. Each year in the USA, 200 million thefts in shops (of goods in sum $11.6 billion) occurs.

According to the Internal Revenue Service of the USA in 1990, the federal government got only 4/5 of all taxes. This underpay of taxes made for $100 billion in arrears. More than 660 ways of evasion from payment of taxes were disclosed.

In the USA, Russia, and other countries, swindle has become one of the main problems of economic safety of the state.

2.2 Errors of personnel

It is notable that not only failures of technical elements of a system result in accidents. A cause of interruption of normal operation may be a single unintended wrong action of an operator or a single omission of a needed action (so-called errors of the personnel), it can also be a combination of technical failures and errors of the personnel.

In history of atomic power stations, more than 50% potentially dangerous incidents (small infringements in work) occurred due to mistakes of the personnel [65]. It should be taken into account that nuclear stations follow the program of quality maintenance, and the required degree of quality of the equipment, the personnel, and auxiliary devices of the "person–machine" dialogue is achieved. The concept of auxiliary devices of the "person–machine" dialogue includes the necessary and sufficient operator's devices for reliable and safe control of the power unit.

Failures of refuel systems of launching rocket systems can be classified as follows: 70% of failures were caused by aging and deterioration of the equipment as a result of long exploitation, 11% of failures were due to mistakes of the personnel, 5% occurred because of constructive defects, 11% were by exploitation factors, and origin of other 3% is unknown [37].

Because the human being is often "a weak component," state of CS and its safety quite often cannot be estimated without taking into account quality of the personnel and working conditions in CS.

2.3 Asymmetric actions of terrorists

The sense of asymmetric actions of terrorists consists in making the greatest harm with the least expense (amidst the expenses terrorists count their own lives, too). It is sad but today acts of terrorism on potentially dangerous objects and in places of mass accumulation of people are feasible.

Now and in the foreseeable future, as acts of terrorism in the USA have shown (attacks on the World Trade Center in New York and the Pentagon in Washington), in Russia (explosions of buildings by the Chechen terrorists), in Israel (explosions in public places by Palestinian terrorists), etc., mankind is vulnerable to the small radical groups, who are ready to play "not fair."

2.4 Hacker attacks on informational networks

Now we cannot imagine all gaps in protection of our civilization. For example, the level of safety of global computer networks with occurrence of computer viruses changed drastically. The racing of more perfect viruses against more effective antivirus programs is going on. As the role of information infrastructure grows, the given class of risks can become more important. Dangers and risks can proceed from a person — hacker, not being stipulated by any technological necessity.

2.5 Personnel in modern civilization

Here we present the results of the analysis of personnel work in the modern civilization, given by Sato Susumu and Kumamoto Hiromitsu in their book "Re-engineering the environment" [64].

Black boxes. Personnel are also becoming system components like robots for process un-automated on economic grounds. For example, the chemical plants are automated; all physical and chemical reaction processes are divided into unit operations or processes. Each unit operation is considered as a black box, automated, and all unit operations are then integrated and controlled. Of interest are input and output relations for each unit operation, and the internal mechanisms of the operations are often neglected. The control over a unit operation is performed on the basis of various measurement variables such as temperature, pressure, rates of heat generation, and stream flow rates. The unit operation looks like a black box to the human operators. Automation has increased the number of black box systems. This inevitably increases the risk of accidents, due to incomplete understanding of processes inside the black boxes or of the ways of interaction between the black boxes.

Automation is fully based on modern rationalism that

- subdivides the whole into elements,
- neglects qualitative aspects of objects,
- recognizes objects by quantities.

Each element thus becomes a target for investigation, the elements are integrated to form the whole, and the result is controlled by computer. Real objects, however, have qualitative and quantitative aspects, and the automation cannot fully represent real processes.

Human errors. Automated manufactures require control and monitoring from the control center as well as daily inspection and maintenance of each elementary process. The automated systems are designed in such a way as to monitor each process by control panels in the control room. However, the machine and process may sometimes cause abnormal events that cannot be monitored from the control center. When these events are overlooked, serious accidents may occur.

Consider a chemical plant where unit processes are connected by pipes. Assume that high temperature and high pressure fluids (or gasses) flow through the piping network. Such a chemical plant has a high risk of small leakage of fluids. The leaked fluids may accumulate, and a spark can cause explosions. This type of leakage cannot be detected by indicators on the control panel, and daily inspections are required.

Operator errors are inevitable for current automated systems. Monitoring tasks are monotonous, boring, and leading to loss of concentration. Humans are not good at this type of monotonous work. They find more satisfaction in tasks that require judgments to adapt themselves to changing environments because such tasks lead to learning by experience. Monitoring tasks with such lack of stimulation are confidence destroying and error prone. The human errors do occur frequently in modern automated systems. And the errors symbolize unacceptance of the monotonous monitoring tasks.

A system or a subsystem is shut down by safety devices when a stable technological process is disturbed by operator errors. Failed components and other consequences of the accident must then be repaired to resume operation. Human errors are also characteristic for the processes of shutdown, repair, and resumption. In automated manufactures, experience and expertise are minimized, types of labor are standardized, and the number of expert workers is decreased. Thus, it is difficult to find people to cope with failures and malfunctions. Engineers have less experience in preventing unexpected chain-initiating event from developing into a large accident because each engineer is engaged in desk designs of small portions of the automation system. This fragmentation of knowledge may also be imposed by management, so that an engineer or scientist cannot go off on his own and start a competing design or construction company as he only knows a small part of the complete process.

Automation and intelligence. Some people suppose that automation increases the ratio of scientific or intelligent labor to manual labor. Others claim that blue-collar labor comes closer to white-collar labor by automation; blue-collar workers are replaced by gray-collar workers who are engaged in monitoring tasks; white-collar workers have risen to manage personnel and materials. It is said that automation requires intellectual labor that can only be performed by people with education levels higher than high school graduates.

An opposite view claims that gray-collar labor is literally gray because a stimulus challenging the labor disappeared. It is difficult to improve the human capabilities through gray-collar labor. The monitoring tasks make the nerves atrophy, causing a new form of fatigue unbearable for human beings. Modern labor-related medicine has pointed out that:

- optic nerves cannot sustain long periods of focusing on flat monitor surfaces,
- extensive periods of monitoring may yield diseases such as autonomic ataxia (loss of muscle coordination).

Therefore, monotonous labor typically observed in modern automated manufactures is no less inhuman than severe physical labor. The transition from blue to gray-collar labor does not imply a transition toward more intelligent or more humane labor. The increase of workers with higher education has nothing to do with the ability or the level of intelligence of labor. The tendency of common higher education is a fashion induced by a longer life span, rather than a result of a conversion from heavy-type industries to a light-thin-short-small type of production.

It may seem that system programmers have the best work, as they are the brain and the center of automation of any manufacture. It is the case at the stages of development and implementation of new projects of automation. But after the project is finished, they are forced to leave the manufacture or remain for support of automation system and to perform boring routine work, and may kill time by writing viruses, or by another hacker activity.

Management intensification. As meaningless, inhumane, and isolated labor increases, management is being intensified. In the traditional steel production, management lines were not separated from the technological lines. These two types of lines were united into a technology/management system. Technological skills were important in these factories, and management was performed by various types of technological experts.

Clear separation of managerial and subordinate work is observed in recent reports on the steel industry. In Japan in the steel industry, the shift supervisor is a key person. He, as a "steel man" by definition, has to manage shift members not only at the factory but also at their homes. Monotonous monitoring tasks granted by only the nervous tension, subordinate tasks controlled by a time-table under the mask of scientific management, and increasingly intensive labor drive the shift workers to despair. Worker's feelings are summarized by representative comments, like "It turns out that I am now working three times harder than before." Shift workers are being eroded by the labor intensification; their family life disintegrates, which in its turn causes harmful influences on the worker's performance. Scientific management by the shift supervisor is no longer sufficient. He controls the lifestyles of subordinates after working hours by making the excuse that he is taking care of their families. This style of management is required to push workers to work under conditions that make them lose their stimulus to work.

Increasing routine workers. Automation in steel industries has created various types of routine labor while retaining some types of routine physical labor. The total number of workers has been decreased by automation. However, the number of routine workers increases considerably in subcontract factories. Automation results in increase of the percentage of routine workers. Similar situations are observed in other industries. Rapid automation is in progress in car industries where the number of routine workers increases in assembly lines that are difficult to automate.

Some people predict that in the future, every process will be automated; they consider the current automation as a transition stage. It should be noted here automation replaces routine tasks by machine operations only when such replacements are cost-effective. Some tasks are still difficult to automate. Besides, automation itself creates new types of routine tasks around human-machine interfaces. Computerization increases data input tasks at the input side, and data monitoring increases tasks at the output side. Automation results in reduction of old types of routine tasks and growth of new types of such labor. It is notable that in automation, the total number of workers decreases, but the percentage of routine workers increases.

Third and fourth levels of industry. The reduction of the labor population in secondary industries (mining, construction, manufacturing, etc.) increases the number of laborers in the tertiary industries. The development process follows the transition of ascent from:

- primary industry (agriculture, forestry, and fishing industry) to
- secondary (mining, construction, manufacturing), then to
- tertiary (commerce, distribution, transportation, communication, public relations, education, services, and finally to
- the fourth level (banking, insurance, real estate).

The expansion of third and fourth level industries is not a social needs but a result of oversaturation of the second level industry with labor population. The expansion of third and fourth level industries is evidenced by the flood of various types of advertisement, persistent and irrelevant enticements to buy goods, and excessive numbers of shops, banks, and insurance companies. This inflation yields a transition of worker types from blue to gray and then to white collar. Some people claim that human labor has become more intellectual and less physical due to this transition.

Consider as a typical example a Japanese city bank that is a center of the money market. Today, the city bank is a leading company, but the labor in the bank is not challenging. Many white-collar workers are engaged in the counter services. The cashier at the counter continuously counts money received from customers. The money is handed on until eventually a final worker at a cash desk receives it. At some point of this process, the amount of the money received is printed on a bankbook, relevant data are sent to a host computer via communication link, and the data is processed and stored in the computer. Money withdrawal follows a reverse process. Most bankers are thus doing routine jobs around the computer. Other bankers repeat routine home public relations (advertising) visits. The bank workers seem to be a bright group of white collars, but their jobs are unattractive, and many bank workers have resigned from their companies.

The third and fourth level industries require many “key punchers.” This job requires physical labor because it involves data entry via keyboards. The job uses mental labor because it reads a computer program list. However, such a physical or mental job is restricted to an extremely narrow domain. Such job of “key punchers” results in inflammation of sheaths of tendon of wrist and autonomic ataxia and proves the inhumanity of this job.

3

Principles of Risk Management in Design

The knowledge of some principles quite often compensates ignorance of some facts.

Gelvitsiy

Occurring incidents are failures, non-successes, accidents, and catastrophes. We appreciate the risk as the non-success probability, damage, and the admitted risk. Principles of risk management, which shall be considered below, at the design stage are applicable with some success for technical, economic, and organizational systems.

At the design stage, the system project in the form of the appropriate documentation is created; problems of risk management at stages of development and operational tests are solved, the corresponding programs of tests and the monitoring system for the operation stage are developed.

At the design stage, the scenarios of danger of the whole system and its components are developed and analyzed, and structural, logical, and probabilistic models of risk are constructed. The graph of dangerous states is built from the top — the final failure event or accident. Derivative events and initiating events are introduced. The possibility of localization of dangerous conditions at their occurrence is taken into account. At the given or chosen probabilities of initiating events, the risk of failures and accidents is estimated. It allows us, as a result of modeling and risk analysis, to choose constructive, technological, and structural decisions for achievement of acceptable risk.

3.1 Style, concepts, and methods of designers

Let us consider style, concepts, and methods of the chief designer in providing safety and the minimal risk of a created new product. As an example, we shall describe the style, concepts, and methods of work of the well-known aircraft

E.D. Solojentsev, *Scenario Logic and Probabilistic Management of Risk in Business and Engineering*, DOI 10.1007/978-0-387-77946-1_3,

designer A. N. Tupolev [66], the founder of known airplanes: ANT, Tu-2, Tu-16, Tu-104, Tu-114, Tu-134, Tu-144, Tu-154, etc.

Style. The thinking method by A. N. Tupolev always corresponded with the level of the chief designer. He did not lose the common view, did not change into the narrow expert, but also did not miss those details that are determining ones. He thought that the chief or main designer is appointed to realize the main idea. The head position obliges him to protect this idea from encroachments, from uncountable "corrections" under which pressure the idea can be simply buried.

The chief designer who does not make great demands on people cannot make a machine quickly and with a good quality. A. N. Tupolev strongly influenced the origin and development of the aviation science. First of all it was revealed in his exclusive strictness to authenticity of test materials, and also in irrepressible aspiration to understand the essence of considered processes and the physical sense of the investigated phenomena.

A. N. Tupolev accepted and realized only those new ideas that had strong scientific and technological basing. "I am not interested, that you think. I am interested, that you have learned, seen, understood, that you have made."

Concepts. First of all the skills is to find and use among the set of the new ideas, providing progress, those that can be realized at this moment. It is well-known that the great Leonardo da Vinci put forward a big number of new and progressive ideas including the helicopter, parachute, and ball-bearing, however for their realization mankind required five centuries.

In the project of a new experimental plane, Tupolev had to make a choice between two new engines: one of them was easier, more economic, and less overall; another was worse on all these parameters. However Tupolev chose the second engine. The reason was as follows: the first engine would demand reorganization of large number of oil refineries, and it would take time and require huge expenses.

The best bomber of World War II, Tu-2 had a number of new features, yet it was possible to realize during war. The experimental plane by designer V. M. Mjasischev was constructed at the same time. It was a very good plane but it contained new solutions that could not be realized at the existing technological level of production, therefore, the plane remained experimental.

For some years, Tupolev searched for a solution to what a heavy jet bomber should be. Two interesting ideas allowed the project to be realized: A. A. Tupolev proposed to arrange engines behind the wing closer to the fuselage, and A. A. Judin proposed to retract undercarriages in special gondolas on a wing.

Young experts were often amazed with his apparent inertness in questions of introduction of new, progressive proposals. The skilled people also understood that refusal of introduction of innovations on the final design stage helped to speed up development and to implement the new experimental object. There are examples when groundless use of innovations resulted in creation of "crude" designs; their debugging strongly delayed introduction in mass production and in exploitation. A. Tupolev did not simply refuse innovations, but placed them in "portfolio" and used them in the following development.

His variant of a passenger airplane borrowed elements of design of a previous military airplane comprehensively checked up in exploitation.

Methods. In the work of the chief designer, most time is taken by the organization of obtaining information on failures, by analysis and elimination of failures at all stages of life cycle of the plane.

Even the construction stage of the experimental plane and its units already brought the new information, including negative one: technological non-efficiency, excess of the given weight, insufficient strength, defects and failures of the equipment and mechanisms on test benches and input control.

He thought that only by full-scale tests and checks of parts and elements of the future airplanes on test benches is it possible to find the confidence in reliability of the decisions made. Full flight airplane characteristics, characteristics of stability, controllability, maneuverability; and fighting qualities of airplanes were determined at the stage of the state tests. The following order of classification of defects by results of the state tests was established. All defects were divided into four groups. The first group includes defects that are dangerous to flights and unconditionally needing immediate elimination. The second group is the defects that are harmless for flights but complicating the job and not allowing normally to perform tests. The third group is the defects allowing to perform tests, but needing unconditional elimination on the tested plane. The fourth group is the defects requiring elimination on prototype serial plane, with obligatory checking the performed actions at the following flight tests of the prototype serial plane as a standard for series.

At exploitation phase, there is new information about the airplane such as statistics of failures and destructions due to industrial or constructive defects and defects on unknown earlier reasons. The defect might be new kinds of resonant fluctuations in structures and systems, fatigue failures, unexpected increase of forces acting on levers of control or interaction in electric circuits and hydraulic systems and much others, including problems in products of suppliers. Failures and accidents during mass exploitation, as a rule, are caused by a combination of defects of engineering (construction or manufacture) and operation (errors of personnel or flight crew). A. N. Tupolev made "uncompromising fight" for correctness of reason analysis of accident and the defects requiring elimination by the producer and the customer.

3.2 Axioms for construction of technology of risk management

General knowledge in risk area is the basis for designing and management of risk and safety of complex technical, financial, and organizational systems

[3, 30, 67, 68]. The scientific paradigm proceeds from inevitability of failures and accidents and determining the necessity to use the concept of acceptable risk. To general scientific knowledge in the area we shall refer the following principles:

- Errors in projects of CS are inevitable;
- If there are stocks of nuclear, chemical, biological, and other energy, there are also the ways for their leakage to the environment and for occurrence of accidents;
- If there are money, material, stocks, they can be stolen;
- There is no profit (in business) without risk;
- Everyone can swindle under pressure of circumstances, if valuables are badly guarded and it is possible to hide the trickery for some time;
- It is impossible to manage the risk without quantitative measurement and analysis of risk;
- Designers of a system should think on its normal functioning, but they should also consider it from positions of the saboteurs looking for ways of its destruction (I. Ryabinin's rule);
- Chief designer of a system should not apply all possible innovations at once; debugging the system in this case is practically impossible (Tupolev's rule);
- It is necessary to borrow reliable and well tried elements and decisions from other products, companies, and countries (Guderian's rule);
- It is necessary to minimize the variety of accepted decisions because it is impossible to provide high reliability for considerably heterogeneous systems (Solojentsev's unification rule).

If we do not acknowledge these rules as scientific axioms, it is impossible to construct appropriate technologies for risk management. We note many of these positions were rejected and not fixed in standards and laws of Russia. It was a common opinion it is possible to ensure zero risk in any system; besides, faultlessness and usefulness of activity of conscious people was supposed.

Let us make some comments on Guderian's rule. It appeared during World War II. When the tank T-34 had proved its power in the actions, the special commission headed by G. Guderian made a conclusion that the German industry could not create a similar tank for two years because for this time it was impossible to create the fuel equipment for the diesel aluminum engine. The decisions and technologies could only be borrowed from the enemy.

3.3 Models and rules

The models that are necessary for risk management at designing complex systems, as a rule, are not described by differential equations. Mainly, the following types of models are used: [3, 7, 30]:

- scenario model of risk;
- structural or graph-models of risk;
- logical risk models;
- probabilistic risk models;
- models of trouble forecasting as critical predicates;
- models of testing technology, consisting of procedures and operations;
- models of objects description in the form of requirements of technical specifications;
- models of expenses of means for decisions and possible damage at absence of decisions;
- models of organizational management in the form of frame networks, providing support of technology and protocols, release of reports, and notifications;
- models of programs and test reports, represented by tables;
- models of states of system in exploitation in the form of the table "Conditions and parameters."

An information technology of the system designing should provide a convenient representation of these models and their communications through a database. Risk management at designing is also provided by expert systems with rules "if–then" and "by analogy," semantic and frame networks. For example, the technology of designing automated debugging and operational tests have procedures of forecasting, modeling, planning, and decision making that contain about 100 elementary operations. From them about 1/3 operations are operations of documenting, about 1/3 operations are operations of calculations on models, and about 1/3 operations require intellectual support for decision making with use of rules.

3.4 Occam's razor

In conditions of uncertainty or incomplete information for description of complex natural and social phenomena, it is necessary to choose one of two or several theories (methods). For resolving such questions, scientists and experts should know the so-called rule "Occam's razor" named after philosopher William Occam [3, 55]. The meaning of this rule is that more simple explanations of some phenomenon is correct with high probability than are more complex hypotheses. In other words, if we have two hypotheses, explaining the same phenomenon, it is necessary to choose that from them which includes the least number of assumptions and difficult calculations.

The sense of the metaphor, used to name the specified rule, is in cutting off superfluous principles and in constructing the model with minimal possible number of assumptions. "Occam's razor" is an extremely useful but rarely used methodological tool. Nevertheless, it is necessary to notice, that it is a philosophical principle, which is not true in all cases and, therefore, it should be applied with some care.

For example, consider the problem of approximation of ten points, solved to tare a spring. Should we use a straight line or any curve from a practically infinite set? Application of Occam's razor results in the choice of the most beautiful and economical decision: the straight line is simpler than a curve.

Similar problems are usual in science and its applications, including modeling and analysis of risk of accidents and failures. The majority of models of accidents and failures, published now in scientific journals, are described by systems of differential equations, which origin is not clear, or use the catastrophe theory or the chaos theory, or enter "abstruse" distributions of probabilities of casual events and "special" descriptions of "fat tails" of such distributions, etc. Above mentioned approaches, along with demonstration of mathematical methods, usually do not give anything practically useful and it is impossible to check these models.

At the same time, there is the simplest solution or "straight line," which consists in constructing logical scenario of connection of events and the writing down on basis of the scenarios the logical and probabilistic risk functions of accident that with the large simplicity and clarity will show as all the processes can occur and at which values of probabilities. It is necessary to compare all other proposed models with "straight line," but often it is not done. The outstanding modern mathematician von Neumann, one of the founders of cybernetics, discrete mathematics, and the theory of computers, claimed that the mathematical methods, which would be developed for their application in social sciences, would be based on the logic, the combinatorial theory, and the set theory rather than on differential equations.

3.5 The transparency of risk models in business

During the evolution of any bank system, it is possible to trace the influence of two differing direction tendencies: the aspiration to the maximal profit on the part of commercial bank owners and the aspiration to the maximal stability on the part of the state. The functioning stability is the main requirement of the society to the banking system. It distinguishes banks from any other branch of the economy. The special regulation of the banking sector by the state is caused as the bank specificity, connected to services of the special sort, so the different negative consequences, which the banking crises carry for the national economy and social stability [69].

The banks take the special place among others specialized financial intermediaries because of the unique duality of the carried functions: the passive one (the attraction of investors money) and the active one (their accommodation in loans).

There are techniques and *software* that are important for banks, companies, business, and economy. They are intended for the estimation of credit risks, ratings, quality of functioning companies, and the risk of investment portfolios. However these techniques and software in many respects are not transparent and accurate [4].

Each of the named sectors has the big multi-billion business with hundreds of business schools, the expensive courses and seminars and, naturally, the corresponding *software* for marketing. Periodically the techniques "are improved" and they begin the new boom for training clients and sales.

The input in the market is free and generates competitors and the unfair competition. Practically all banks and companies are found in nets of this prospering business because the market obliges to receive and periodically to update the credit ratings, the quality certificates, etc. Obviously, it is possible to find agencies and professionals that are ready for the compensation to overestimate the rating of a bank or company.

The techniques, technologies, and *software* are necessary in the named sectors of business for increasing efficiency, profit, and competitiveness. The trouble is in that, despite the global activity of the numerous rating agencies and the quality centers, ruins and bankruptcies occur often with banks and companies having the high ratings and parameters.

Attempts to explain this circumstance do not have success. It is found out that the information on banks and companies is confidential. The corresponding techniques, technologies, and *software* are the intellectual property of the rating agencies and the quality centers and consequently cannot be presented. These techniques have some expert estimations of factors and consequently are not transparent.

Problems of the transparency and the risk are closely connected. The business loves the non-transparency as the basis of the super profit: and for leaving from taxes (20 offshore branches of *YUKOS* company), and following to the motto "without risk there is not the profit." The state controllers, blinded fetishism of money, sciences and democracies, do not manage with their functions on the regulation of the stability of banks and market.

The open society by Soros and the transparency of business are different things. They enter in business with the new effective transparent technique and technology is not simply: the market is occupied, and techniques can appear unnecessary (the other example is the models for the revelation of bribes).

The problem of the techniques *transparency* for the estimation of the credit risks and ratings is considered in Chapter 7 and in Chapters 12–21. We show that the use of the risk LP-models with the GIE essentially raises the transparency of results and analysis.

3.6 The admitted values of parameters

The physical approach to debugging and operational tests, monitoring and diagnostics, risk management and safety of the CS consists in the estimation of functional system abilities and their elements based on measurement of parameters and calculation of risk [3, 7, 30].

The measurement of parameters and indicators is applied to physical destruction (durability and wears), behavior of personnel, economic parameters, ecological parameters, and the accident and failure risk. The physical approach provides rapid estimation of functionalities of systems, and the approach is alternative and complementary to the accelerated and durable tests resulting in failure or accident.

The physical approach provides an information communication of tasks of designing, testing, monitoring, risk, and safety of the CS on the basis of the parameter measurement.

At the construction of the non-success risk scenarios, at the risk calculation and at the risk management, we will use the concept "the admitted value of the parameter." Parameters having the admitted values are, as rule, random variables and have the distributions of probabilities. It is possible to calculate the probability that the parameter is less or more than the admitted value. Below we give the list of some parameters having the admitted values.

The admitted values of parameters (for a man and an animal):

- the temperature of the air: $T_{min} < T < T_{max}$;
- the atmospheric pressure: $P_{min} < P < P_{max}$;
- the content of oxygen in the air: $A_{min} < A < A_{max}$;
- the feed, calories;
- the water for drinking, liters;
- the duration of dream, hours;
- the noise, decibel;
- the radiation level, roentgen.

The admitted values of parameters for personnel:

- the cognitive resource (training and re-training);
- the psychomatical resource (the support of capacity for work);
- the motivational resource (the formation of motivations to work).

The admitted values of parameters for business:

- the minimally admitted yields;
- the profit rate;
- the liquidity;
- the reservation of resources under the risk, etc.

The admitted values of parameters for population of the country:

- the living standard;
- the educational level;
- the level of health services;
- the safety level;
- the openness level of society and democracy.

The admitted values of parameters for machines and details:

- the amplitude and frequency of vibration (millimeters and hertz);
- the surface cleanliness (class of cleanliness);
- the size accuracy (class of accuracy);
- the geometry of circles (ovality, obliquity);
- the hardness of materials (Brinell's or Rockwell's hardness);
- the specific gravity of materials;
- the weight of machines;
- the dimensions of machines;
- the reliability;
- the safety;
- the deterioration;
- the toxicity (exhaust gases in engines);
- the factor of the surplus of the air for the combustion (in engines);
- the factor of the surplus of the blowing-off air (in engines);
- the first and second space speeds.

3.7 Scheme of complex object management

Management of state and development of complex system and also its testing and operation will be performed by us as complex object control of the control theory [3, 70]. Such management consists in control of movement on the chosen program trajectory and correction at the deviation from it (Fig. 3.1). As the parameter, specifying the trajectory, the risk can also be chosen.

Thus, the complex system is moved from the initial condition A to the given final condition B following the chosen program trajectory A–B divided into some stages $j = 1, 2, \ldots, n$. The correction is performed in case of deviation of system from the program trajectory.

Proceeding from this interpretation, the following basic concepts are introduced:

$Y(Y_1, Y_2, \ldots)$ are controllable parameters;

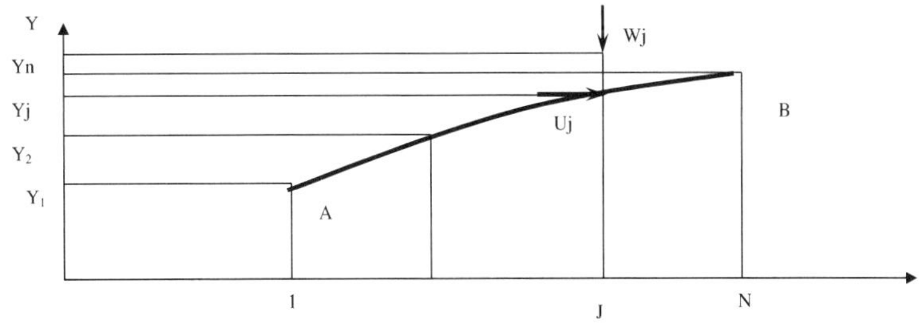

Fig. 3.1. The scheme of control of complex object: Y is controlled parameters, U is control actions, W is corrective actions

$H(H_1, H_2, \ldots)$ are stages;
$U(U_1, U_2, \ldots)$ are controlling influences for organization of stage;
$W\ (W_1, W_2, \ldots)$ are adjusting influences during stages.
The named parameters are vector values.

Controlled parameters Y are measured or observed parameters, which we use to judge about the system capacity to work. Leading at the first stage (point A) is chosen to be minimal in order to not destroy the system, the last stage (point B) is done on nominal or maximal functioning mode (mode with the maximal loading). The system is moved from initial condition to the final one through a finite number of discrete stages with progressive increasing parameters.

During development of the management program, designers beforehand prepare for possible accident by providing W-corrections, which are certain variants of constructive or technological decisions or resources. In creating development program (debugging), it is necessary to determine values Y, W, U for the stages of debugging H.

For complex system vectors, Y, W, U have big length and their realization can demand excessive means. For optimum choice of components of these vectors, it is necessary to know the expenses:
$Q_y(Q_{y1}, Q_{y2}, \ldots)$ – on measurements and management;
$Q_u(Q_{u1}, Q_{u2}, \ldots)$ – on control influences;
$Q_w(Q_{w1}, Q_{w2}, \ldots)$ – on adjusting influences;
$Q_h(Q_{h1}, Q_{h2}, \ldots)$ – on stages;
and also the following possible damages if the actions are not made:
$R_y(R_{y1}, R_{y2}, \ldots)$ – at absence of measurements and controls;
$R_u(R_{u1}, R_{u2}, \ldots)$ – at absence of control influences;
$R_w(R_{w1}, R_{w2}, \ldots)$ – at absence of adjusting influences;
$R_h(R_{h1}, R_{h2}, \ldots)$ – at the absence of stages.

The scheme of management of a complex object is invariant concerning any object. This scheme could even be used for management of market transformations of economics in Russia on the program "500 days" by G. A. Yavlinsky.

3.8 Minimization of the number of decisions

Accepting as few as possible variants constructive decisions, it is possible to provide (at manufacturing and in exploitation) higher reliability of each accepted decision. We shall consider the designing problem with use as the criterion function "the minimal number of different decisions or the maximal unification of accepted decisions." The problem is of interest for a developing company, aspiring to satisfy needs of different consumers with more reliable products and smaller expenses, or for a company having high development and wishing to reduce too wide range of manufactured products and, accordingly, to reduce expenses for manufacture and increase the product reliability.

The formulation of such problem assumes the existence and possibility to find in the mathematical model of the product designing the set of admitted solutions [3, 7, 67], exactly, the admitted discrete solution set (constituted for example by values belonging to series of sizes of the basic detail, established by the state or branch standard). The problem is solved in the dynamic optimization of standardized series of productions during performance of new projects. Beforehand, the time of an order receipt for products with new parameters and sizes is unknown.

Let us explain the problem statement using Fig. 3.2. Here, the abscissa axis is the number (the basic size) of the product of standard series and the ordinate axis is the number of the project in a sequence of its performance within, for example, two years under orders of the customer. The fat points on the horizontal lines are the admitted solutions for the project from the considered series.

There are four allowable decisions for the first project, so, the probability of the choice of the correct decision is equal to $P_1 = 1/4$. It is possible that the chief designer will choose decision 3 because it provides the minimal dimensions for the first project. Then making decision for the second project he will regret because the decision he has chosen and started in manufacture for the first project is bad from the viewpoint of unification. There are five admitted decisions for the second project; the probability of the choice of the correct decision is again small $P_2 = 1/5$, etc. Thus, depending on "luck" of the designer, for six projects he could obtain from six different constructive decisions $(1 \to 3, 2 \to 5, 3 \to 9, 4 \to 7, 5 \to 6, 6 \to 2)$ up to two $(2,3,4,5) \to 8; (1,6) \to 3$.

Let us state an algorithm of the solving of this problem of dynamic optimization of series with the criterion "of small possible number of different standard sizes." Elements of established series of standard sizes $D_1, D_2, \ldots, D_n$

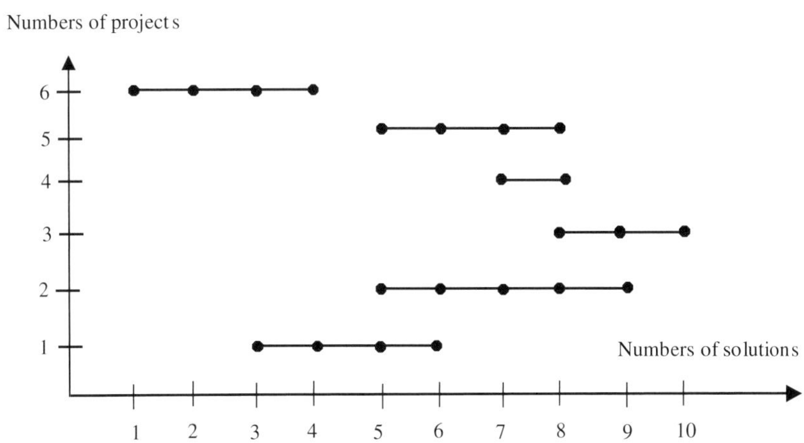

Fig. 3.2. The scheme of dynamic training dimension-type row

are given by normalized weights $C_1, C_2, \ldots, C_i, \ldots, C_n$ ($\sum Ci = 1$). If the company only begins development and manufacture of projects, these weights should be fixed $C_1 = C_2 = \ldots = C_n = 1/n$. If the company has already developed projects, the normalized weights can be chosen, for example, proportional to the quantity of product release with the given standard size. As a criterion function, we shall use the following expression for entropy of series

$$I = -\sum Ci \cdot \ln Ci, \quad i = 1, 2, \ldots, n \tag{3.1}$$

the meaning of the latter can be explained with example of information entropy. Suppose that information either D_1, or D_2, or ... or D_n is transferred with probability of these events $C_1, C_2, \ldots, C_n$. If probabilities $C_1, C_2, \ldots, C_n$ are approximately equal, it is impossible to give preference to any of the events. In this case one says that information entropy is small. If probabilities $C_1, C_2, ..., C_n$ differ appreciably, it is presumed that the information having the greatest *a'priori* probability was transferred.

The entropy of series (3.1), as well as the entropy of information or thermodynamic system serves for an estimation of a measure of "disorder" of series members. Maximization (increasing) of entropy of series corresponds with the increasing distinctions in weights between members of series, it means increasing manufacture of members of series with the greatest weights.

Thus, during designing new products, it is necessary to choose from the admitted decisions one that as much as possible increases the entropy of series (3.1) or, similarly, to choose as the decision the element of series that has the maximal weight. Naturally, the weight of this series member needs to be increased proportionally increasing serial productions, and it is also necessary to make normalization of weights of all members of series (3.1) and, by that, to obtain *a posteriori* probabilities (weights).

The stated approach was used for designing cylinders (characteristic size is the diameter) of piston compressors in the industry. The series of diameters (according to the standard for diameters of piston rings) consists of 108 members, and the admissible decisions for the project were selected by solving the problems of linear programming, first on *min*, and then on *max* [3, 7].

The group of the projects having a place actually for the last three years was presented to the weighed series of cylinder diameters on serial productions. The group of compressors of one of the companies could have, basically, 32 cylinders with different diameters. During consecutive development of projects, the designer chose cylinders with 24 different diameters; 13 different diameters were chosen by the considered method.

3.9 Structural design

Structural designing serves the important purpose of designing systematization and increase of object reliability. Structural designing is the method of modular designing objects at which process of designing is represented as

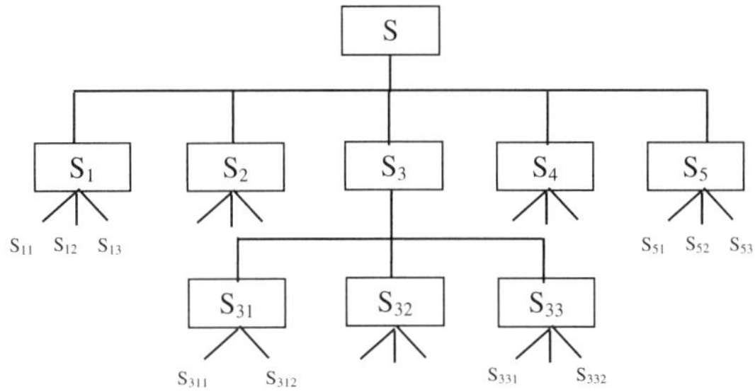

Fig. 3.3. Scheme of structured design

hierarchy of levels of comprehension of the object [3, 7, 67]. Thus, each certain level is completely isolated from details of the lower levels (Fig. 3.3).

The method of structural designing assumes that at the first stage of design, the project S is expressed in terms of its essential concepts. At this level of abstraction we fix some objects (elements) of the second level $S_1, S_2, \ldots$. These components are considered further as the object components, which will be decomposed at the following level. The process of definition proceeds up to the level where elements of the object become elementary and indivisible. Moving deeper into essence of the problem and dealing with the most complex object-elements, the designer abstracts from details of the lower level. Therefore, it is possible to provide for modification of the object. The objects of lower functional level can have some alternative realizations. It leads to the problem of optimal choice or unification of decisions with use of existing wares of the world market.

Such sequence of the designing, named "from top to down," is one of the basic ideas of the structural designing. The second important idea of the structural designing is the use of simple and evident schemes of management of the project, which becomes foreseeable and "controllable."

The object development process begins from comprehension of requirements of consumer and market on creation of original elements or borrowed elements from former designs, or from realized elements by specialized firms.

Certainly, during "from top to down" designing, the returns to higher levels are possible, if there is no effective decisions for elements of the lower level. However, it is not a reason to reject the basic ideas of the structural designing method.

3.10 Concept of the acceptable risk

The central philosophical question in the safety problem is the choice between the concept of "absolute" safety and the concept of "acceptable" risk. For the first time, the concept of the acceptable risk was stated in I. A. Ryabinin's papers. We shall describe this idea following one of his monographs [2].

Substantiation of the concept. At first, more humanistic concept (at first sight) of the absolute safety was accepted. It was a basis for definition of the appropriate standards in the nuclear power and in other branches of industry. The detriment of the zero risk concept is by the presumption that it is possible to exclude any danger for population and environment if we do not spare efforts and means for creation of safety engineering systems and the serious organizational acts, providing high level of discipline.

However, even use of most effective safety systems and advanced methods of monitoring in technological processes does not provide, and cannot provide in principle, the absolute reliability of a system operation, excluding any accidents. The zero probability of catastrophes is reached only in systems with no reserved energy, chemically and biologically active components. On other objects the catastrophes are possible; they are not excluded even with the most expensive engineering acts. The concept of the absolute safety is contrary to the internal laws of nature, which have probabilistic character.

The concept of the acceptable risk has many opponents. They consider it as immoral, saying that this concept gives designers the right to plan failures with probability less than acceptable one. However, it is more immoral to mislead ourselves with hopes on the unattainable absolute safety.

Probabilistic risk analysis having been used outside Russia for many years, has allowed us to accept a set of new actions to increase safety of operation of nuclear stations and other potentially dangerous industries. The concept of the acceptable risk allows us more reasonably (with open eyes) to concentrate and distribute means not only for accident prevention, but also for preliminary preparing for emergency actions in extreme conditions.

Having agreed with the acceptable risk concept and necessity of probability calculation of danger of technical systems, it is necessary to choose suitable mathematical tools. Such tools, as a rule, are the probability theory, mathematical statistics, and mathematical logic. The development of the logical and probabilistic safety theory (LP) of CS seems to have perspective.

As the safety LP-theory we understand the basic knowledge on calculations of the risk of failures and accidents in the structural complex systems. It is based on the logical presentation of development of dangerous conditions and mathematical methods for calculation of the validity of functions of the logic algebra. The LP-methods of safety research allow us to reveal objectively the most dangerous places, the reasons and initiating conditions; the methods form another ideology of developers and induce experts to concentrate their efforts to the decision of principal problems.

Economic choice of the acceptable risk. The most general and universal methods of calculation of the risk value [71, 72] is the approach based on the economic analysis of safety. According to this method, the criterion of optimum of safety level is the minimum value Q that is the sum of two components: $Q_1(r)$ is the given charges for safety security with the risk r, and $Q_2(r)$ is the direct damage, caused by the risk r. Thus,

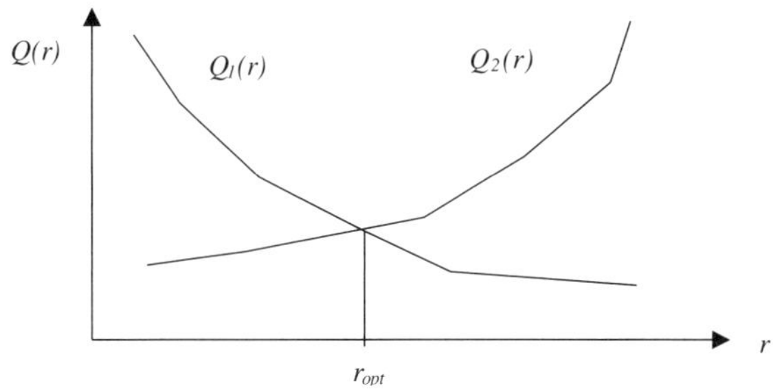

Fig. 3.4. Finding the best value of risk

$$r_{opt} = \arg\,\min Q(r) = \arg\,\min[Q_1(r) + Q_2(r)]. \tag{3.2}$$

The value r_{opt} can be accepted as the acceptable risk value. The graphic illustration of the above mentioned expression is given in Fig. 3.4.

The acceptable risk value depends on the national economics level. The higher is the economics level, production relations, and safety culture, the higher is the level of requirements, made by society, to the safety of potentially dangerous objects, i.e., to the lower value of the acceptable risk. In the process of economics development, the requirement to safety should increase, and the value of the acceptable risk should reduce. Introduction of the failure risk as the universal characteristic of the safety meant, in some sense, revolution in theory of safety management.

3.11 Markowitz's and VaR-approach to investment risk

Investments are the basis of the market economy in developed countries. The security portfolio theory is the most widespread modern theory of investments. It makes it possible to optimize, simulate, analyze the risk and operate by the security portfolio risk. It solves the problems of forecasting and optimization of yield and risk.

In Markowitz's theory and VaR-approach (Value-at-Risk), "models of averages and dispersions" are used [17, 18]. For each security in a portfolio, the yield, as the mean of distribution, and the risk, as the mean square deviation and measure of uncertainty of yield, are taken into account. Such concepts as diversification, curves of indifference of the investor, available and efficient sets of portfolios are used. The normal distribution laws of yield both for each security and for total portfolio are used.

Problem of selection of a portfolio. The investor has a fixed sum of money for investment and wants to invest this money on a certain time interval. The beginning the period is designated $t = 0$ and $t = 1$ corresponds with the period end. At the period end, the investor sells securities that were bought. As the portfolio has some various securities, making this decision is equivalent to selection of the optimal portfolio from a set of possible portfolios.

Making a decision in the moment $t = 0$, the investor should take into account that yields of securities and the portfolio in forthcoming period are unknown. However, the investor can estimate expected (or average) yields of various securities, being based on some assumptions and then to invest money in securities with the greatest expected yields. Markowitz notes that it will be, in general, an unreasonable decision. The typical investor wishes the highest yield but simultaneously wants the yield to be so determined, as far as possible. It means that the investor, aspiring simultaneously to maximize expected yields and minimize uncertainty (the risk), has two purposes, contradicting each other, which should be balanced at purchase decision making about at the moment $t = 0$. Markowitz's approach for decision making makes it possible to take into account both these purposes adequately. The approach implies diversification, i.e., purchasing not one, but several securities.

The yield of security j for one period can be calculated using the formula

$$Z_j = (Z_{j\ t=1} - Z_{j\ t=0})/Z_{j\ t=0}, \tag{3.3}$$

where

$Z_{j\ t=0}$ is the security yield at the moment $t = 0$;
$Z_{j\ t=1}$ is the security yield at the moment $t = 1$.

As the portfolio is a set of securities, its yield can be calculated analogously:

$$Y = (Y_1 - Y_0)/Y_0, \tag{3.4}$$

where

Y_0 is the portfolio yield at the moment $t = 0$;
Y_1 is the portfolio yield at the moment $t = 1$.

At the moment $t = 0$, an investor cannot know the yield for all portfolios. Hence, the investor should consider the yield, connected to any of these portfolios, as a random variable. Such variables have characteristics, one of them is the expected value $\overline{r_p}$, and another is the standard deviation σ_p.

The investor should base the decision of the portfolio selection exclusively on the expected yield and the standard deviation. It means that the investor should estimate the expected yield and the standard deviation for each portfolio and then choose "the best portfolio," basing on the ratio of these two parameters.

Expected yield. As it was marked above, the portfolio represents some set of various securities. Thus, the expected yield and the standard deviation of portfolio should depend on expected yield and standard deviation of each security, included in the portfolio. Besides, obviously, it should be taken into account what part of money is invested in each security. The expected yield of the security portfolio is as follows:

$$\overline{Y} = \sum_{j=1}^{n} x_j\ \overline{Z_j}, \tag{3.5}$$

where:

x_j is a part of money, invested in the security j;

$\overline{Z_j}$ is an expected yield of the security j;

n is a number of securities.

Standard deviation. The measure of the risk should estimate deviations of the achieved result from the expected one. The standard deviation is the measure, allowing us to do it, as it is an estimation of the real yield deviation from the expected one. In the case when the yield distribution of a portfolio can be approximated by a curve of normal distribution, the standard deviation really is a very good measure of uncertainty degree for estimation of portfolio trend. The approximation is often considered as the plausible assumption at the yield analysis of diversified portfolios when the investigated period of holdings of securities is short (for example, quarter or less).

The formula for calculation of the standard deviation of a portfolio is

$$\sigma_p = \left[\sum_{i=1}^{N} \sum_{j=1}^{N} X_i X_j \sigma_{ij} \right]^{1/2} \tag{3.6}$$

where σ_{ij} is a covariance of security yields i and j.

Analysis of portfolio. In Markowitz's approach to the decision of the problem, an investor should estimate alternative portfolios from the viewpoint of their expected yields and standard deviations, using indifference curves. In the case when the purpose is to avoid the risk, the portfolio, laying on the indifference curve, which is located higher and more to the left than other curves, will be chosen for investment.

From a set n securities it is possible to combine the infinite number of portfolios. Fortunately, the investor should only consider a subset of the set of all possible portfolios, belonging to the so-called efficient set. The investor will choose the optimal portfolio from the portfolio set, where every portfolio provides the maximal expected yield for some risk level or provides the minimal risk for some value of the expected yield. The set of portfolios, satisfying these two conditions, is the efficient set or the efficient border.

The VaR-approach for portfolio selection (Value-at-Risk). The VaR-approach for the portfolio selection by criterion of allowable losses *(draw down criteria)* is an alternative to Markowitz's approach. We shall consider the typical case of an investor who is willing to avoid risk. The optimal portfolio choice is made by the condition of maximization of the admitted yield

$$Y_{ad} = \overline{Y} - h_\alpha \cdot \sigma \to \max; \tag{3.7}$$

taking into account that $VaR = h_\alpha \cdot \sigma$. The latter formula can be written:

$$Y_{ad} = \overline{Y} - VaR \to \max. \tag{3.8}$$

Here: $\overline{Y}, Y_{ad}$ are the expected and minimal admitted yields of security portfolio, respectively, h_α is the number of standard deviations in quantile of order

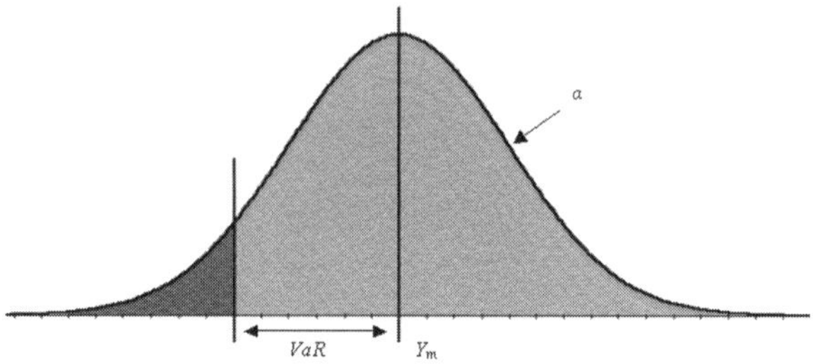

Fig. 3.5. Distribution of portfolio yields

α (level of trust); for example, for trust level $\alpha = 95\%$, the value of h_α is 1.65; and for $\alpha = 99\%$, the value of h_α is 2.33; σ is the standard deviation of the portfolio yield, and VaR is a probable loss (Fig. 3.5).

In conclusion, we note that the assumption about the normal law of distribution of every security and portfolio yield (according to Markowitz's theory and VaR-approach) is very strong and not always justified.

As it will be shown in Chapter 15, the non-success risk LP-theory with GIE successfully solves the choice problem of the optimal security portfolio unifying Markowitz's and VaR techniques. It allows us to remove the essential assumption in the portfolio theory about the normal distribution law of the yield of each security and the whole portfolio and to solve new problems for the analysis and forecasting of the portfolio risk.

3.12 Active and passive management of risk

Let us formulate the concept of safe deterioration of the material resource part [3, 30, 36]. If a material resource in the start of usage has a value exceeding the necessary resource for operation of the object, then the process of resource deterioration has two stages. At the first stage, the remained not-depreciated part of resource completely provides trouble-free operation of the object. The probability of a failure, caused by deterioration of the given resource, does not differ from zero. At the second stage, the material resource is depreciated much, so that the probability of failure of the object because of deterioration of the given resource accepts some non-zero. The time corresponding with the moment of transition from the first stage to the second stage is the so-called threshold time. It is possible to control the threshold time: the influence of any material resource on non-failure operation of an object can be investigated.

The material resource can be considered as a set of resources, where each resource is wearing out in the course of time. For each resource there is some function $Rs(t, x_1, \ldots, x_m)$, which represents dependence on time (t) and on conditions of operation ($x_1, \ldots, x_m$). This function is usually investigated beforehand by experiment with material resource i. In the range $t_0 \div t_1$,

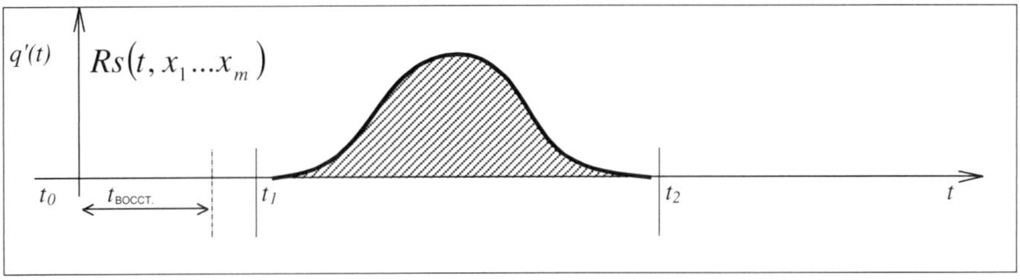

Fig. 3.6. Frequency distribution of failures as a result of wearing material resource

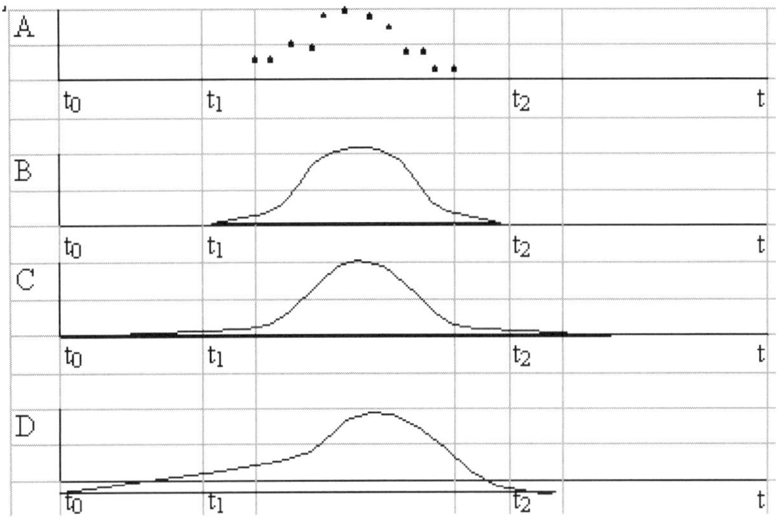

Fig. 3.7. Different interpretations of frequency distribution of refusals as a result of wearing material resource

the function is not defined (Figs. 3.6 and 3.7), which corresponds with the fact that frequency of failures as a result of deterioration of the given resource in the range $t_0 \div t_1$ is equal to zero. Because generally the material resource consists of a set of internal resources, always there is a danger of existence of the deterioration function of an unexplored internal resource $Rs(t, x_1, \ldots, x_m)$ that has some non-zero finite values in the range $t_0 \div t_1$. A single fact of failure because of deterioration of an internal resource $Rs(t, x_1, \ldots, x_m)$ leads to necessity of research that will let change exploitation conditions in such a way that the repeated occurrence of the given event in the range $t_0 \div t_1$ becomes impossible. Because to the left of the point t_1 the probability of breakdown is equal to an infinitesimal value, and to the right of the point the probability has finite values, it is reasonable to suppose that the threshold time is near the point t_1. The traditional approach to selection of the distribution function of failure does not assume existence of the threshold time.

Experimentally obtained data on the object breakdown, which happen because of deterioration of any controllable material resource, represent the operating time, laying in a limited range of time. The life of a material resource can be divided into three qualitatively different periods:

$t_0 \div t_1$ is the period when the material resource is sufficient and there is some reserve of the resource;

$t_1 \div t_2$ is the critical period when the material resource has no reserve and continues to wear out;

$t_2 \div t_\infty$ is the period when the material resource is already exhausted.

The moment t_1 is the threshold time. If work with the material resource is correctly organized, that is beforehand, before the threshold time, the material resource is regularly restored, it is possible to claim that it is possible to control the threshold time. The value of the reserve factor for the time between two procedures of restoration of the material resource following one after another can be found on the basis of the investigated material resource and the value of a possible damage in the case of failure, caused by deterioration of the given resource. Hence, the opportunity to control the threshold time directly depends on research; that is, legitimacy of concept of the safe deterioration of the material resources depends on realization of timely and sufficiently wide program of researches of material resources.

In the case when the material resource is widely used, information on the resource deterioration is usually sufficient and available. Research should be directed to revealing qualitatively new features of the resource. If the material resource is used seldom or unique, research can be productive only when the intensity of the material resource deterioration that can be achieved in researches is higher than one in the real operation process.

Practical use of the concept of safe deterioration of the material resource consists in the following: before the threshold time, it is necessary to manage the risk actively, raising reliability of the weakest and most dangerous elements of the system. It is necessary to use results of monitoring conditions of these elements. After achievement of the threshold time, it is necessary to replace dangerous elements or if possible and reasonable to insure the elements or the system as the whole.

3.13 Algorithmic calculations

It is not a secret that most mathematicians and mechanics are still demonstrating their skill by analytical solutions and transforms and are using a computer for printing results of their intellectual refined excises [30,63]. However, at quite "strong" assumptions, their problems are come to analytical calculations for points, strings, plates, cases, etc.

Problems of logic and probabilistic estimations and analysis of risk are always connected with complex algorithmic calculations, which are so labor-intensive, that the problem of estimation of the complex algorithm and decreasing laboriousness of calculations arises.

Algorithmic calculations are the following stages of constructing and using the LP-models:

- constructing the risk L-function and the risk P-function;
- solution of the optimization problems with calculation of criteria and characteristics of risk models, of risk elements, risk objects, and the system in the whole;
- fulfillment of combinatoric and logical and probabilistic analysis of risk.

3.14 Arithmetical and logical addition

Initiating factors (signs) influencing the final event of the system can be added arithmetically or logically [3, 27]. The number of such added factors can be from several units to several tens. Below we shall study the dependence of the final event probability from values of probabilities of factors and their numbers, and also we shall compare results of arithmetical and logical addition of probabilities of sign-events.

The logical function for addition of events $Z_1, Z_2, \ldots, Z_n$ is as follows:

$$Y = Z_1 \vee Z_2 \vee \ldots \vee Z_j \vee \ldots \vee Z_n. \tag{3.9}$$

In words, it means that the failure occurs, if any one, any two ... or all initiating events occur. After orthogonalization of the logical function (3.9), the following probabilistic function can be written (probabilistic polynomial):

$$P = p_1 + p_2 q_1 + p_3 q_1 q_2 + \ldots, \tag{3.10}$$

where

$p_1, p_2, \ldots$ are probabilities of events $Z_1, Z_2, \ldots$;

$q_1 = 1 - p_1, q_2 = 1 - p_2, \ldots$.

The arithmetical function for addition of events is as follows:

$$P = P_1 + P_2 + P_3 + \ldots + P_j + \ldots + P_n, \tag{3.11}$$

where $P_1, P_2, \ldots, P_j, \ldots, P_n$ are weights of factors $Z_1, Z_2, \ldots, Z_n$.

The value P of the probabilistic polynomial (3.10) always belongs to the interval [0, 1] at any values of probabilities of initiating events $0 \leq P_j \leq 1$; $j = 1, 2, \ldots, n$.

If there is one sign-event ($n = 1$), the probability of the final event P in logical addition (3.10) will linearly depend on the probability of this sign-event P_1 (Fig. 3.8). If there are two initiating sign-events ($n = 2$) in the logical addition (3.10), the probability of the final event P will have S-type dependence on probabilities of sign-events P_j, $j = 1, 2$ (which are given with identical values). If there are three and more sign-events, the probability of the

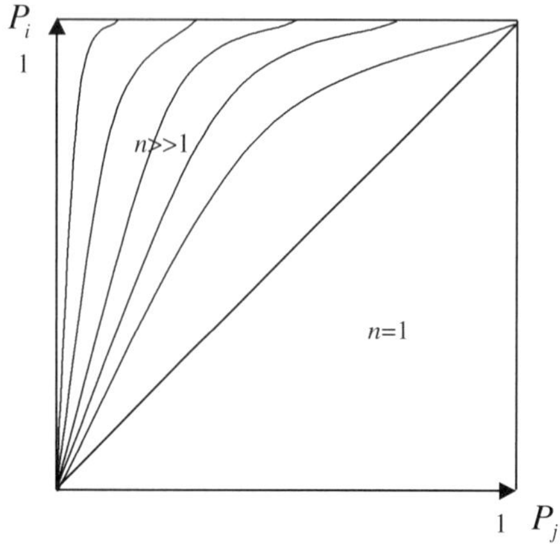

Fig. 3.8. Risk in function of number and probabilities of initiating events

final event P also will have S-type dependence on probabilities of sign-events P_j, $j = 1, 2, \ldots$ (which are given with identical values, too). The steepness of S-curve will increase with increasing n.

The probability of the final event in the logical addition (3.10) depends both on the number of sign-events and on their probabilities. The saturation of probabilities ($P = 1$) also depends on these factors. We note that only low probabilities of initiating events (sign-events) provide the small total risk ($P = 0.02 \div 0.04$ for $P_i = 0.001$).

Comparison results of logical and arithmetical addition of probabilities of sign-events are shown in Fig. 3.9 for the number of sign-events 41, 20, 5, 1.

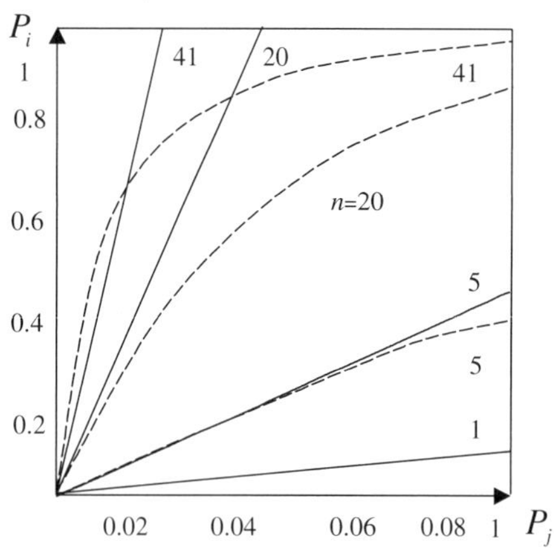

Fig. 3.9. Risk in function of number and probabilities of initiating events

For the big values of weights of signs P_j, $j = 1, 2, \ldots, n$, and for the big number n, the final event probability, calculated as the arithmetical sum of probabilities, becomes absurdly large ($P > 1$). The arithmetical and logical sums are close only for small values of probabilities of initiating events and the small number of the events. Therefore, the technique based on arithmetical addition has satisfactory accuracy only for small number of signs $n = 1 \div 3$ and their small weights $P_j = 0.001 \div 0.0001$, $j = 1, 2, \ldots, n$.

Comparison of polynomials for arithmetical and logical addition (3.10) and (3.11) shows that the logical and probabilistic polynomial (3.10) has more complex structure and, consequently, better possibilities for the adequate description of the final event risk.

It is also notable that the polynomial (3.10) can replace with success a neural network (NN) with arithmetical addition of edge weights. The logical function (3.9) for the probabilistic polynomial (3.10) also has quite clear sense in comparison with NN. Formulas on basis of NN are deprived of the physical and logical sense.

4

Risk Management at Debugging Tests

Debugging and operational tests is a search of mistakes in the project.

Robert Stevens

In this chapter, we shall discuss results of analysis of debugging processes of some complex objects. The meaning of debugging tests is to find and to remove errors in a project. We shall state principles and schemes of the debugging management and describe procedures of debugging technology. Examples of failure scenarios and building structural, logic, and probabilistic models of non-success risk of debugging tests of machines are given. We shall propose two quantitative criteria for the management of debugging process: the coefficient of debugging difficulty and the non-success risk of debugging tests. An example of development of a program for debugging tests is given. The obtained results can be used for management of debugging test of complex engineering, technological, and organizational systems.

4.1 Definition of debugging tests

Bad quality of the debugging technology of complex objects results in large losses of time and means in the debugging and risk of failures and accidents in operation. Though debugging always was a part of any technology, there is a small number of publications on the theory of debugging. Mainly it is explained by difficulty in formalization of debugging process. It is necessary to note works [3, 7, 70] in which the questions of automation of debugging tests on stands were considered, and works [73, 74] in which questions of debugging tests in real conditions of operation with real personnel are stated.

E.D. Solojentsev, *Scenario Logic and Probabilistic Management of Risk in Business and Engineering*, DOI 10.1007/978-0-387-77946-1_4,

Unlike the mentioned works, we shall pay basic attention to questions of formalization of debugging and optimal control of debugging process on the basis of construction of scenarios of failures and structural, logic, and probabilistic models of debugging non-success risk.

Examples of complex and yet not solved problems in design and debugging are nuclear fusion, nanotechnologies, airships, free-piston machines, aeroplanes Sukhoi SuperJet-100, An-148-100, Tu-334-100; aircraft engines Rolls-Royce, Pratt & Whitney, SaM146, 117C, AL-31M1; surface and subsurface ships of the navy fleet of class of "Svetlyak," "Zubr," "Tuchkov bridge," "Kapitan Nikolaev," "Oden," "Akula," etc.

Complex objects, for example engines, are characterized by the following attributes [3, 7, 75, 76]: structure complexity and a large number of systems and mechanisms, complex thermodynamic, gas, and hydrodynamic processes, high thermal and mechanical loading basic details, interrelation of processes and influence of their parameters on reliability of units and details; discrepancy of various properties (profitability, reliability, weight, cost) and consequently complexity of the selection of optimal solutions; variety of modes of operations and conditions of an environment; stochastic behavior caused by random factors, by evolution of the object, and by drift of parameters.

For a complex object, along with stages of design and manufacturing, of importance is the stage of debugging. The concept of "debugging" is identical to concepts of improvement and operational development. In debugging it is necessary to obtain parameters of processes, level of reliability, and quality of the object given by the technical specification and project. In debugging one manages a non-stationary stochastic object, and debugs, as a rule, individual samples. The basic purpose of operational debugging is searching errors in the project and making decisions on their elimination. The information on errors in the project is obtained at tests by measurement of parameters, failure-breakdowns, and failure-restrictions consisting in exceeding of admitted values by some parameters.

The object condition during debugging at the moment t is the random event. It is either failure-breakdown, or failure-restriction, or expected functioning. The object condition Y is defined by random conditions $Y_1, Y_2, \dots, Y_n$ of their subsystems and units. In their turn, the failure-events $Y_1, \dots, Y_n$ are caused by random initiating factor-events $Z_1, Z_2, \dots, Z_m$. The random events $Y, Y_1, Y_2, \dots, Y_n$ and $Z_1, Z_2, \dots, Z_m$ are logically connected. The graph-model of non-success risk of debugging (or the structural model) is constructed following the scenario of failures, written by experts being well aware of the object. Using the structural model, we build the logic non-success risk function the orthogonal logic non-success risk functions, and the probabilistic non-success risk polynomial [3]. Furthermore, we use the polynomial for the quantitative estimation of debugging non-success risk at development of the debugging test program and the debugging process management.

4.2 Analysis of debugging process

Loss at debugging. The absence of rigorously justified technology of debugging, non-optimal decisions, and the intuitive approach in conditions of the large uncertainty result in significant losses of means and time both in debugging and in operation of poor debugged objects. We can give some examples.

The development process of engines and compressors lasts for 4–8 years, thus, expenses of time and means for debugging take up to 80% of all the project cost. Two thirds of general charges on the space program "Apollo" was spent for systems of ground tests. The cost of the development program of a 5th generation destroyer is evaluated by experts at $10 billion. Half of costs is the development of an engine. Some projects of civilian objects have by high complexity and cost, but they cannot hope for large investment and have to compensate by knowledge and intellect of experts.

Normative documents. The standards on the product development do not provide "debugging" in the life cycle of products. The factory and inspection tests cannot replace debugging the product prototype, as the tests establish only conformity of parameters of the object to the technical project and documentation. Therefore the customer does not have the legal basis to require the realization of qualitative debugging tests, which cost is rather significant in comparison with the project cost. Now debugging is carried out on the basis of intuition of experts and normative documentations of companies.

Analysis of debugging processes. Let us present results of analysis of debugging processes performed for more than 20 different objects [3, 70].

Debugging stuffing-box of the 6th step of the compressor VSH-2,3/630. The complexity of the work was determined by the high pressure of very penetrative gas hydrogen (63 MPa), the small diameter of the plunger, the absence of its cooling and limited greasing stuffing-box. The debugging took two years.

Mobile compressor device MCD-30/120. The time of debugging tests was only 360 hours for 6 years, due to serious failures and breakdown. In the debugging, the autonomous debugging systems and mechanisms and special means were not used.

Ethylene compressor of super high pressure 4M40-250/2500. The complexity of the work was due to high pressure of the gas (250 MPa). The debugging proceeded for 4 years and finished successfully. In the debugging a unique stand was used.

Free-piston machines (FPM). Some organizations carried out the development of free-piston gas generators and compressors for more than 15 years. The work was stopped. In the debugging of FPM, the highest failure intensity took place in the debugging beginning, and the FPM collapsed after the first start.

Diesel locomotive engine DLE 26/26. Debugging successfully finished for 7 years. However, later large volume of work to increase resource and reliability

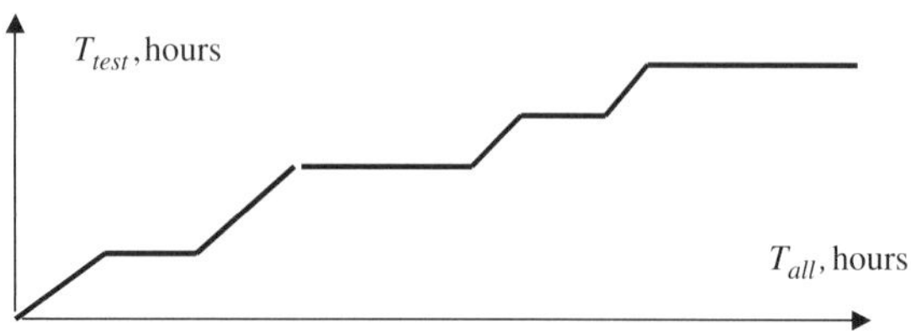

Fig. 4.1. The graph of useful usage of time at debugging complex object: horizontal lines are downtime

was carried out. The debugging locomotive diesel engine DLE 32/32 was performed on the basis of achievement in development of the diesel engine DLE 26/26, but yet the process passed with large expenses of means and time.

Debugging analysis of these and other machines (engines, compressor stations, ships, airplanes) shows that efficiency coefficient of debugging (Fig. 4.1), equaling relation of the time of testing T_{test} in hours to calendar time T_{all} in hours, is equals to $5 \div 12\%$ only:

$$\eta = T_{test}/T_{all}. \tag{4.1}$$

The rest of time is spent for restoration of the object after breakdown, preparation of measurements to figure out reasons of failures, manufacturing details with the new constructive and technological decisions, etc.

For the above described objects, statistical models of failure intensity in debugging were constructed. For this purpose, the time of continuous work of the object during the debugging was broken into intervals. The normalized cost of expenses on failure elimination was attributed to the interval middle. The failure data are best approximated by Waybull's distribution with the greater failure intensity in the beginning of the debugging process. For example, if the debugging time of FPM for a period equals to approximately 1000 hours, 40% failures occur for the first 30 hours, 70% happened for the first 100 hours, etc. In view of such failure intensity, it is difficult to make correct decision, and often the effective solution does not exist, because realization of the decision requires large expenses and long detention, and also development of optimum strategy of debugging.

4.3 Management of debugging process

Principles. Now we observe change of perception of debugging as an art to understanding that the debugging is organization. The approach "as it happens" is replaced by the approach with well-defined plan. It means that the developer should proceed not from optimistic expectations of debugging tests being easy and short in time, but from knowledge that a large number of failures will happen caused by errors of designing and manufacturing. Therefore, it is necessary to plan voluble resources for development of methods and means of debugging.

In organization of debugging process, the following principles [70, 73] are used:

- the physical approach to search for restriction-failures by measurement of parameters and their comparison with design and admitted values;
- decision-making for elimination of failures in mathematical models trained on measurement data;
- forecasting failures with the help of the trained mathematical models;
- consecutive complication of debugging stages to avoid destruction-failures at the first stages of debugging and to provide time to measure parameters;
- forecasting and planning debugging at the design stage;
- replacement of long-term tests to destruction-failures by the program of the short tests, intended for revealing restriction-failures;
- making the first test in the regime, where the probability of object destruction is small;
- maintenance to provide massive revealing restriction-failures and decision-making for their elimination;
- maintenance of continuous debugging by automation of search of restriction-failures, decision-making, and forecasting failures;
- performing autonomous debugging systems and details on special stands.

Management. The process of debugging object tests is interpreted as a process of management of complex object [75, 78] with movement from the initial condition to the given final one following a chosen trajectory and with correction in the case of deviation from the trajectory (Fig. 4.2). Based on this interpretation of the process of debugging tests, the following notations are introduced:

$H(H_1, H_2, \ldots)$ are stages of tests;

$Y(Y_1, Y_2, \ldots)$ are controlled parameters;

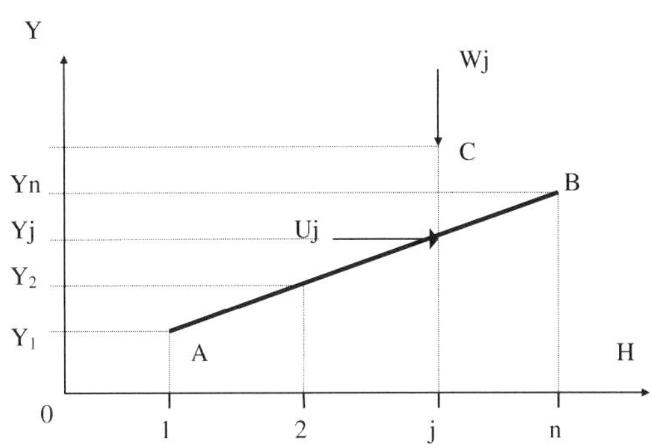

Fig. 4.2. Scheme of management of debugging process: Y are controlled parameters, U are control actions, W are corrective actions

$U(U_1, U_2, \ldots)$ are managing influences for test stages;
$W(W_1, W_2, \ldots)$ are adjusting influences for test stages.
The named parameters are vectors.

The controlled parameters Y are measured or observed parameters, which inform us on serviceability of the object. In the debugging, the object moves from the initial condition A to the final condition B on the program trajectory AB. The first stage (from the point A) is chosen to not destroy the object, the last stage (to the point B) is carried out on the nominal regime or regime with the maximal loading. The object is moved from the initial condition to the final through some number of discrete steps.

The object debugging complexity is generally characterized by all or by only chosen parameters of processes. The ordered set of the final number of parameters is represented by the vector Y. During debugging, this vector describes the trajectory in multi-dimensional space as a function of time. Because each parameter has the upper and lower admitted values at each debugging stage, the trajectory passes inside a plait in the multi-dimensional space. The debugging problem is to keep the trajectory inside the limits and to ensure that at the debugging end the plait would be strapped into a point.

The vector Y may have large dimension, and this is extremely inconvenient for debugging management. Bellow we shall apply the operation of convolution and use the scalar value as the debugging criteria. We shall introduce two criteria of debugging:

(1) The parameter of difficulty of debugging K_{diff};
(2) The non-success risk of debugging P_{risk}.

At development of the debugging tests program, we should be prepared for troubles, by providing W-corrections, which represent variants of constructive and technological decisions. Besides, at development of the debugging program, the values Y, W, U are determined for each stage H. Vectors Y, W, U have large dimension and dealing with them may require excessive expenses. For optimal distribution of the limited allocated resources for debugging, it is necessary to choose some components of these vectors. For this purpose, it is necessary to know component costs and possible damages due to their absence.

The proposed debugging diagram and basic concepts can be used for debugging any complex machines and objects and complex technical, technological, and organizational systems. As measured and observed parameters, we can use some parameters, and managing and adjusting influences, such as financial resources and actions.

4.4 Technology of debugging tests

The object debugging technology is presented by the logically closed sequence of procedures of the knowledge diagram. The following logical procedures [3, 67, 70] are used:

- forecasting failures,
- modeling expenses and damages,
- planning the test program,
- realization of tests and decision-making on correction,
- passing protocol and closer definition of models.

Forecasting. At forecasting stage, we make the lists of controlled parameters Y, managing influences U, adjusting influences W, and stages H of debugging tests. Based on the object parameters and restrictions of specifications and opinions of experts, the possible failures and damages at stages of debugging tests and dangerous values of the controlled parameters $Y_1, \ldots, Y_k$ are defined.

Modeling. At modeling we determine the ratio of expenses to damages for each component of vectors Y, W, U, and H. Starting from the values of the ratios, we distribute funds allocated for debugging tests. It is necessary to take into account not only material damages, but also losses of time in the case of destruction-failures and restriction-failures. After solving the problem of optimal distribution of resources, we determine components of vectors Y, W, U, H that will be used. Thus, we find the structure of the debugging test program.

Planning. In planning at stages H, we determine numerical values for the controlled parameters Y and managing ones U and for the adjusting W influences usually, for eliminating possible destruction-failures and restriction-failures. Three to five values for each adjusting and managing influence are provided.

Testing. At debugging tests, we apply the adjusting influences W, if destruction-failures or restriction-failures appear. The records in the testing protocol with indication of losses of time and expenses for elimination of failures are made.

Processing test protocols. At processing the test protocols, comparing results of forecasting, modeling, planning debugging, and real tests, we specify the following knowledge by the identification methods: the models of the object and its elements; the models for estimation of expenses and damages; the knowledge on parameters Y, U, W, and H (if new components and rules for decision-making were introduced).

4.5 Non-success risk scenarios of debugging

Object of debugging. Let us consider scenarios of failures at debugging tests of free-piston machines. The free-piston generator of gas (Fig. 4.3, b) produces the gas for a gas turbine. The compressed air of the buffer carries out the reverse motion of pistons.

From compressors, the air arrives at the receiver and further at the diesel engine, where it is heated up and arrives at the turbine transmitting the

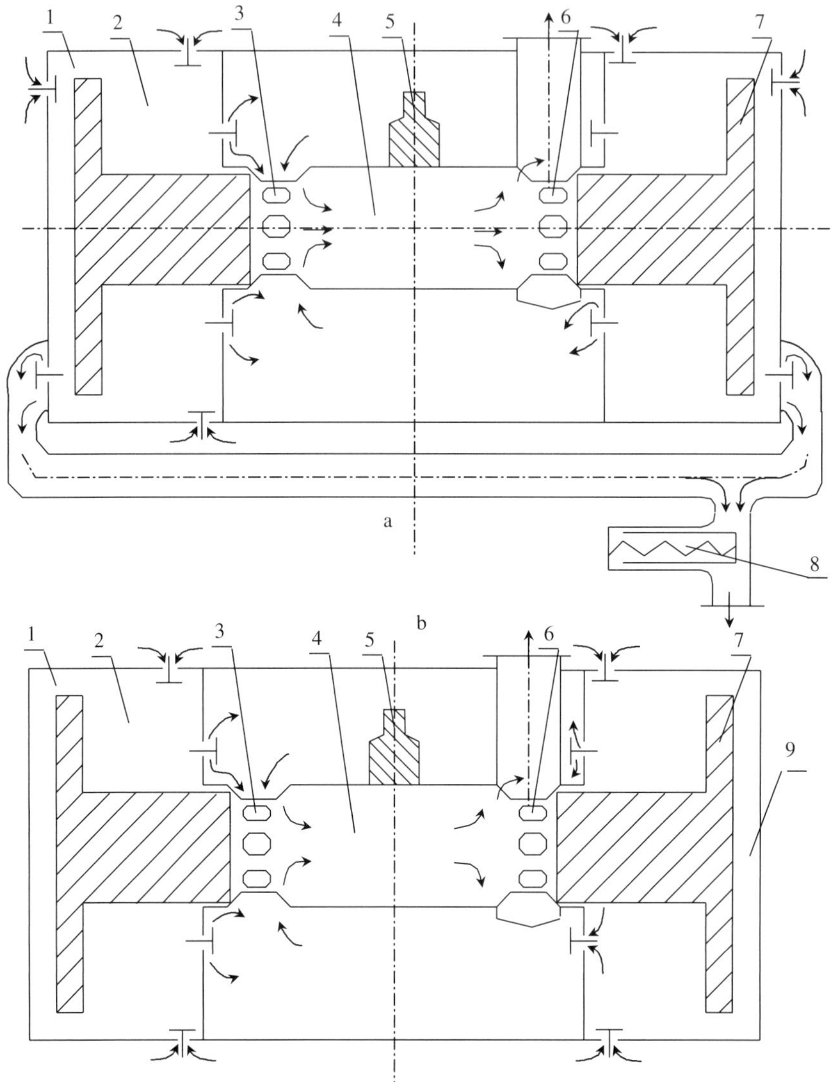

Fig. 4.3. The principal schemes of free-plunger compressor (a) and free-plunger gas-generator (b): 1 - plunger; 2 - compressor; 3 - windows; 4 - diesel; 5 - sprayer; 6 - windows; 7 - plunger block; 8 - pressure keeping valve; 9 - buffer

power to a consumer. The free-piston compressor (Fig. 4.3, a) produces the compressed air. The compressor can have one, two, or four stages. Distinction of FPM from crank engines and compressors is the connection of sizes of engines, compressors, and other cavities and their processes with dynamics of movement of pistons in the absence of fixing position of the outer (o.p) and the internal (i.p.) dead points.

Non-success scenario of debugging. The feature of debugging process of FPM is the impossibility of opportunity of separate debugging processes in the engine and in other cavities, constructive decisions, and control systems. It is also not yet known whether one can obtain the necessary parameters of processes in cavities.

Complexity of debugging is also explained by the fact that the object elements (piston and cylinder of engine, etc.) are not debugged up yet, as well as the system of their cooling. Most first tests can result in breakages of these elements. On the other hand, debugging these elements and systems of their cooling should be carried out later with satisfactory process in the engine in regime closely approximating the nominal one.

At debugging there is multi-dimensional space of dependent parameters: managing and adjusting influences, geometrical and working parameters, deviations from the norm of cavities air-tightness, coaxiality of cylinders and guide of synchronizing mechanisms; insufficient check out of fuel equipment, automatic valves of compressors, details of cylinders and pistons, lack of run-in of the mechanism of movement, labyrinth of packing valves; technological and constructive defect of details and units. Under action of these random factors, it is difficult to find the true reason of breakage and to make the correct decision for elimination of defects.

Because of features of piston movement, at the debugging beginning the compressor valves may have gas-dynamic losses, considerably distinguished from expected ones, or the limited resource. In view of this reason, the fuel equipment, in spite of testing on the special stand, may have characteristics that essentially differ from expected ones.

At debugging, random factors essentially influence organization of working processes in the object. These factors are the following: non-coaxiality of cylinders of the engine, compressors and directing rack-cutting the synchronizing mechanism; non-air-tightness of cavities, etc. The numerical characteristics of these factors are random variables.

At debugging, the fuel equipment condition is also a random variable. When it is installed, the air may throw into the system. By fastening the fuel injector, it is possible to cause deformation of its body, which will result in the jam of plunger or the fuel injector needle at heating-up. Change of sprayer characteristics after some operating time is possible, too. For the combustion chamber the sprays are not designed yet even for the given i.p. and o.p. dead points. The combustion chamber may be narrow or wide, and the spring of the fuel injector needle, determining fuel injection and the dispersion quality, may not correspond with the pressure of combustion P_z

During debugging, the friction force may exceed the design value because the movement group was not run-in. Therefore, the fuel feed will increase and the excess air coefficient for combustion will decrease. The gaseous exchange may be unsatisfactory because of small value of the piston pass.

Thus, the final event Y will take place (destruction of the object), if any one, or any two,..., or all from the following derivative events happen: Y_1 is the breakage of rings and crosspieces between flutes of rings in the engine piston, Y_2 is the burning of the piston and the sprays, Y_3 is the coking rings and sprays, Y_4 is the failure of the fuel equipment, T_5 are fins of cylinders and guides, Y_6 is the breakage of the compressor valves, Y_7 are other defects.

Non-success risk scenarios of system debugging. The failure of the complex object is caused by failures of its systems, mechanisms, units, and details, which in their turn are caused by some factors. Let us describe the action scenarios of these factors.

Z_1 *is the excess air coefficient for combustion* (α). The more is the value α, the more is the probability of normal realizing combustion process, even if the fuel equipment is badly adjusted and the air leakage is great between cavities. Naturally, the blow off air factor φ must not be less than 1.3–1.4 even for large values of the factor α.

Z_2 *is the cycle dynamic parameter (N)*. The less is the number of cycle N, the more is the reliability of compressor valves, the less are inertia force and deterioration, the better are conditions of work of the fuel equipment and synchronizing mechanism.

Z_3 *is the mechanical intensity parameter* (P_z). The less is the value of maximal combustion pressure P_z, the less is the probability of breakage of rings and straight arch between flutes in the piston, the less is the intensity of details of the fuel equipment.

Z_4 *is the thermal stress criterion* (P_t). The less is the value P_t, the less is the probability of failure of cylinders, pistons, and fuel equipment. The research of temperature fields and thermal stress of pistons and cylinders with purpose of their debugging become possible, when the processes in FPM have satisfactory condition. The value P_t is determined by the function of the average piston speed C_m, the engine cylinder diameter D, pressure and temperature of air P_s, T_s in the blowing receiver, the display pressure P_{id}, and fuel rate G_i [7, 70].

Z_5 *is the fuel equipment faultiness.* In FPM all the fuel is injected and burned up to i.p. Because of the large part of the lost piston stroke, slow movement of pistons in o.p., and change of the form of the combustion chamber, the problem of fuel mixing is difficult, despite the high values $\alpha = 2$–2.3. The faultiness of the fuel equipment results in the following defects: engine overheating, pistons and cylinder burning, sprayers firing, rings and sprayers gumming-up, engine ring elasticity loss, start-up absence, bad work of the stabilization system of places i.p. and o.p., sprayer needle lagging.

Z_6 *is the non-air-tightness.* In FPM the working gas is the air passing consistently from one cavity into another one. The air passes the compressor, the blowing receiver, the engine, the gas receiver, and the gas turbine. The cavities are isolated from each other with the help of valve boards, the stuffing-box.

The flow-over from one cavity into another are serious barriers for the debugging. At bad air-tightness the following defects are revealed: the increased gas temperature in the turbine, the overheating cylinders and pistons, the process deterioration in the engine because of the small value α, pistons and sprays burning; engine ring elasticity loss; the bad work of stabilization system of places i.p. and o.p. The loss of tightness occurs also because of breakages of valves. The standards of tightness are difficult to fix beforehand; they vary in process of running-in of valves, rings, and other details. We can estimate a degree of tightness by measuring speed of filling in cavities by air before start-up using indications of a manometer. For check of tightness of cavities, the air from extraneous sources may be used.

Z_7 is the non-coaxiality. When the engine cylinders, the compressors, the buffers, and directing the synchronizing mechanism are non-coaxial, the following failures can happen: the scuffing of these elements, the deterioration of stuffing-box; shift of the engine cylinder; shift of guides of the synchronizing mechanism with blocking channels for greasing; the difficulties in start-up because of the large forces of friction; the breakage of system of start-up and impacts of pistons to the valve board because of high pressure of the starting air. The defects elimination requires big expenses. If checking of coaxiality is not performed, then wrong decisions are selected. They consist in excessive increase of the durability of details and units or changing their design.

Z_8 is the i.p. and o.p checking. The checking limit positions of pistons is one of the basic problems at debugging FPM. It is carried out by means of the rotary barrel and the rod, on which end the pencil is mounted. The piston stroke change from cycle to cycle is recorded. It allows us to establish the i.p. distance, at which occurs stably the first flare; to determine pressure of starting air, to find out the reason of a stop of the machine after several cycles (for large i.p. distance this is a small distance of compression, for small o.p. distance this is the non-sufficient opening blow off windows); to establish the place of pistons after flare; to adjust value of fuel feed; to uncover scuffing in mechanisms of movement in the first 15–40 seconds after start (then the positions of limit points are unstable); to adjust the system of regulation to work with low flare pressure P_z. At the initial stage of debugging, the stabilization system is not adjusted because of considerable air leakage, losses on friction, and bad combustion. The stable work occurs at the large degrees of compression and the big pressures $P_z = 15$–21 MPa. Thus, the combustion chamber has the adverse form and the fuel flames get to the bottoms of the pistons. The sprayers and pistons are burned, the crosspieces between flutes in pistons and rings are broken. At the first stages of the debugging, it is forbidden to work with P_z higher than 11.5–12 MPa.

Z_9 – good design of the machine.

Z_{10} – productability of the machine.

Z_{11} – regularity of geometric of the machine (diameters and outer and internal dead points).

4.6 Construction of the criterion of debugging difficulty

Let us define the technical and economic weights of conditions $Y_j, j = 1, 2, \ldots, 7$ and factors $Z_i, i = 1, 2, \ldots, 11$ (Table 4.1). The total cost of the jth condition for debugging process is determined by the formula:

$$S_j = \sum_{t=1}^{N} (C_t \cdot T_t + Q_t), \tag{4.2}$$

where t is the number of a condition occurrence; T_t is the time of the break of debugging process; C_t is the cost of losses because of the break; Q_t is the cost of restoration.

Table 4.1. Weights of conditions and factors

Name of conditions, Y_j	Weight conditions, P_{ij}	Factors, Z_i			
		α	n	P_z	P_t
		Z_1	Z_2	Z_3	Z_4
Breakage of rings and crosspieces	0.11	0.05	0	0.35	0
Burning the piston and the sprays	0.25	0.2	0	0.05	0.02
Coking rings and sprays	0.08	0.25	0	0	0.25
Failure of the fuel equipment	0.18	0	0.2	0.27	0
Scuffing cylinders and guides	0.13	0	0.15	0	0.1
Breakage of the compressor automatic valves	0.10	0	0.8	0	0
Other defects	0.15	0	0	0	0
Weight of initiating factors, P_i	-	0.0755	0.1355	0.0996	0.083
Group weight of factors, P	-	$P_{1-4} = 0.3935$			

Factors, Z_i						
Fuel equipment	Air-tightness	Coaxi-ality	Checking i.p. & o.p.	Const-ruction	Tech-nology	Geometry FPM
Z_5	Z_6	Z_7	z_8	Z_9	Z_{10}	Z_{11}
0	0.05	0	0.35	0.1	0.05	0.05
0.15	0.095	0.025	0.05	0.11	0.05	0.07
0.175	0.175	0.025	0	0.05	0	0.075
0.2	0.1	0	0.075	0.08	0	0.075
0	0	0.7	0	0.05	0	0
0	0	0	0	0.1	0.1	0
0	0	0	0	0.7	0.3	0
0.0875	0.0623	0.09925	0.0645	0.1884	0.073	0.0425
$P_{5-8} = 0.313$				$P_{9-10} = 0.261$		$P_{11} = 0.042$

The weights of the event S_j are determined from the following expression:

$$P_j = S_j \Big/ \sum_1^7 S_j \ ; \quad \sum_1^7 P_j = 1. \tag{4.3}$$

The weight P_{ij} (Table 4.1) of the factor Z_i regarding the event Y_j are given by the method of expert estimation. The weight of the factor with regard to all conditions are determined by the formula:

$$P_i = \sum P_j \ P_{ji}; \quad \sum P_i = 1. \tag{4.4}$$

The factor-events Z_5, Z_6, Z_7, Z_8 are random events; their importance is significant: the group weight of these factor-events is equal to $P_{5-8} = 0.3135$ (Table 4.1). It is necessary to reduce essentially the group weight of these factor-events by development and realization of a complex of special actions on the stand. The factors Z_5, Z_6, Z_7, Z_8, thus, get to the category of strictly controlled ones.

We cannot actively influence the factor-events Z_9, Z_{10}, Z_{11}; they are found and eliminated during debugging. The factor-events Z_1, Z_2, Z_3, Z_4 have the high group weight $P_{1-4} = 0.3935$.

It is possible to weaken influence of these factors choosing the appropriate strategy of debugging. Namely, one needs to work in the most favorable regimes, in order to get the best conditions for organization of working process, to reduce the tension of details, consecutively to pass from one test stage to another one.

The criterion of the object debugging difficulty on the factor-events Z_1, Z_2, Z_3, Z_4 is equal to [7, 70, 75]:

$$K_{diff} = a_1 \overline{Z}_1^{-1} + a_2 \overline{Z}_2 + a_3 \overline{Z}_3 + a_4 \overline{Z}_4, \tag{4.5}$$

where: a_1, a_2, a_3, a_4 are the normalized weights of the factors Z_1, Z_2, Z_3, Z_4, taken from Table 4.1; $\overline{Z_1}^{-1}, \overline{Z_2}, \overline{Z_3}, \overline{Z_4}$ are values of the factors Z_1, Z_2, Z_3, Z_4 regarding nominal regime.

The criterion of debugging difficulty K_{diff} depends on managing and adjusting influences at stages of debugging, and it is determined (by calculated or measured factors $\overline{Z_1}^{-1}, \overline{Z_2}, \overline{Z_3}, \overline{Z_4}$). The criterion (4.5) demonstrates the arithmetic addition of actions of the factor-events with their weights.

The reduction of the thermal and dynamic intensity of FPM is achieved by reduction of the cyclic and increasing the coefficient of excess air for combustion in the engine [3, 7].

The reduction of cyclicity is achieved by increasing the piston weight, reduction of compression degrees in the engine, buffer, and compressor. Thus, the thermal intensity of the engine, the inertial forces in the moving mechanism and the fuel pump are brought down and the resource of the compressor valves is extended. In order to increase α, it is necessary to reduce the dead space of the compressor or the pressure of compression, to cool the air in the receiver, to add the air to the blowing receiver from an auxiliary source.

4.7 Construction of the logic and probabilistic model of debugging non-success risk

Scenarios of an object's failures, systems, units, mechanisms, and details allow us to construct the structural model or graph-model of debugging non-success risk. An example is shown in Fig. 4.4; the random events correspond with the object condition Y, with the element conditions $Y_1, Y_2, Y_3, Y_4, Y_5, Y_6, Y_7$ and with the factors $Z_1, Z_2, \ldots, Z_{11}$. We designate these events as logic variables using the same letters. The final event Y (failure) occurs, if any one, or any two, ..., or all events $Y_1, Y_2, Y_3, Y_4, Y_5, Y_6, Y_7$ occur. In their turn, these derivative events are caused by the factors or initiating events $Z_1, Z_2, \ldots, Z_{11}$. Let us construct the graph-model of the debugging non-success risk (Fig. 4.4). The risk graph-model has logic connections OR, each of initiating events $Z_1, Z_2, \ldots, Z_{11}$ causes some derivative events $Y_1, Y_2, \ldots, Y_7$. The non-success risk graph-model may be more complex and had logic connections AND, OR, NOT and cycles [2].

Using the risk graph-model, we write down the non-success risk logic model of debugging non-success risk

$$Y = Y_1(Z_1, \ldots, Z_{11}) \vee Y_2(Z_1, \ldots, Z_{11}) \vee \ldots \vee Y_7(Z_1, \ldots, Z_{11}). \qquad (4.6)$$

This risk logic model may be written down in the orthogonal form [2, 3]

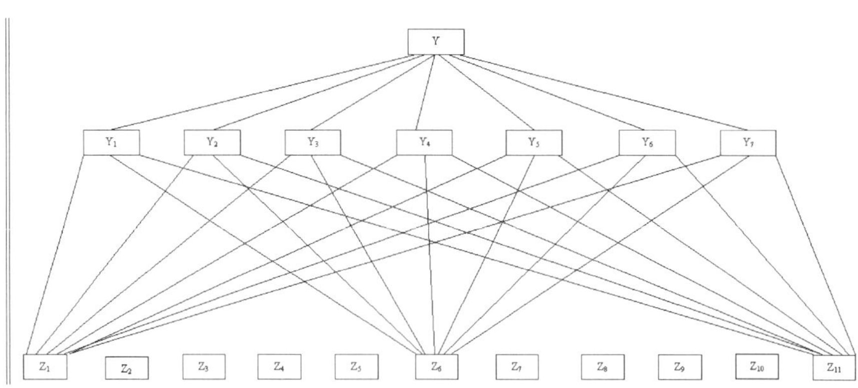

Fig. 4.4. Graph-model of the failure risk of the complex debugging object: free-plunger compressor

$$Y = Y_1 \vee Y_2\overline{Y_1} \vee Y_3\overline{Y_2Y_1} \vee \ldots, \tag{4.7}$$

and as the non-success risk probabilistic model of debugging

$$P_{risk} = P\{Y = 1\} = p_1 + p_2(1 - p_1) + p_3(1 - p_2)(1 - p_1) + \ldots \tag{4.8}$$

where $p_1, p_2, \ldots, p_7$ are probabilities of events $Y_1, Y_2, \ldots, Y_7$.

In order to use (4.8) for quantitative estimations of the debugging non-success risk at development of the debugging program and object debugging management, it is necessary to know probabilities of the initiating events $Z_1, Z_2, \ldots, Z_{11}$ for derivative events $Y_1, Y_2, \ldots, Y_7$. Let us consider approaches and methods for determination of the named probabilities using the statistical data from the protocols of previous debugging tests of similar objects.

Let us also use as the debugging criterion the non-success risk of debugging (4.8), constructed with the risk structural model (Fig. 4.4). This criterion demonstrates the logic addition of events.

We recall that because of using estimation by experts and in view of formula (4.5) features, weights in Table 4.1 have the following properties: the weights P_j, $j = 1, 2, \ldots, 7$, in the sum are equal to 1; the weights P_{ij}, $j = 1, 2, \ldots, 7$; $i = 1, 2, \ldots, 11$, in the sum are equal to 1 for any $j = 1, 2, \ldots, 7$; the weights P_i, $i = 1, \ldots, 11$ in the sum are also equal to 1.

Now, for risk calculation, as probability of non-success of debugging, we abandon the artificial conditions of normalization of the weights P_j, P_{ij}, P_i in Table 4.1 and pass from these weights to the corresponding probabilities p_j, p_{ij}, p_i.

We shall take into account only the factor-events Z_1, Z_2, Z_3, Z_4 because the factor-events Z_5, Z_6, Z_7, Z_8 are considered as strictly controlled ones and we cannot actively influence the factor-events Z_9, Z_{10}, Z_{11} — they are to be found and eliminated during debugging.

Earlier it was already pointed out that efficiency coefficient of the debugging process (4.1) is equal to $\eta = 0.075$. Therefore, we accept, for definiteness of statement, that the non-success risk of debugging is equal to $P_{risk} = P\{Y = 1\} = 1 - \eta = 0.925$ under action of all factor-events $Z_1, Z_2, \ldots, Z_{11}$. If we take into account only the controlled factor-events Z_1, Z_2, Z_3, Z_4, then the non-success risk of debugging is equal or proportional to their group weight $P_{1-4} = 0.3995$ in Table 4.1. Thus, the non-success risk of debugging only from action of the factor-events Z_1, Z_2, Z_3, Z_4 is equal to $p_{1-4} = P_{risk} \cdot B_{1-4} = 0.354$.

Let us construct Table 4.2 with the probabilities of condition-events $Y_1, Y_2, Y_3, Y_4, Y_5, Y_6$, factor-events Z_1, Z_2, Z_3, Z_4, and the factor-events with regard to the condition-events. At that, we do not consider the condition Y_7, which means other defects, as it does not depend on the factors Z_1, Z_2, Z_3, Z_4. The probabilities of the condition-events $p_1 \div p_6$ are calculated proportionally to their weights $P_1 \div P_6$ and taking into account the following expressions:

Table 4.2. Probabilities of condition-events and factor-events (the last stage of debugging, P_{risk}=0.354)

Name of conditions events, Y_j	Probabilities of event-conditions, P_j	Factor-events			
		α	N	P_z	P_t
		Z_1	Z_2	Z_3	Z_4
Breakage of rings and crosspieces in the piston, Y_1	0.0542	0.0068	0	0.0476	0
Burning the piston, cylinder and sprays, Y_2	0.1232	0.094	0	0.0235	0.0094
Coking rings and sprays, Y_3	0.0394	0.0197	0	0	0.0197
Failure of the fuel equipment, Y_4	0.0887	0	0.0386	0.0521	0
Scuffing cylinders and guides, Y_5	0.0641	0	0.0391	0	0.026
Breakage of the compressor automatic valves, Y_6	0.0493	0	0.0493	0	0

$$Y = Y_1 \vee Y_2 \vee Y_3 \vee Y_4 \vee Y_5 \vee Y_6; \tag{4.9}$$

$$P_{risk} = p_1 + p_2 \cdot (1 - p_1) + p_3 \cdot (1 - p_1)(1 - p_2) + \ldots = 0.354. \tag{4.10}$$

Probabilities of the factor-events p_{ij}, $j = 1, 2, \ldots, 6$; $i = 1, 2, \ldots, 4$ are calculated proportionally to their weights P_{ij}, multiplied by the coefficients for every condition j, to get values of probabilities, calculated from (4.10). At that we use the risk LP-models for every condition-event:

$$Y_j = Z_1 \vee Z_2 \vee Z_3 \vee Z_4;$$

$$p_j = p_{z1} + p_{z2}(1 - p_{z1} + pz3(1 - p_{z2})(1 - p_{z1}) + \ldots, j = 1, 2, \ldots 6.$$

The non-success at first and other stages of debugging is calculated as follows

$$P_{risk} = P\{Y = 1\} = p_1 + p_2(1 - p_1) + \ldots +$$
$$+ p_6(1 - p_1)(1 - p_2)(1 - p_3)(1 - p_4)(1 - p_5),$$

where on debugging stages probabilities p_{ij} are corrected, as for (4.5), corresponding values Z_1, Z_2, Z_3, Z_4 relative to the nominal regime (Table 4.2).

4.8 Example of development of the debugging program

The debugging stages of the free piston gas generator were carried out at the gas pressure in the turbine $P_g = 0.415$ MPa. Their complexity was estimated by both the debugging difficulty criterion (4.5) and the debugging non-success risk criterion (4.8). As the measurable managing influences U were chosen the compression degree in the engine E_d, the piston stroke S_{max}, the linear value of the dead space of the compressor C_k, the cooling blow off air $\triangle T_{cool}$. These influences are done respectively: by changing the initial pressure in the buffer and fuel feed cycle, by changing the dead space of compressors, and cooling the blow off air.

The managing influences have the upper and lower limits of change, defined by the condition of stability of work or practical possibility of influence realization. The test regimes were calculated by numerical modeling on a computer [7, 70] with determination of parameters of working processes in all cavities, dynamics of movement of pistons, and thermal loading cylinders and pistons.

The working processes in the cavities and pistons dynamics are described by a system of ordinary differential equations with variable coefficients, which is solved by numerical integration cycle-by-cycle up to convergence. The equations are obtained on the basis of the laws of conservation of energy, mass, and impulse and take into account processes of fuel combustion, heat exchange, intake and discharge of the gas and the air through the valves and the engine windows, and leakage through rings and stuffing-boxes. The total number of the equations in the system depends on the object design and can reach one hundred; the time of the calculation of one regime of object operation takes 1–2 hours on a PC of the Pentium type.

The control parameters for 16 regimes are changed in intervals: $E_d = 7 \div 10$; $_{max} = 0.15 \div 0.176$ m; $T_{cool} = 0 \div 150$ grad.K; $C_k = 0.01 \div 0.027$ m. Moreover the basic checking parameters are checked in intervals: $\alpha = 1.6 \div 3.1; \varphi = 1.5 \div 2.5$; $N = 1370 \div 1616$ cycles/min; $P_z = 8.0 - 13.8$ MPa; $P_t = 4.4 \div -8.65$; $P_{id} = 0.55 \div 0.99$ MPa; $T_g = 605 \div 820 grad.K$; $K_{diff} = 0.65 \div 1.0$.

Taking into account the computed values of the debugging difficulty criterion K_{diff} and the expedient sequence of exception of constructive changes at the debugging stages, connected with cooling the blow off air and increase of the dead space of the compressor, the debugging stages are obtained and they are given in Table 4.3. The last n-th stage of debugging is the nominal mode. Ten debugging stages are chosen; they are placed with gradual increase of the debugging difficulty criterion K_{diff}. The calculated values of the debugging non-success risk criterion P_{risk} are given in the same table in the last column.

The management of debugging process by criteria K_{diff} and P_{risk} allows us almost to halve the complexity of the first debugging stage in comparison with the last debugging stage. The ranges of change of the logic debugging non-success risk criterion P_{risk} is $0.235 \div 0.354$. Naturally, it is appreciably lower than ranges of change of the arithmetic debugging difficulty criterion $K_{diff} = 0.488 \div 1.0$.

The risk LP-model and the risk of the debugging process as a whole from 10 stages:

Table 4.3. Stages of debugging tests

Stages debugging	Values of controlled actions				Criterion of difficulty debugging K_{diff}	Criterion of non-success risk debugging P_{risk}
	E_d	S_{max}, m	$\triangle T_{cool}$, grad K	C_k, m		
10	10	0.1765	0	0.0105	1.0	0.354
9	9	0.1765	0	0.0105	0.952	0.3410
8	8	0.1765	0	0.0105	0.881	0.3284
7	7	0.1765	0	0.0105	0.814	0.3152
6	7	0.164	0	0.0105	0.786	0.3027
5	7	0.15	0	0.0105	0.733	0.2893
4	7	0.15	0	0.0147	0.6925	0.2764
3	7	0.15	0	0.0273	0.5785	0.2632
2	7	0.15	50	0.0273	0.54	0.2501
1	7	0.15	150	0.0273	0.488	0.2352

$$Y = Y_1 \vee Y_2 \vee Z_3 \vee Y_{10}; \tag{4.11}$$

$$P_{risk} = P_1 + P_2(1 - P_1) + \ldots + P_{10}(1 - P_9)(1 - P_8) \cdot \ldots \cdot (1 - P_1) = 0.948, \tag{4.12}$$

where $P1, P2, \ldots, P_{10}$ are non-success risks of debugging on stages $1, 2, 3, \ldots 10$. By this means, the probability of failures at debugging equals to approximately 1.

4.9 Operating tests

Usually there are a lot of annoying surprises after starting operation of a complex system (CS), despite the fact that it was tested in a laboratory environment. The developers of systems and equipment become more and more aware of this fact. And they are inclined to the idea that the test programs of new complex systems should be more detailed and the tests should be performed in the actual operation environment. They wish that tests are carried out by staff whose qualification is the same as that of a real user [3,73].

Management by operating tests is implemented under the same scheme, which we described above for debugging tests. We also forecast failures and develop non-success scenarios, distribute resource for tests, and generate the test program. Therefore below we will only consider the problems that are specific for operating tests.

Forecasting non-success and scenarios. The debugging operating tests and estimation (OTE) include processes of prediction and planning tests [30, 73]. There are critical questions (CQs), for which answers can be found at operating tests.

In the past, operating tests started with preparation of detailed descriptions, in which each characteristic of the system was detailed and criteria were established, which the given system should meet. The final stage of this methodology was the formal test of the system for determination of its compliance to the required characteristics.

However, soon it became apparent that this type of test program is unsatisfactory. Products often corresponded with all characteristics at tests, at the same time not functioning as expected. A car, for example, could correspond with any of hundreds of detailed technical conditions, at that being not demanded in the market, because the buyers could find that it is inconvenient or complicated in control. On the contrary, some product could not comply with some characteristics, but it completely satisfied its functional destination.

Now the necessity of new type tests becomes evident; in these tests, a given system, including equipments and personnel, is tested in conditions to the maximum approximating real operating conditions. It allows us to define if the system corresponds with its destination.

For the first time, the necessity in OTE became obvious when it was necessary to deal with complex military systems.

The operating tests and estimation have four main purposes:

1. To define whether a system together with its operators, staff, and auxiliary equipment executes current tasks and corresponds with given purpose.
2. To develop methods and to find modes of optimal exploitation of a new system or new usage of old system for solution of new problems or appropriate matching of old system to new equipments.
3. To establish limitations, characteristics, and capabilities of a new system to make easier its being introduced in existing structure of management and to establish requirements to staff, material, and technical supply.
4. To obtain information that will help in research and development of new systems, by documenting solutions for functional improvement of functions, and by documenting all detected defects of system operation in the testing period.

Debugging many objects at one time is rather labor-consuming. For example, at design of cars in the USA and Japan, up to 20 thousand parameters are debugged and at that it is necessary to distribute optimally funds and time for debugging. The concept of CQ is an approach for forecasting possible errors in the project and for controlling uncertainties in functional abilities of the complex system. There are critical questions such as:

- How well does the complex system CS execute the tasks that it is intended to do?
- Can the CS be provided and supported in exploitation ?
- How sound is the criticism of CS opponents, doubting in its capabilities?

Elimination of the critical questions is essential for success of the CS. Many CS fail after start of operating because not all critical questions, connected with their functions, were realized and answered.

Methodology of critical questions. The critical questions are expressed in the form of questions to CS, which take into account argumentation and uncertainty of capabilities of operational efficiency, practicality, effect on environment, etc. It is always necessary to clear up why the CQs emerge and what they mean in fact.

When the CQs are posed, then defects of the CS can be uncovered and corrected, and it is possible to optimize modes of exploitation (to train operators, to organize manufacture of service tools, to organize material and technical supply).

The critical questions create the basis for structural analysis methods, establishing lists of necessary computational and experimental researches at design and necessary measurements at tests, and the base for detail planning design activities and tests.

The critical questions are developed in two stages:

- Definition of a circle of questions, which can be critical ones;
- Discussion of each of these questions for estimation of its real criticality.

Outcome of these stages is the CQ list, which demands obtaining extra information for answering the questions.

Sources of critical questions. The sources of critical questions are

- Retracing the CS activity when carrying out given operations, following the stream of information and CS functioning in the typical operation conditions. At each step it is asked: what requirements are made to the corresponding operation at this point?
- Discussion of the CS with the people who are well familiar with design of the system or usage of similar systems. There are common problems for all objects and their test programs in the given field, and these problems are known to professionals.
- Studding all documents corresponding with the project to tabulate CS functions, requirements to it, and its characteristics. Putting questions to the CS or critical remarks will allow us to consider all these questions and to collect facts for solution of discussed questions.

Not each question is the CQ for tests. To clear it up, it is necessary to answer the following questions:

- Whether this question is still unsolved? If it has already been solved finally by research or administration, then it is not critical.
- Whether this question is really critical? The best way to evaluate criticality of the question is to find out all possible answers, which can give tests of the system behavior as the whole. If differences in behavior of the CS are little for all the set of answers, this question is not critical.

- Whether tests can give an answer to this problem? The question can be important, but tests might not be able to answer on it and consequently it is not critical. For example, the question whether means will be available for operational service of the system in the time of its activity cannot be answered at tests.
- Whether tests are the best way of solution of the question? If it is possible to use analysis, simulation, modeling at the design stage, then this question should not be included in the list of the CQs for tests.

The critical questions can appear during the whole project stage. This results in correction of the test program. Thus, CQ definition is an evolutional process, which continues during all the time of OTE program. Tests and discussions bring up new CQs.

Usage of critical questions. If CQs are determined, it is necessary to create a test program with the purpose to collect information for answering each critical question. Each test can answer one or more critical questions. The more questions a test can answer, the higher is the efficiency of the test. The time of solution of CQs depends on the questions. Some CQs demand solution at the earliest stages of the creation process of the CS.

Test tools can be very diverse, including special test stands, multi-factor experiments, special tests in conditions of high noise and vibration, simulation, modeling, and different kinds of analysis. There are many ways for using these various means for obtaining desirable information as correct and reliable as possible [7, 75–78].

The CQ usually cannot be answered by a single measurement during tests. On the contrary, separate measurements give answers to different subquestions. All obtained answers need to be grouped together for the full solution of each CQ. Therefore, each CQ should be broken into a few in-depth questions, each of which in its turn can be broken into some subquestions again. Finally all questions should be sufficiently simple to answer them with the help of certain measurements.

Scenarios and critical questions. Using analysis results of CS non-successes in the past, we shall describe some precedents to not stating CQs timely.

1. A car corresponded with one hundred detailed technical specifications, but it did not have a big demand on the market, as the customers found it difficult to control. The tested and debugged prototype was heavier than a production sample and therefore had different controllability.
2. Carbine M-16 of the USA passed all tests and was put into production; but in combat conditions of Vietnam it appeared to be unreliable because:
   (a) it was used as a crutch during combing marshes and in jungles (hot weather, stuffiness, serpents; mosquito, heavy rucksacks, etc.),
   (b) it was greased by instructions for the old carbine,

(c) after coming out of the jungle, soldiers shot using not short, but long burst of fire as required by the instructions to show they worked well. In these conditions the carbine often jammed; after modification these defects were eliminated.

3. The airplane "Concorde" did not find wide application, as during solution of complicated problems, designers forgot noise standards in airports and ozone layer problems.
4. At design of free-piston machines, experts thought that pistons had only the reciprocal motion in cylinders. However, it was found that the pistons had also rotatory motion, which results in breaking the synchronizing mechanism.
5. The light aluminum alloy was used for the pistons of the engine. Bridges of piston-rings are guaranteed to resist the pressure less than $P_z \leq 8.0$ MPa. However, the pressure P_z is a random variable. In 1% of cycles, the combustion pressure exceeds $P_z > 9.0$ MPa, and pistons after short activity are disrupted.

Selection of conditions for testing. In the past, the main reason of failures of OTE was not taking into account effects of different operating conditions on system operation. Exotic conditions of weather region, forest cover, geography, temperature, illumination, etc., and combinations of the conditions were the cause of the unique effects on the system during its tests. The test results in one environmental condition are completely unsuitable for other conditions. Sometimes this circumstance is apparent. It is not astonishing if in the jeep, which passed tests satisfactorily at the environments of the continental zone of the USA, the oil freezes and it cannot run in temperature conditions of Alaska.

It is less apparent why the laboratory measurements of reliability is nearly two times larger, than ones the system will show in natural conditions, even when the environmental conditions do not reach extreme values. Even more elusive is the fact that a human body can be a severe environment; so severe that electronic stimulation of the heart, which has service life up to 5 years in laboratory conditions, results in only 18 months of life when it is implanted into a human body. These examples demonstrate that special factors of environment, which appreciably influences efficiency of work of the system, is not always easy to find out. Thus, at tests it is necessary to use a full set of conditions, in which the system should work after the start of exploitation.

If there are no limitations in the test, then the section of the document reg- arding the operation conditions should be simple; in this section it is offered to test the system at all conceivable combinations of operating conditions. Unfortunately, it is impracticable. Taking into account different available limitations, it is necessary to formulate an idea of test in operating conditions.

4.10 Evolutional tests (ETs)

The debugging and operating tests can use the idea of evolutional tests. It is similar to the scheme of management of complex objects (Fig. 4.2).

The complex system is often designed in the best way, when efforts are concentrated on solution of one main problem from the full list of problems. When this problem is solved, it is possible to begin solution of following problems until the system is able to solve all the problems.

This method has advantages. Already at the earliest stage of tests, the system is able to execute some functions. The old rule of system architects wisely states: "The system should work before it will be able to work well". *ET*s can be planned so that tests have evolutional characters and are carried out with parallel evolution development of the system. In this case, we have the extra advantage — during the process of creation of the system, we can take into account results of tests. In this case, improvements of the system are cheap. The realization of *ET*s in such a way demands extremely close cooperation between system developers and testers.

First of all, the schedule of *ET*s should be aimed at testing the main function of the system or a set of functions. Then the aim is development, evolution, and modification of the system. The new stage is realization of more detailed debugging tests, etc., until the full completion of works on the system is achieved.

4.11 Conclusions

1. The poor-quality technology of the debugging of a complex object results in the large losses of means and time at the debugging and opportunity of failures and accidents in operation.
2. The essence of debugging tests is defined as the search and elimination of errors in the object project.
3. The principles of debugging management, the scheme of debugging management as the complex object, and the procedures of debugging management technology are considered.
4. The examples of failure scenario and construction on their base of structural, logic, and probabilistic models of the debugging non-success risk are given.
5. The debugging difficulty criteria and the debugging non-success risk are suggested; the example of development of the program of debugging test is stated.
6. Operating tests should fulfill real conditions of operating and with using personnel who have qualification equatable with real users; test programs should make up on the base of methodology of "critical questions."

7. The fulfillment of evolutional tests calls for the close cooperation between system developers and investigators. They are conducted on the scheme of management of the complex object.
8. The suggested approach and models can be used for management of debugging or development of technical, economic, and organizational systems.
9. For complex objects, systems and technologies should construct non-success scenarios and non-success risk LP-models on statistical data protocols before debugging objects or like analogs.

5

Risk Management in Operation on Basis of Monitoring

Monitoring — an integral part of control systems of safety and risk in complex systems.
V. E. Prohorovich, O. V. Krasnov

At operation stage an estimation of complex system (CS) accident risk is made on the basis of scenarios by using results of monitoring. The monitoring allow us to estimate values of element deterioration, real loads and vibrations, operation features, readiness and conditions of safe operation. The quantitative estimation of accident risk allows us:

- to analyze the CS risk;
- to accept the reasoned decision on prolongation of safe operation and service life of CS;
- to develop proposals on maintenance of safe operation;
- to organize the process of personal training for safe operation;
- to plan acting in dangerous situations.

Monitoring is the essential part of safety system management of complex technical, economic, and organizational systems.

5.1 Destruction, wearing, and deterioration of equipment in operation

For construction and buildings, being in operation for a long time, the cause of accidents can be degradation of material properties, beyond-limit levels of stored damages, appearance of uncontrollable development of cracks, cavitation wearing, etc.

During long-time operation, CS elements are worn out and age. We observe corrosion, active degradation of equipment components, effects of corrosion-active substances, effects of such dangerous and harmful factors as higher and lower temperature, humidity, vibrations of different spectrum and amplitude,

E.D. Solojentsev, *Scenario Logic and Probabilistic Management of Risk in Business and Engineering*, DOI 10.1007/978-0-387-77946-1_5,

Table 5.1. Characteristics of using design resource of objects

Level of resource usage	Relative number of objects
More than 0.50	60%
Up to 0.75	20%
Up to 1.0	15%
More than 1.0	5%

etc. Combined simultaneous actions of the named factors results in accumulation of rust, appearance of cracks in elements of construction and in welds, breakdown of air-tightness of flange packing, reduction of insulation resistance of cable lines because of aging polymeric covers etc.

Potentially dangerous objects and manufactures, as a rule, have considerable usage of resource (Table 5.1). In the most crucial branches (power, petrochemical industry, gas chemical industry), the potentially dangerous objects have used resource at a level of 75–90% [39]. In the first years of the 21st century, the amount of potentially dangerous objects with the indicated levels of used resource will increase by approximately 10%. Thus, the required expenses for liquidation of consequences of extraordinary situations, modernization, renovation, and withdrawal of this equipment will result in the reduction of gross national product as much as 5–10%.

Today, many atomic power stations have high, above 65%, level of wearing of basic production equipment. Insufficient attention is paid to modernizing, repair, and preventive maintenance of equipment. By social causes, the industrial and technological discipline drops. In the chemical complex, wearing production equipment comes to more than 80%, and about half of long distance pipes have been in operation for more than 20 years. Repairing and replacement of worn-out equipment are much behind the needs.

5.2 Monitoring in engineering

In engineering, a typical situation is the conflict between reached values of operating time of complex technical systems and rigid requirements to quality of their operation during operation. First of all, it concerns the CS with high cost, unique ones, with a long time of building (nuclear reactors of power stations and submarines, launcher of rocket systems, power equipment, etc.). Therefore, it is necessary to create new operation-saving technologies [37, 79–83]. The monitoring, as information technology, is intended for evaluation of technical condition of the CS and its resource, for decision-making on prolonging resource, and maintaining safe CS operation with the prolonged resource.

The essence of the new approach to safe operation consists in a large scale application in industry of monitoring technical condition of the exploited CS in order to obtain timely information about their actual condition and to make decision about the order of their further exploitation. Thus, the safety of exploitation of CS is determined both by its technical condition and by readiness of personnel to supply successful and safe operation.

Monitoring CS exploitation is the process of systematic obtaining and initial processing information about conditions of CS elements, affecting factors of environment and the operational processes realized in the CS. Further processing information is used for numerical estimation of non-success risk of CS with prolonged service life.

The theoretic basis of CS risk estimation with the help of monitoring consists in the following. First, we write scenarios and the risk LP-models of accidents, which allow us to compute the risk if probabilities of initiating events are known. The properties of CS elements change with time as they wear out, age, and come to ruin. Monitoring fixes these changes. For prediction of change of element properties, equations of mechanics can be used.

At this stage we also build models connecting properties of elements and probability of their failure (changing properties of elements results in changing probabilities of their failures). These models can be physical-statistical, statistical, and expert ones. Examples of construction and usage of such models are given in [37, 83] for components of refueling systems of rocket launchers. Models are built for corrosion damages of pipelines, wear of movable components, aging polymer and rubber products, and errors of staff at localization of dangerous condition of the CS.

Thus, we can calculate accident risk for each CS condition by using monitoring results, and make a decision on the possibility of its exploitation, prolongation of resource, required repair and replacement of components, or on impossibility of CS exploitation because of high inadmissible risk.

5.3 Monitoring infrastructure of rocket launcher

The world experience of space activity testifies to the fact that problems of risk estimation and safe operation risk analysis of elements of the ground-based space infrastructure (GSI) are urgent. It is also necessary to create well-founded methods of achieving required risk level [37, 81–83]. Basic GSI elements are refueling stations, technical facilities, launcher, measuring stations, and other similar objects.

GSI is a complex system including a set of objects of higher danger. Thus, the majority of GSI elements is now exploited on the basis of decision on prolongation of resource and service life. The features of exploitation of GSI elements at the stage of prolonged service life consist of the following. First, the GSI exploitation occurs in conditions of degradation of components of the equipment, and it is difficult to meet the requirements on reliability and safety. Secondly, in a number of cases the GSI exploitation occurs in conditions of poor technical readiness of tools of safety supply. All this results in decrease of reliability and safety of GSI operation, which confirms the urgency of developments, directed to creation of methods of risk analysis, safety maintenance, and risk management of GSI exploitation. The methods should take into account the actual technical condition of equipment and functional condition of personnel.

Scenarios of accident appearance. For providing safe GSI operation the following problems are solved:

1. Selection of the risk parameter that allows us to estimate the safety level quantitatively;
2. Normalization of requirements to safe GSI operation in the selected parameter;
3. Development of models for calculation of the risk parameter value of safe GSI operation;
4. Development of technology for obtaining initial data for the models;
5. Development of methods for analysis and management of safe GSI operation.

For quantitative risk estimation of safe GSI operation, a vector parameter can be proposed. It is the vector of probabilities of appearance of possible accidents:

$$P_{<L>} = (p_1, p_2, \ldots, p_l, \ldots, p_L). \tag{5.1}$$

Application of such parameter enables us to formulate the requirements to safe GSI operation taking into account the level of possible damage by various accidents. The GSI application has as its final goal (the condition S_k) obtaining some useful effect C_p (for example, profit). At the same time, during operation of GSI, there always exists the risk of accident appearance and damage to people, environment, equipment of the GSI, and the space rocket. The condition S_p corresponds with the accident event and it is characterized by the probability p_n and the damage W_p. The graph of dangerous GSI conditions is shown in Fig. 5.1.

Following basic principles of the concept "the admitted risk" [2], it is possible to state that the risk of safe GSI operation can be considered as an admitted one in the case when the positive effect from GSI operation is more than the effect by accident, that is the inequality holds:

$$(1 - p_n) \cdot C_n > p_n \cdot W_n. \tag{5.2}$$

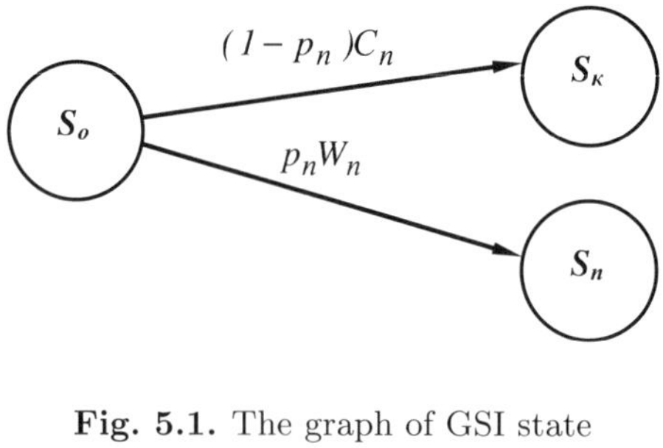

Fig. 5.1. The graph of GSI state

The inequality (5.2) reflects only the technical and economic aspect of maintenance of safe GSI operation and does not take into account the current condition of social relations or, speaking in other words, the level of "admitted risk" currently accepted in society. For elimination of this flow it seems appropriate to introduce the coefficient of admissible risk k_{ad}. Transformation of inequality (5.2) with the coefficient k_{ad} allows us to obtain the maximum admitted probability of appearance of accident in the GSI starting from possible damage and expected useful effect from proper GSI application:

$$p_l^d = \frac{k_{ad} C_n}{k_{ad} C_n + W_{n_l}}, \tag{5.3}$$

where W_{n_l} is the possible damage by appearance of l-type of accident in GSI.

The full list of possible accidents on the GSI is found by the method of morphological analysis. For example, for the space-rocket launcher "Proton" the full list of possible accidents during preparation and launch of the rocket includes 66 items.

The quantitative estimation of the risk of safe GSI operation is made by the binary scheme (if the criterion is realized, then the demanded level of safety is ensured) and consists in realization of criterion of suitability:

$$G : P_{<L>} \in (P^d_{<L>}), \tag{5.4}$$

where $P^d_{<L>}$ is the vector of acceptable values $(p_1^d, p_2^d, \ldots, p_L^d)$ of GSI accident appearance probabilities (calculated by (5.3)).

The criterion (5.4) formally means:

$$G : \{ p_1 \le p_1^d, \; \ldots \; p_L \le p_L^d. \tag{5.5}$$

The components of the parameter $P_{<L>}$ of safe GSI operation are probabilities $p_1, p_2, \ldots, p_L$ of appearance of accidents. They are calculated with the help of models of accident appearances in GSI. The basis of these models are so-called "scenarios of accident appearances." They are prepared on the basis of models of the GSI exploitation process and formally described by logical and probabilistic functions of accident appearances.

The essence of the approach for construction of models of accident appearances in GSI is the following. For the stage i of GSI operation, we determine all possible accidents, and for each of them we construct the scenario of accident appearances. On the basis of this scenario, a logic function Y_i of accident appearances is constructed. It permits, by using algorithms described in [2], to determine the probabilistic function of accident appearances $P_l = P\{Y_l = 1\}$ expressed in terms of probabilities of initiating events (initiating events and conditions) of the scenario of accident appearances.

The probabilities of initial events in scenarios of accident appearance are calculated as probabilities of their appearance at the considered stage of operation. In these models, the process of accident appearance is considered as

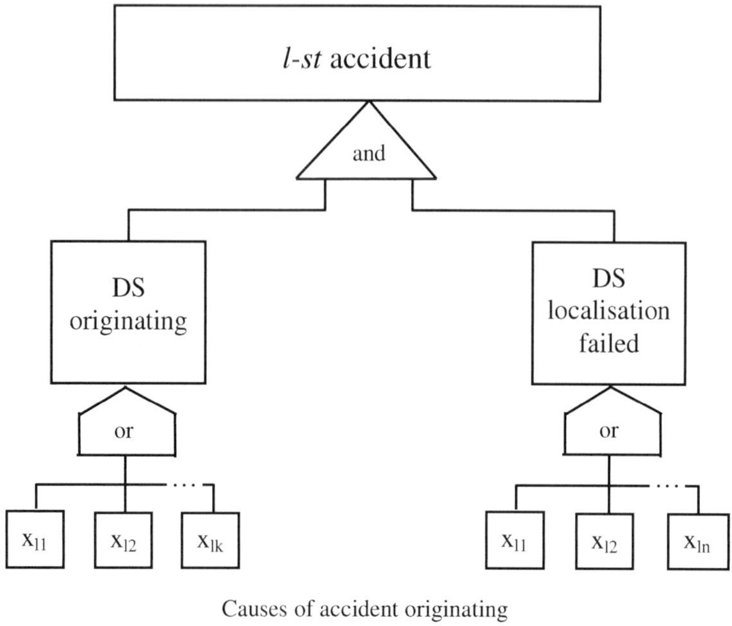

Fig. 5.2. The scenario of incident originating

consisting of two stages: appearance of the dangerous situation and its development as accident. Thus, we take into account the possibility of localization of the dangerous situation and prevention of damage. An example of the scenario of incident appearance is shown in Fig. 5.2.

The main feature of the scenarios of accident appearances is the presence of internal "horizontal" connections. This dictates necessity of application of algebra of logic for their formal description [2].

The causes of the accident l, possible at fulfillment of the considered operation stage, can be presented as events consisting in failure of an equipment element (or an operator error) during operation of GSI at the considered operation stage. In the formalized form, these events can be written as follows:

$$\hat{x}_{lk} = \langle \hat{u}_{lk} < \hat{z}_{lk} \rangle \,, \tag{5.6}$$

where u_{lk} is the conditional (only within the considered operation stage operating) mean time between failures of the element k or mean time between errors of the operator; z_{lk} is the required mean time between failures of element k within the considered operation stage for its completion.

For determination of probabilities of the events $\hat{x}_{lk}$, it is necessary to find distribution functions of random variables u_{lk} and z_{lk} described with the help of physical-statistical, statistical, and expert models (depending on the structure and quality of the data).

The application of the logic and probabilistic methods for description of accident appearance scenario essentially simplifies estimation and analysis of risk of safe GSI operation. The scheme of such analysis with application to one accident is described in detail in [2, 37]. The meaning and features of the risk analysis of safe operation with application to GSI and taking into account the selected parameter can be described as follows. First, a list of possible incidents of GSI is prepared. Further, the structural models of their appearance are designed, and the structural weights of each of the incident causes are determined [2, 37]. This allows us to form the ranged lists of causes for each accident. The latter reflect the parameters suitable for safety management. Besides, the LP-method allows us to determine the minimum cross-sections of the safe operation (or paths, which guarantee safety).

The possible accidents causes are divided into four groups: failures of GSI equipment elements, unrepaired errors of operators, non-success at localization of dangerous situations, and initiating events. For each cause group, the standard models and recommendations on their application are proposed within the considered problem. For failures of equipment elements of GSI and non-success at localization of dangerous situations, depending on structure and quality of initial data, one can use the physical-statistical, statistical, and expert models. Thus, for physical-statistical models the given data are the results of measuring main parameters, for the statistical models the data are the operation time until failure or the time of censoring elements, for expert models the data are the expert interval estimations of corresponding parameters. For modeling unrepaired eliminated errors of operators, it is proposed to use the known experimental models [2, 37]. The initiating events of incidents are proposed to model with the help of the function with values 0 or 1, that is the probability of appearance of initiating events can be equal to 0, or to 1, depending on the capability to manage them.

System of monitoring. The basic tool for maintenance of the demanded safe GSI operation is the monitoring parameters determining safe operation of systems during regular operation. Monitoring is understood as the process of regular obtaining and primary processing information about parameters of technical condition of the system, which change influences safe operation of the system.

For achievement of the purpose of monitoring within its realization, it is necessary to solve step by step the following problems:

1. To carry out the operation of safety analysis of GSI means with the purpose of definition of parameters that require control;
2. To determine needed structure and volume of information about the checked parameters of elements;
3. To develop methods of control for the observed parameters and parameter monitoring strategy, which include formation of the optimum schedule of estimation and analysis of safety by given criteria;

4. To conduct collecting statistical and expert information on parameters determining safe operation of GSI systems and their checking by methods of nondestructive control;
5. To estimate and analyze the current safe GSI operation on the basis of obtained information. To forecast safe operation parameter values in the planned span of time with taking into account results of checking statistical and expert information on parameters determining safe operation of technical systems;
6. To elaborate actions on maintenance of demanded safe operation of the considered system;
7. To update models of incident appearances in systems and the list of checked parameters and requirements of safe operation.

The system of parameters permitting one to estimate safety of operation of technical systems is introduced. The two-level system of parameters is proposed for quantitative estimation and analysis of safe operation of technical systems. Vector of probabilities of accident appearances $Q_{\langle L \rangle} = \{Q_1, \ldots, Q_L\}$ is introduced. Here the subindex designates the type of accident. The components of the vector $Q_1, \ldots, Q_L$ are parameters of the first level. Appearance probabilities of incident causes (which, generally, can be element failures of a technical system, personnel errors, non-successes at localization of dangerous situation, and initiating events) are denoted by $\{q_1^1, \ldots, q_\rho^1\}$, $\{q_1^2, \ldots, q_\eta^2\}, \ldots, \{q_1^L, \ldots, q_\xi^L\}$, and these are parameters of the second level. The probability values $Q_1, \ldots, Q_L$ of incident appearances at technical system operations will be determined by parameter values of this level. We shall now consider in detail problems solved at monitoring.

The first problem is the analysis of safe operation of technical systems with the purpose of determination of parameters that require checking. In solving this problem, we investigate the operation process and the structure of technical systems with the help of the morphological analysis method and determine the full list of possible accidents at operation.

Then the models of incident appearances are constructed on the basis of models of operating processes of technical systems. The basic of these models are scenarios of incident appearances, described formally by logical and probabilistic functions of incident appearances. It is obvious that the probabilistic function of accident appearances can be expressed in terms of probabilities of the incidents causes, which are components of the vector $Q_\Sigma = \{\{q_1^1, \ldots, q_\rho^1\}, \{q_1^2, \ldots, q_\eta^2\}, \ldots, \{q_1^L, \ldots, q_\xi^L\}\}$. After that, ranking causes with regard to the risk factor is carried out, and the structural weights of the causes are determined. We assume that all causes have the same probabilities. Then we find out the structural significance of accident causes possible at GSI operation. This is necessary to determinate influence of the causes to the safety of operation of the technical system.

The second problem of monitoring is determination of necessary amount of information on checked parameters, sufficient for their estimation, and development of instrumental base, necessary for obtaining this information. For this purpose, it is necessary to define the possibility to measure value of the checked parameter or to define a list of indirect diagnostic characteristics permitting one to determine its value. We use it for determining the structure of instrumental base needed for obtaining the demanded information.

The subsequent problem of monitoring parameters, which determines safety, is the development of methods of checking the observed parameters. The problem solution requires implementation of the following steps:

- Grouping elements by the type of checked parameters;
- Definition of sets of checking methods permitting one to control corresponding groups of elements;
- Development of methods for checking corresponding parameters for each group of controlled elements.

At realization of the first step, it is expedient to divide all sets of controlled elements of technical systems into groups with controlled parameters of the same type. For each of the element groups, it is possible to compose a list of methods of control and a list of devices for realization of these methods. Selection of a control method should be based on knowledge about conditions of an element's operation, its geometrical sizes, physical characteristics of an element's material, suitability of elements for control, and sensitivity of existing methods of control. Definition of parameter monitoring strategy consists in formation of optimum schedule of estimation and analysis of safety of operation of technical systems by the given criterion.

The quality of solution to the problem of collecting statistical and expert information and control by methods of nondestructive testing parameters determine the quality of input data needed for estimation of safety.

The problem of risk estimation and analysis of current safety of operation of technical systems based on monitoring results supposes step-by-step fulfillment of stages of the same algorithm.

The following problem of monitoring is the development of a plan of action on maintenance of demanded safety of operation of the technical system on the basis of results of estimation. Besides, in solving this problem we can correct models of accident appearances, the list of controlled parameters, and requirements to safe operation of GSI. At this, we update both the initiating data for models of accident appearances at the GSI operation and the model structure. The necessity of correction of the list of controlled parameters, determining safe operation of technical systems, is necessary in view of the changes of accident weights during monitoring. The correction of requirements to safety is made at each stage of monitoring.

The considered approach to the risk estimation and analysis of safe operation of elements of the ground-based space infrastructure enables us to estimate quantitatively the risk of safe operation with taking into account results of monitoring technical condition parameters of objects, to determine parameters that are the most effective for the risk management, and to substantiate methods of safety management (in the concept of “acceptable risk”).

The experience of usage of such system of monitoring for the space-rocket system “Proton” proves the high efficiency as the information technology for solving problems of estimation and forecasting technical condition and residual technical resources.

6

Risk Management of Dangerous Plant

The Bernoulli approach assumes that the risk does not give in to regulation or meaning is not adjusted. Columbus's approach assumes that the risk should be reduced up to admissible due to corresponding adjustment.

N. K. Pechenin

The quote above is from N. K. Pechenin's works [36]. The choice of passive or active risk management for each component of a dangerous plant depends on possible damage. The passive risk management is performed on the insurance basis, whence the active one is performed on the basis of regulation of restoration procedures of material resources and components of personnel reliability.

6.1 Difficult problems

In the theory and practice of risk estimation, there are some difficult problems. The problem of "the human factor" is most perceptible. The detailed analysis of reasons of failures and accidents in complex systems shows that often the technological and organizational documentation direct individuals to measures for prevention of failures, but the measures are not realized by personnel (Fig. 6.1). Most reasons of failures can be removed by paying more attention to personnel or by training by personnel.

Attempts to approach to the human reliability problem with the same criteria as for the engineering reliability problem reveal their inconsistency. The human failure probability can be determined precisely only for the specific person, social conditions, and short time period. Generalization of obtained data to different peoples, social conditions, and large time periods results in the growth of the result uncertainty.

E.D. Solojentsev, *Scenario Logic and Probabilistic Management of Risk in Business and Engineering*, DOI 10.1007/978-0-387-77946-1_6,

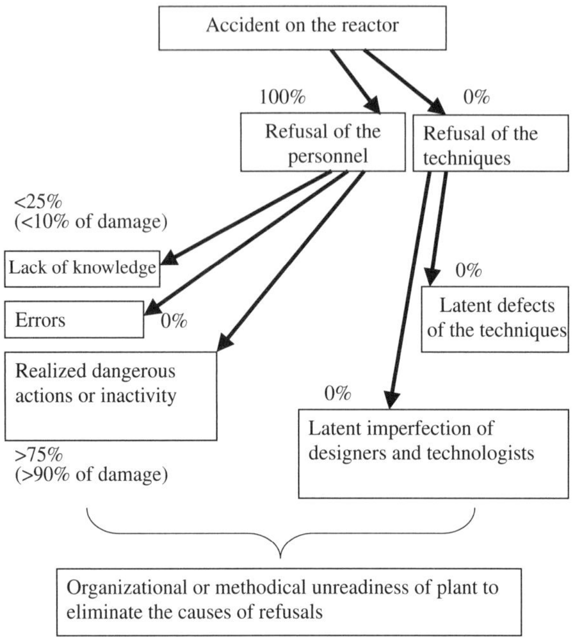

Fig. 6.1. Systematization of causes of failures of nuclear reactors

Among other problems, attention should be paid to principles underlying the bases of existing ways of failure probability estimation. In many respects they determine the lay efficiency of risk management: the syncretic reason of failure and extrapolation of probability density function outside area limits of values obtained in experiment.

The syncretic reason principle means characterizing the object by a uniform parameter of failure probability, when all or a part of failure reasons are not investigated, that is, the influence of the following factors on failure probability is not considered: by wearing each material object resource separately; by each reason external to the object; by each action of a man influencing object reliability.

The principle extrapolation consists in transferring a selected function describing failure probability density, outside the limits of values, obtained in experiment. This principle allows us to fix finite, distinct from zero, values of failure probability density at all points belonging to the range, for which there are not experimental data. The extrapolation principle is applied both in relation to the object and in relation to each of its material resources. The last plays the special role in practice of definition of object failure probability, because the operation time of the material resource is usually out of the range of values obtained in experiment. That is, for probability estimation of the failure, occurring as a result of wearing the material resource during operation, one uses values obtained by extrapolation of experimental data instead of the data itself.

Actually, the failure probability is set as a result of interpretation of experimental data by enough arbitrary chosen function of failure probability density in the field of experimental data and extrapolation of this function in the area of failures, not containing experimental data.

The hypothesis on existence of distinct from zero values of failure probability density at any operating time is not based on anything. The correctly described real situation with wearing material resource demonstrates the legitimacy of other hypothesis: before wearing material resource results in the object failure, there is a period of time when the resource wears out, but part of the resource, which is not worn, completely provides non-failure operation of the object.

6.2 Management of risk

Period of safe wearing resource. If the operation beginning the material resource has a volume, exceeding the one necessary for the object work, then the process of resource wearing passes two stages: at the first stage, the remainder, not worn out resource part completely provides the object trouble-free operation, and the failure probability, caused by wearing the given material resource, is indistinguishable from zero; at the second stage, the material resource is worn out to the extent such that the object failure probability due to wearing the given resource takes finite non-zero values. The moment of transition from the first stage to the second one is named the threshold time.

The material resource can be presented as a set of resources, each of which wears out in due course. For each resource there is some function $Rs(t, x_1, \ldots, x_m)$, representing dependence on time (t) and conditions of operation ($x_1, \ldots, x_m$). In the range $t_0 \div t_1$, the function is not defined (see Figs. 3.6 and 3.7). Actually, the failure frequency, as the result of wearing the given resource, is equal to zero in the range $t_0 \div t_1$. A single fact of failure due to wearing an unexplored internal resource $Rs \cdot (t, x_1, \ldots, x_m)$ makes necessary researche and the subsequent change of conditions of operation in such a way that the repeated occurrence of the given event becomes impossible in the range $t_0 \div t_1$.

Because on the left of the point (t) the failure probability is equal to an infinitesimal value, and on the right of it the probability has finite values, it is reasonable to suppose that in the vicinity of the point (t) there is the threshold time. The traditional approach to selection of failure probability distribution function does not assume the existence of the threshold time.

Let the moment (t) be threshold time. If the work with the material resource is correctly organized, that is, beforehand, before the threshold time, the material resource is regularly restored, it may be thought that it is possible to check the threshold time. The value of the safety factor in defining time between two subsequent procedures of restoration of the material resource is determined taking into account investigations on the material resource and values of possible damage from the failure, caused by wearing the given resource. Hence, the opportunity of the check of the threshold time directly depends on the opportunity of realization of timely and sufficient researches of the material resource.

It has already been mentioned briefly in Chapter 1 that *Columbus' approach* to risk management should be applied to processes, such that the losses from non-success considerably exceed expenses necessary for regulation of the process. It is a troublesome approach, but the expenses for its realization grow linearly depending on complexity and danger of the process, and the losses from non-success of complex and dangerous processes grow in the geometrical progression. At some complexity and danger of the process, *Columbus' approach* economically justifies itself.

We shall recall the main distinction of the two approaches to risk estimation.

The feature of Bernoulli's approach consists not in using the Law of Large Numbers by *Bernoulli* and not in using the mathematical probability theory on the basis of this law, but in refusal to regulate the process, in which non-success risk is estimated.

The feature of Columbus' approach consists in detailed regulation of those parts of the process in which failure results in catastrophic damage of all the process. The consecutive realization of this approach leads to decrease of failure occurrence risk on the adjustable part of the process to infinitesimal values.

The effective application of the economical mechanism of insurance of dangerous and expensive objects is possible only after transition from the risk insurance in the mode of *Bernoulli* to work in the mixed mode, when the catastrophic damage occurrence risk on dangerous parts is reduced to infinitesimal values with the help of *Columbus' approach.*

Bernoulli's (without regulation) estimations of failure probability, obtained by interpretation of the experiment data, are put into the basis of risk calculation with the help of the probability theory. Both the faults of engineering, its non-appropriate usage, infringement of technology, and erroneous actions of personnel, all these reasons of failure at *Bernoulli's* approach lose individuality and are transformed to an abstract parameter.

Columbus' approach to the risk estimation requires another operating with facts. On the basis of *Columbus'* interpretation of facts, the following principle lies: each material resource, each element, each component of process should be considered separately so that it could be regulated.

Risk systematization and classification of problems. If all the variety of reasons, raising risk of failure or accident of a dangerous object, is located on a straight line (Fig. 6.2), placing on the left-hand side the hidden defects of technical products, used on the dangerous object, and on the right-hand side the obvious destructive actions by a man, then it is possible to

Technics			*Human*
4	3	2	1
4- *Lack of know ledge of staff*	3 - *Staff faults*	2 - *diverse changed motivation of an act*	1 - *diversion , sabotage*

Fig. 6.2. Areas of method applicability of risk estimation

represent the applicability area of any risk estimation method as follows. The Shaded Area 1 in Fig. 6.2 includes the purposeful destructive actions. Area 2 includes failures due to realized infringements of service regulations, not directed on destruction, but causing failures and accidents. The general name of the reasons belonging to the Areas 1 and 2 is changed motivation of action. The Area 3 includes failures caused by mistakes of personnel (by random actions, not made after the time period or ineptly made actions, psychophysical or emotional unreadiness of the man to work). Area 4 includes failures caused by lack of professional knowledge both of workers and manufacturing organizers. The structure of risks is given in Fig. 6.3.

In the widespread concept "the human factor," the failure reasons related to human actions lose their individuality. The division of reliability of the man into three components is necessary for organization of effective work to guarantee the safety of dangerous manufactures. It is necessary to take into account separately the influence of people directly connected with manufacture, and indirectly, through engineering and engineering specifications. The people, directly influencing the safety, are the personnel of dangerous manufacture. The people, indirectly influencing the safety, are taken into account during reliability estimation through parameters of reliability of engineering and authenticity of documentation. The personnel of dangerous manufacture have complex structure: operative, repair, service, administrative, operational, and auxiliary personnel. Functionality of some sorts of the personnel overlay some types of personnel complement each other. One thing unites them: personnel directly influence safety of manufacture.

The given systematization of risks requires selection of four types of problems on the basis of risk estimation and maintenance of reliability of object work (Figs. 6.4 and 6.5).

1. Problems of the failure risk estimation of either materials or equipment or designs (MED) because of their hidden defects. If the failure of MED does

Risk of technical refusal because of a latent defect.	Risk of specialist's absence of knowledge, in the given moment.	Risk of under taking of an error by the person.	Risk of motivation change of the executor actions.

Fig. 6.3. Structure of the risk

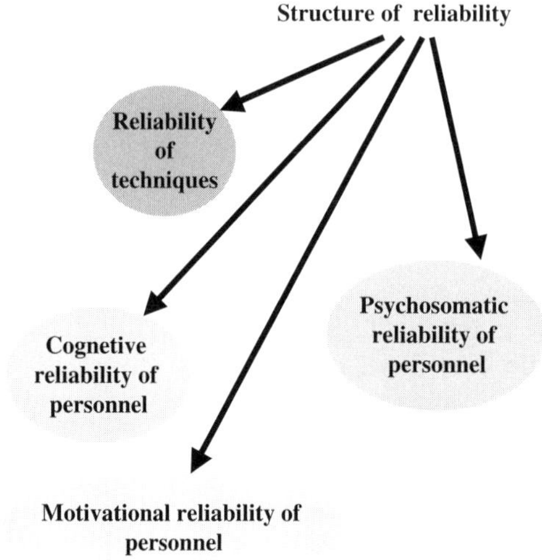

Fig. 6.4. Structure of reliability

not result in large damage, then *Bernoulli's* approach to risk estimation is sufficient. Considerable warehouse stock, systematic delivery of spare parts and equipment, and also risk insurance, proportional to damage, are capable to ensure normal functioning of any technical object. If the failure of the MED leads to significant damage, which appreciably exceeds cost of timely replacement of the misfunctioning material, equipment, or design, it is necessary to

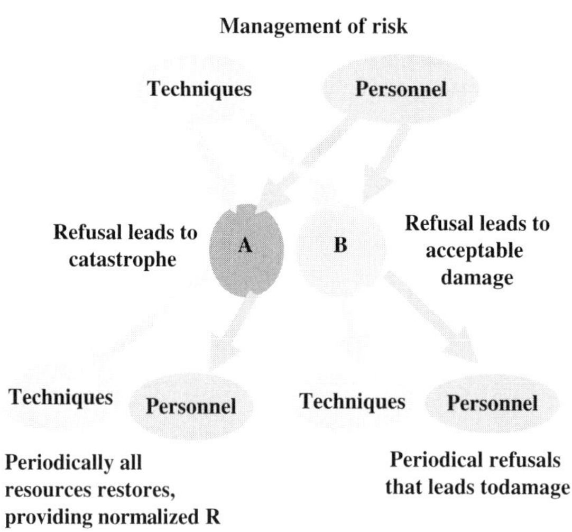

Fig. 6.5. Management of the risk

apply *Columbus' approach* to the risk estimation. In this case for each potential source of damage, one should determine the time T, after which the moment failure of the MED becomes very probable. The obtained value of the time $T^{0.5}$ should be divided by the reliability factor R. As the result the value of time is obtained, after which the MED should be replaced by new ones. If the MED includes materials or elements where failure leads to failure of equipment or design as a whole, it is necessary to treat this material or element with the same *Columbus' criterion.* The time of replacement is equal to $\Delta T_{rep} = T^{0.5}/R$, where R is the safety factor of the admitted risk.

The time after which the failure of the MED is probable can be considered as the mathematical expectation of failure, providing the limits of integration include all area, in which the failure probability density distribution function $f(x)$ is defined. If integral is taken not over the whole area, then the obtained value of the average operating time until failure cannot be applied in computation as the time after which the failure of MED is most probable, because usage of the value leads to incomplete development of the given resource, the reserve of serviceability of elements or components, i.e., reduces the production efficiency. The time of the most probable failure can also be determined as the median or 50% fractile or as the mean time of the operating time until failure, (50% of life length until failure).

2. Problems of estimation of risk that the expert does not have knowledge necessary at the given moment. In all cases, when the failure of the equipment, caused by lack of knowledge of the worker, does not result in large damage, it is sufficient to apply *Bernoulli's* approach to the risk estimation. Thus, it is enough to create the professional emergency and repair services, and also to carry out the insurance of damage due to lack of knowledge of the experts. The given actions are capable to ensure normal functioning of the object. If the failure of the equipment, caused by the lack of knowledge of the expert, leads to significant damage, which appreciably exceeds cost of professional training and retraining the expert, it is necessary to apply *Columbus' approach* to the risk estimation. In this case for each expert, it should be determined the time after which the absence of necessary knowledge of the expert is most probable. The value of this time is determined by:

- Intensity of renewal of information and realization of innovations in this area;
- Quality of knowledge obtained by the expert in the educational institution (this includes category of the educational institution and expert's grades);
- Individuality of the man on forgetting special information.

The obtained time is divided by the reliability factor: $\Delta T_{train} = T^{0.5}/R$. After expiration of this time, it is necessary to retrain the expert.

3. Problems of the mistakes risk estimation. If the failure of the equipment, caused by a mistake of the personnel, does not result in large damage, it is sufficient to apply *Bernoulli's* approach to the risk estimation. One should insure the risk of damage by making the mistake. This measure is capable to guarantee normal functioning of the object. If the failure of the equipment, caused by the mistake of the personnel, leads to the significant damage, which appreciably exceeds cost on duly psychophysical and emotional preparation of the personnel, it is necessary to use *Columbus' approach* to the risk estimation. For each man from the operative, repair, and administrative personnel, it should be determined the time after which making mistake is most probable. The value of the time of making first mistake is determined by:

- Intensity and monotony of working actions, intensity of attention, levels of illumination, noise and noise information, convenience of control panels, volume and distinguishability of the working information, and other ergonomic characteristics of the workplace and the process;
- Individual psychophysical features of the man;
- Intensity of application of special psychophysical technologies for increase of emotional stability and readiness of the man to work in all possible (design and emergency) regimes.

The obtained time is divided by the reliability factor. After expiration of this time $\Delta T_{rest} = T^{0.5}/R$, it is necessary to make restoration of missed abilities of the worker. The restoration time should be determined both for the man as a whole, and for his or her separate functions and qualities. The appropriate procedures, allowing one to restore some temporarily missed qualities (for example, attentiveness), can be applied more often than others.

4. Problems of the risk estimation of conscious non-fulfillment of the required actions. If the failure of the equipment, caused by conscious non-fulfillment of required actions, does not result in large damage, it is sufficient to apply *Bernoulli's* approach to the risk estimation. One should insure the risk of damage due to conscious non-fulfillment of the required actions.

This measure can be effective and capable to ensure normal functioning of the object. If the failure of the equipment, caused by conscious non-fulfillment of required actions, leads to the significant damage that appreciably exceeds cost of formation of motivational readiness of the personnel to make their work, it is necessary to use *Columbus' approach* to the risk estimation. It is necessary to determine the time after which the conscious non-fulfillment of the required actions is most probable. The value of the time of the most probable conscious non-fulfillment of the required actions is determined by:

- State of structure of the public relations in the world and the country;
- Readiness of the structure of the public relations, generated in the labor collective, to withstand the expansion of moral norms from the overactive social institutes;
- Internal coordination of the industrial structure of the enterprise;
- Features of connections of small groups (working groups, brigades, shifts) with the structure of the public relations of the collective.

The obtained time is divided by the reliability factor. After expiration of this time $\Delta T_{social} = T^{0.5}/R$, it is necessary to apply the social technologies, forming motivational readiness to the safe qualitative work.

The special block of problems. It is not enough to have four problem blocks for transition from *Bernoulli's* risk estimation to maintenance of reliability by *Columbus'* approach. It is necessary to have a problem block for definition of the reliability parameter R:

- Investigation of data on failures and accidents for parameter estimation of reliability R, which characterizes the system and the personnel at the moment, preceding the accident;
- Development and statement of the normalized values of the reliability parameter R_{norm}, admitted for dangerous objects (the parameter should depend on plant types, on product damage, and on sources of danger).

The failure occurrence risk calculus on the dangerous plants includes the following works:

1. Definition of the object list, for which *Bernoulli's* approach to the risk estimation of damage is sufficient. For these objects, one should apply the technique of the probabilistic analysis of safety *(PAS)* or the *LP*-modeling for risk estimation and analysis, or to use the practical results — the primarily overestimated resources are specified by real development of events. This provides the reliable work of the objects. The problem of the manufacture organizers is to ensure supply, and fulfillment of the restoration and repair work. For definition of periodicity of repairs, volume and periodicity of deliveries, volume of the emergency stock, the number of the operational, repair, and emergency personnel, one should apply concepts and ways of definition of values, developed in the reliability theory. In *Bernoulli's* approach, both hidden defects of equipment and influence of the personnel are depersonalized.

The insurance of *Bernoulli's* objects is traditional procedure of definition of the insurance tariff, taking into account the practice of insurance or results of calculation by techniques of *PAS* or by *LP*-models.

2. The list of objects, plants, or their parts is defined that need application of *Columbus' approach*, that is, the application of all kinds of regulation allowing one to avoid failures. The following problems should be solved:
(1) Definition of the normalized value of the parameter of reliability R_{norm}.
(2) Preparation of specifications:

- Regulation works for designs and equipment, their elements and materials, which failure does not result in significant damage; for maintenance of their normal operation, it is sufficient to apply *Bernoulli's* approach to the risk estimation,
- Construction and equipment, their elements and materials, such that their failure leads to significant damage;
- Regulation works, which, if they are not done, result in failures.

(3) Definition for each construction, equipment, element, and material the time $T^{0.5}$ after which failure is most possible.
(4) Calculation of the time $T_{change(repair)} = T^{0.5}/R_{norm}$ after which materials, equipment, designs, or elements should be replaced by new or repaired ones. After repair the probable time until failure $T^{0.5}$ should be recalculated.
(5) Definition of the following parameters of professional readiness of the experts to the technology requirements:

- Intensity of updating the information in those fields of knowledge in which experts work;
- Quality of knowledge obtained by each expert in educational institutions;
- Individual features of each expert on forgetting information.

On the basis of the obtained parameters for each expert, it is possible to determine time after which the absence of necessary knowledge is most possible. The obtained values of time should be divided by factor of reliability R. In result, for each expert the value of time will be obtained after which the expert should pass retraining.

(6) Definition of parameters of the personnel is predisposition to make mistakes:

- Parameters of ergonomic features of each workplace (intensity of actions, monotony, pressure of attention, illumination of the workplaces, noise level, convenience of the control panel and equipment on them, level of the information noise, volume and distinguishability of the obtained information, and a number of other ergonomic characteristics);
- Parameters of individual psychophysical features of each person from the object service personnel (temperament, ability for mobilization of attention and efforts, influence of fatigue toward mistakes, susceptibility of large volume of the information).

(7) Definition of time (on the basis of the obtained parameters) for each of the personnel after which making mistakes is most probable.

(8) The obtained values of time should be divided by the factor of reliability R. Thus, for each expert the time value is obtained after which the person should restore the readiness of organism to reliable performance of work.

(9) Definition of parameters of readiness of the labor collective to resist formation of extraneous motivation on the workplace because of the public relations in the world, country, and in collective, behavior of small groups of workers, brigades, changes and their connection with the public relations in collective.

(10) On the basis of the obtained parameters for each group of the labor collective and for all labor collectives, the time should be determined after which expiration of the formation of extraneous motivation on the workplace is most probable.

(11) The obtained values of time should be divided by the factor of reliability R. For each labor collective and group, the time value will be found after

which it is necessary to carry out measures on restoration of motivational readiness for safe productive work.

The basis of *Columbus' approach* to the risk estimation problem is the maintenance of total regulation of production. For this purpose, complex and dangerous production should be considered in parts. The serviceability of each part is concerned with development of one or a group of material resources, for each of which the function of probability distribution density of failure is known. The diagram of this function consists of two qualitatively distinguishable parts: in the first part the material resource wears out, but it is sufficient for maintenance of non-failure operation of the object. In the second part, the material resource is worn out so that in some cases it can lack for non-failure operation of the object, and there is a finite value of failure probability due to development of this resource. Some material resources can have unusual form. For example, the weakening screwed connections during work is necessary to consider as wearing the material resource, which value depends on vibration, character and value of loading, from character and quality of the groove surface, from application of greasing and from the effort applied at connection. For known values of these parameters, the screwed connection will be characterized by the diagram of the failure probability distribution density function, on which there are two qualitatively distinguishable parts: the time when the connection works trouble-free and the time when the probability of failure has a finite value. For this function, we compute the value of the most probable time of operating before failure. Using this time, with the help of the reliability parameter we should determine the periodicity $\Delta T_{rep} T^{0.5}/R_{norm}$ of necessary strengthening screwed connection.

The part of production in *Columbus' approach* is also the personnel. This approach considers engineering as a set of the large number of material resources, each of which is necessary to regulate. The personnel, as more complex than the engineering component of the process, is considered in three aspects:

- As a carrier of knowledge that is necessary for realization of production (cognitive component);
- As an owner of own body and own mentality, more or less suitable for maintenance of process (psychomatic component);
- As a participant of the public relations, in which the motivation of the person to fulfillment of acts (psychosocial component) is formed.

Retraining the experts on each of cognitive resources should be done with periodicity $\Delta T_{train} = T^{0,5}/R_{norm}$. For each of psychometrical resources, it is necessary to use procedures of resource restoration with periodicity $\Delta T_{rest} = T^{0,5}/R_{norm}$. For each of psychosocial resources, it is necessary to work with collective and small groups with periodicity $\Delta T_{socia} = T^{0,5}/R_{norm}$.

The diagram of failure probability distribution function, as it indicated above, at *Columbus' approach* is characterized by the initial part, on which the basic opportunity of object failure cannot completely be excluded, but any finite value for such failure probability estimation is not defined.

The experience of operation of any object, say gear-wheel, gives another failure distribution than that obtained in carefully prepared experiment. In practice, the failure of gears in operation is possible always. The reasons of earliest failures are connected with deviations from conditions of technical operation. In *Bernoulli's* approach, the distinctions in the reasons causing failure are ignored, therefore it works well for the practical data estimation. For the description of the operating time distribution, this approach recommends functions with finite values of probability in the range of operating time from 0 to ∞. *Columbus' approach* requires separation of the reasons of failures. Along with the technical reasons, by wearing out material resources (for example, fatigue deformation of metal in the berth of gear cogs), other reason groups should be recognized, related to insufficiency of expert knowledge, to mistakes of the personnel, and to the changed motivation of actions. Such division allows us to apply functions with infinitesimal values of probability in the first part of the diagram for risk estimation.

An uncertainty is brought into the logic of *Columbus' approach* by the problem of accuracy of definition of the operating time before failure for each of material resources of engineering. The uncertainty is also by problems with definition of functions of the operating time before failure on each field of expert knowledge, on each of reasons causing mistakes of the personnel, and on each inconsistency between activities of social institutes. In order to neutralize this uncertainty, in relation to engineering, the methods of non-destructive control are applied. The quality of control methods is a basis for definition of the normalized values of reliability R_{norm} for dangerous objects in engineering. The insufficiency of the control forces one to overestimate the parameter of reliability. In order that the normalized reliability parameter could be determined only by the object characteristics (level of danger, complexity, etc.), each of the control methods should be given with a rank. The set of the rank parameters is used to define the periodicity of repairs and service.

The legitimacy of application of one or another function for the description of the operating time before failure due to insufficiently expert knowledge is also related to the method features for control of expert knowledge and skill to use the knowledge.

A similar situation is with techniques of psychophysical readiness of the personnel for performance of work and psychosocial readiness of labor collectives. In all cases, determination of a system of coefficients describing methods of the control is necessary. We have

$$\begin{aligned} \Delta T_{repair} = T^{0.5}_{engin}/(K_{qual} \cdot R_{norm} \cdot \Delta T_{train} = T^{0.5}_{knowl}/(K_{knowl} \cdot R_{norm}), \\ \Delta T_{reconst} = T^{0.5}_{psycho}/(K_{psycho} \cdot R_{norm}), \\ \Delta T_{social} = T^{0.5}_{social}/(K_{social} \cdot R_{norm.}), \end{aligned} \tag{6.1}$$

where $K_{qual}, K_{knowl}, K_{psycho}, K_{social}$ are factors of method quality of the non-destructive control, knowledge control, psychophysical control, and psychosocial control, respectively.

The use of risk computation results in exploitation.

1. With the help of *Bernoulli's* approach, we usually determine the necessary periodicity and volume of equipment and materials for replacement of those leaving out operation because of failures. Thus, as a basis for computations, we accept the failure probability of equipment, constructions, their elements and materials, probability of violation of the requirements of operational instructions, making mistakes, and personnel knowledge insufficiency. The parameters that are related to the reliability of a person can also be taken into account by implicit ways; they are incorporated into the failure probabilities obtained in practice for different kinds of engineering in different conditions of operation. *Bernoulli's* approach allows us to determine for emergency and repair services the required level of technical equipment and prospective volume of work, and also to determine quantitative and qualitative structure of the operational, repair, auxiliary, and emergency personnel.

2. With the help of *Columbus' approach*, it is possible to determine the list of technical objects and their parts, labor collectives, workplaces, and specialists, which do not allow one to apply *Bernoulli's* approach to the risk estimation. Further, for each of the spent material resources or elements of the equipment, the time should be computed after which the resource or the element needs replacing or repairing. If the repair does not involve complete replacement of the resource, then for the resource the repair periodicity should be determined (taking into account that the probable time of failure after repairs is progressively reduced).

On the basis of the defined work volumes, the requirement of experts, number of workplaces, and manufacture organization structure is determined. For each expert the time is determined after which professional knowledge should be updated. For each member of the personnel related to the dangerous object, the time is determined after which the person should restore psychophysical readiness for work performance: for different kinds of works and for different kinds of the human abilities the periodicity is different. For each labor collective and for each small group, the time is determined after which the social technologies ensuring restoration of motivation of readiness of the personnel to safe productive work should be applied. On basis of the data set, the number and professional structure of administrative and auxiliary services are defined.

Principles of work organization for risk decrease.

Principle 1. Setting in the regulatory documentation:

- The requirements to work out for each dangerous object the lists of the equipment, constructions, elements, and materials for which the calculation of periodicity of replacement or repair is necessary on the basis of the normative reliability parameter R_{norm};

- The requirements to work out the lists of professions, workplaces, and labor collectives, functioning reliability of which should also be supported taking into account the parameter of reliability R_{norm};
- Criteria of classification of dangerous and expensive objects by value of the parameter of reliability R_{norm};
- The requirements to take contracts with research organizations on development of restoration methods for each of four components of reliability and quality control methods by each of components, and also development of methods of the risk estimation;
- The requirements to appoint supervising organizations conducting systematic and random inspection of quality of estimation and maintenance of normalized reliability of dangerous objects, and also the quality check of the risk estimation.

Principle 2. Setting in the normative documentation and in safety rules the requirements on periodicity of repairs and replacements of the equipment, building construction, their elements and materials (for determination of the periodicity, normalized parameter of reliability is used), and on periodicity of work on maintenance of reliability of the personnel related to dangerous objects, which has three directions:

- To update knowledge (for providing quality of performance of the technology requirements);
- To support serviceability (for providing the faultlessness of work of the personnel);
- To increase motivation to safe productive work (for providing the timeliness and accuracy of operation).

Principle 3. Organization of continual work of research and supervising organizations in each of the following directions:

- Research of development of material engineering resources (equipment, constructions, their elements and materials), investigation of methods for estimation and control of development and restoration of resources, licensing methods and techniques;
- Research of necessary knowledge structure for the experts of different professions, dynamics of knowledge update in various branches and research of new methods of replenishment and usage of knowledge, methods of estimation and control, licensing techniques;
- Research of psychophysical readiness of the workers, research of methods of estimation and control, licensing methods and techniques;
- Research of the mechanism of formation of motivational readiness of a person for performance of work and methods of estimation and control, licensing techniques;
- Research and development of mathematical methods of risk and damage estimation;

- Quality control of estimation and maintenance of normalized reliability of dangerous objects, and quality control of estimation of risk to get damage.

For work organization of research and supervising organizations, in compliance with the described principles, it is necessary to develop mechanism of financing with the central place of insurance companies.

Principle 4. Organization of insurance of the damage caused by failures of the equipment. The insurance contract should provide that a part of funds, accumulated in the insurance pool, is used for financing research, organizational and technical work ensuring reliability of the dangerous equipment in *Columbus'* sense.

Principle 5. Development of the following techniques and technologies (methodical base):

- Risk management;
- Calculation of reliability of the dangerous object;
- Classifications of dangerous objects;
- Maintenance of a necessary level of reliability of engineering;
- Maintenance of a necessary level of reliability of technology by regulation of forms and periodicity of retraining experts;
- Maintenance of a necessary level of reliability of operation by regulation of forms and periodicity of restoration of motivational readiness of labor collectives for performance of work;
- Maintenance of a necessary level of reliability of the personnel by regulation of periodicity of restoration of psychometrical readiness of people for performance of work.

After the risk value for one dangerous manufacture is determined, the expenses for maintenance of indefinitely low failure probability of the equipment are calculated, and the risk is compared with expenses, it is possible to make generalization of the results for manufactures of the same type. It will make it possible to correct the current expenses. The means for liquidation of consequences of failures and failures can be accumulated in the insurance companies. Hence, there is no necessity to create funds for emergency use at each enterprise. It is enough to add to the current expenses the current insurance payments. The value of payments is determined by the risk value.

6.3 Financing the risk management process

The basic features of risk management financing is the following:

1. Work on reliability maintenance by *Columbus'* method requires increase of financing for:

- Realization of planned-prophylactic repairs and services;
- Maintenance of necessary periodicity and volumes of repairs and services;
- Maintenance of the necessary forms and periodicity of retraining of the experts;

- Maintenance of the necessary forms and periodicity of restoration of motivational readiness of labor collectives for performance of work;
- Maintenance of the necessary forms and periodicity of restoration of psychomatic readiness of people for performance of work;
- Research on features of manufacture of each of material resources;
- Research on the necessary forms and periodicity of retraining of the experts;
- Research on the forms and periodicity of restoration of motivational readiness of labor collectives for performance of work;
- Research on the forms and periodicity of restoration of psychomatic readiness of the people for performance of work;
- Development of methods of the non-destructive control of engineering;
- Development of methods of knowledge control;
- Development of methods of motivational control;
- Development of methods of psychomatic control;
- Research of volume of controlling actions on directions of work.

2. The basis of insurance organization for a dangerous object is the separation of its components, equipment, designs, elements, and materials into two groups. For the first groups, *Bernoulli's* approach to risk estimation should be applied. For the second group, *Columbus' approach* should be applied. The criterion of division is the value of damage that can be covered by insurance organization, that is, the normalized damage.

3. The normalized damage should be determined from the condition of the minimal insurance tariff taking into account features of expense change:

- The more that part of dangerous object in which reliability is provided by *Columbus'* method, the less the insurance tariff (the insurance by *Bernoulli's* principle refers to the equipment not creating large damage at failure);
- The more that part of dangerous object in which reliability is provided by *Columbus'* method, the larger means are necessary to spend for maintenance of required periodicity restoration of material resources of object and maintenance of reliability of the person. These means can be gained by increase of norm of assignments for repair work and increase of the insurance tariff.

The reduction of the insurance tariff by damage reduction has nonlinear dependence on the amount of the equipment insured by *Bernoulli's* method. The function of increase of the insurance tariff by growth of expenses for maintenance of reliability *by Columbus' method* is close to linear one. Therefore, always there is the area of values area of damage by equipment failure, in which the value of the insurance tariff is minimal. One of the values of this area can be accepted as normalized.

4. The reliability maintenance of dangerous objects by *Columbus' method*, at which the costs of the insurance company on payment of the insurance

premiums are limited to the value of the normalized damage is possible only when significant means are spent on research, development of the methodical, regulatory, and normative documentation and control. Taking this into account, in practice of organization of dangerous object insurance, it is possible to refuse formation of insurance pool, focused on covering expenses from failure of Chernobyl's type. Means of the insurance companies should be used primarily for financing researches, documentation development, and inspection.

5. The mode "of partial insurance" can be used by the insurance company for the dangerous enterprise. The mode assumes the insurance premiums are paid only at failures resulting in damage, not exceeding the normalized one. In this case, the insurance tariff paid by the enterprises to the insurance organization should be reduced, and the difference between the existing high insurance tariffs and the reduced one in the mode "of partial insurance" should be directed to financing research, development of the documentation, and inspection realization.

Let us outline the basic rules for economic calculations:

1. For organization of work on risk management at an enterprise the lists of objects, equipment, construction, materials, professions, jobs, labor collectives should be made, if their failure in work can result in: catastrophic damage; certainly large damage; damage compared with expenses for damage prevention; insignificant damage.

2. For the personnel and engineering of the first three categories (the group A may be, with exception for a part of resources of the category C or all resources of the category C and part of resources of the category B), the computation of expenses, which are spent for restoration of material resources and resources of the personnel, should be made. The computation is done providing that the period from the resource work beginning to the moment of its restoration should be R times less than the period from the resource work beginning to the moment when the probability of its failure takes the given finite value.

3. Group B is formed by the engineering and personnel that were not included into the group A.

The system of the equations describing expenses for work organization according to the stated rules looks as follows:
For all resources $i \in A$:

$$Z_A = \sum_{i \in A} Z_i(R_i); Z_B = \sum_{j \in B} Z_j(q_j);$$

$$Z_j = \min\{Z_j(q_j) + U_j q_j\};$$

$$Rs = \max\{q_j U_j + q_{j;k} U_{j;k} + q_{j;k;l} U_{j;k;l} + \ldots + q_{j;k;l,\ldots,n} U_{j;k;l,\ldots,n}\};$$

for all elements and components $j, k, l \ldots n \in B$:

$$S = S(R_s; Z_A). \tag{6.2}$$

Here:

- Z_A are the total expenses on the group A;
- $Z_i(R_i)$ is the expense for restoration of the resource i, which is carried out with periodicity dictated by the reliability parameter R_i for the resource i;
- Z_B are the total expenses on the group B;
- Resulting expenses on j-th element of the group B, including expenses for maintenance of failure probability q_j and expenses for indemnity of damage U_j (from the condition of minimum of Z_j at constant U_j one determines the value of q_j);
- Expenses for maintenance of failure probability q_j of j-th element of the group B, leading damage in the case of failure;
- The maximal risk in the case of failure of any element or combination of components and elements of group B;
- Damage caused by failure of any element or any combination of elements;
- Insurance payments of the enterprises ensuring compensation of damage caused by failures of elements of group B (partially or completely), and, possibly, spent to partial covering of restoration of resources of group A.

The complete expenses of the enterprise, depending on participation of the insurance companies in organization of its economic activity, can be formed as follows:

$$Z_{compl} = Z_A + Z_B + R_s(1 + \beta);$$
$$Z_{compl} = Z_A + Z_B + S(b, R_s) + (1 - b)(1 + \beta)R_s;$$
$$Z_{compl} = (1 - a)Z_A + Z_B + S(R_s, a, Z_A),$$

where:

β is the factor, increasing the deduction for the risk insurance because of necessity to pay credit rate in the case, when the failures with the large damage take place before the enterprise account accumulates sufficient sum from the last assignment for risk, and it is necessary to take a credit at bank on the damage covering;

b is a share of risk from the insured engineering the group B;

a is a share of expenses on work organization according to the concept of managed risk (on research, development of methodical base, organization of introduction, realization of preparatory works, control), which is paid by the insurance company itself, proceeding from the opportunity to insure many enterprises of the some type.

In the first case, the enterprise does not insure the risk. It leads to the higher expenses for works organization with the personnel and engineering the group A and higher deductions for insurance. In the second case, the insurance contributions on risk are reduced in that measure, in which the insurer covers the damage from failures and accidents of the personnel and engineering the group B. In the third case, the enterprise expenses are reduced in that measure, in which the insurer takes the charges on organization of works according to the concept of managed risk.

6.4 Reliability regulation of engineering and a person

The attempts of taking into account the personnel influence on work of an object of the group A are traditionally restricted by some basic circumstances. First, the failure probability of a person is difficult to measure. Some success is achieved at taking into account the error increase on simple operations in dependence on growth of fatigue of the person or on change of conditions of work. In relation to more or less complicated activity, any authentic data about the human failure probability is not present. Secondly, the man work reliability parameters is influenced by many factors. Description of the factors is an independent difficult problem, because there are no simple rules to define whether and how these factors influence small groups of object's personnel. To account for the influence of these factors on the failure probability now is impossible.

The concept of managed risk allows us to avoid the necessity to estimate human failure probability at any moment of time. All that is required to know about resources of the man is the mean value of probability of failure or the most probable time of failure. Finding this information is possible. The concept allows us to make the resource lists that should be given to the personnel for different kinds of activity.

We can give to the human reliability parameter $Rh^{0.5}$ different values for different level of responsibility of the personnel. This ensures that, with a different stock for different levels of responsibility, the achievement of the critical point of transition from the period with infinitesimal probability of exhaustion of the resource to the finite values of probability will be prevented.

The man features allow us to unite all the human resources to three large groups:

- Cognitive resources, which have the reliability parameter $Rhc^{0.5}$;
- Psychomatic resources, which have the reliability parameter $Rhps^{0.5}$;
- Motivational resources, which have the reliability parameter $Rhm^{0.5}$.

For each group of resources, we apply specific technique of restoration. It is necessary to understand that the concept of managed risk is the first theory that gives an opportunity to take into account the man influence on the reliability degree of dangerous manufactures; it is the first that makes it possible to compare expenses for the man resources restoration with expenses for restoration of object resources.

6.5 Consideration of natural and man-caused accidents

The separation of objects, equipment, and materials into groups A and B does not take into account influence of natural and man-caused accidents on the character of development of the material resource (the accidents are the external phenomena for engineering, which are considered according to the concept of acceptable risk). The concept of managed risk does not exclude application of the concept of acceptable risk.

It is possible to be protected from external failures with the help of technical decisions (underground structures, aseismic design, etc). These structures should refer to the group A and the appropriate work on restoration of their resources should be performed. The accidents probability is insignificant, and, depending on the damage value, the risk value can be compared with the damage due to failures with engineering the category B. In this case, additional measures for decreasing damage is unjustified. The damage can also be reduced with the help of special protection of not the whole object, but of the most responsible parts of it only. However, the concept of managed risk does not consider regulation of damage. The risk insurance related to accidents is possible either for a large number of insurance objects or when the state guarantees credits of the insurer in the case of accident.

6.6 Probability of poor organization

The activity of the developers of methodical base is organized as work of the personnel of group A (with maintenance of periodicity of resource restoration of the personnel).

Taking into account the quality of work of designers and developers of technology. The features of designers and developers of technology is that they are not necessarily the object personnel. Meanwhile, full work with them as with the personnel of group A should be done.

Taking into account work quality of the designers of the special engineering. It is not possible to obtain authentic data about the nature of wearing material resources of special engineering. Therefore, in manufacturing special engineering, the work should be organized as with objects of the group A. With the personnel who produce the special engineering, the work should be organized for restoration of all three components of human reliability.

Taking into account the opportunity of terrorist act. The work on prevention of the terrorist act has two large components. It is the protection against external threat and the protection against internal threat. The prevention of the terrorist act made by the object personnel is provided by dealing with motivational component of the personnel resource. It is a problem of social psychology solved with the help of the analysis of activity of social institutes and application of special techniques, which correct the connection of the man with structure of the public relations.

The operative analysis of activity of social institutes makes it possible to predict external threat, too, in particular, to define the most probable social and psychological portrait of the terrorist. For maintenance of reliability of physical protection of object, the design, equipment, and materials used for organization of the physical protection should be referred to the group A.

7

Transparency of Methods for Estimation of Risk

The transparency of a technique is its property to show and estimate the phenomenon not only as a whole, but also in detail.

Author

This chapter is written on the basis of the materials published together with Dr. N. V. Stepanova [4, 93]. At first, the available scoring methods of credit classification and their limitations are considered. Then the requirements to methods of the credit risk estimation on the basis of the LP-approach are formulated.

7.1 Scoring methods of the object classification

Banks' hiring of qualified experts for credit status estimation has some disadvantages. First, their opinion is subjective; second, people cannot process the great volumes of information fast enough; third, good experts demand considerably high fees. Therefore banks are increasingly interested in such systems of credit classification that would allow them to minimize experts' participation and the impact of human factor on decision-making.

Essence of scoring techniques

The scoring of a physical persons's credit is the estimation technique of the borrower's quality, based on various characteristics of the client: income, age, marital status, profession, etc. The analysis of credit variables yields an integrated parameter in points, which estimates the degree of borrower's credit status via the scale of ranks: the borrower is good or bad. The decision about credit and its limits is made depending on the point estimation.

E.D. Solojentsev, *Scenario Logic and Probabilistic Management of Risk in Business and Engineering*, DOI 10.1007/978-0-387-77946-1_7,

The borrower's credit status is his ability to pay back in time under his liabilities. According to such definition, the primary task of scoring is to find out whether or not the client can timely pay the credit back.

Scoring is a mathematical model by means of which the bank tries to determine on the basis of credit histories of its previous clients the probability of the timely payments of its potential borrower. Scoring is the classification method of the target population into various groups, when the characteristic that divides these groups is unknown.

In the Western banking system, when a client applies for a loan, the bank has the following information for the analysis:

- The questionnaire filled out by the borrower;
- The information on this particular borrower from the credit bureau, in which credit histories of a country's adult population are stored;
- The data of borrower's account transactions if he or she is the bank's client.

Credit analysts use the following concepts: clients' "attribute-parameters" and "grade-values," which attributes take. In the client questionnaire, the attribute-characteristics are age, marital status, profession, etc. In the client questionnaire, the grade-values are the answers to these questions. The scoring model provides the integrated total of certain characteristics. As a result, the integrated parameter (score) is obtained. The bigger it is, the higher is the client's reliability. The integrated parameter of each client is compared with the predetermined level of the parameter. If the parameter is higher than the level, the credit is provided. If the parameter is lower than the level, the credit is not provided.

The problem is which attribute-parameters should be included in the model and what "weight" factors should correspond with them. Scoring philosophy does not imply the search for explanations why the client does not pay. Scoring uses characteristics that are most closely connected with the client's unreliability. It is not known whether a certain borrower will return the credit, but it is known that in the previous years people with the same age, profession, educational level, and the number of dependents did not return (or returned) the loan.

Historical development of scoring

Scoring is the method for classification of credit groups under investigation into various categories. In statistics, the idea of the population classification into groups was developed by Fisher for plants in 1936. In 1941, David Djuran was the first to apply this technique to classify credits into bad and good. The time coincided with the Second World War, when almost all credit analysts were called up to the front. The banks faced the necessity to replace these experts. Before leaving, analysts had written the instructions for non-specialists to be followed in decision-making for credit provision/extension. This was the prototype of expert systems, with so-called "scoring cards"being used (see Table 7.1).

The Western banks have been using scoring systems for a long time. The leader in the field of scoring systems development was American consulting

Table 7.1. **Scoring card**

Parameter	Value	Ball
Age,	20–25	100
year	25–30	107
	...	...
Income,	1000–3000	130
dollar	3001–5000	145
	...	...

company Fair Isaac Corporation (with system FICO). The company was set up in the early 1950s in San Francisco. It provides services to 7 out of the 10 world's largest banks, 97 out of 100 of America's largest banks, and the 50 largest emitters of credit cards.

Scoring systems became widespread with the massive introduction of credit cards. As this banking service is much in demand, it was necessary to automate the decision-making system for loan provision.

In 1974, the USA passed the law on granting equal opportunities on credit provision, which forbade to refuse a person a loan on the basis of his or her race, skin color, nationality, age, sex, marital status, religion, physical disability. The credit legislation in the USA and the consumer credit legislation in Great Britain attached great value to the establishment of the credit bureau service. These bureaus store credit histories of all people who at some time applied for loans to any credit organization of the country. Credit bureaus store the following data:

- social-demographic characteristics;
- court rulings (in case of claiming debt through court);
- insolvency information;
- data on borrowers received from other banks.

There are transnational commercial companies, like Experian, Equifax, TransUnion, Scorex, which use scoring systems and sell clients information as an integrated parameter that is entered in the automated system of the credit organization. Credit bureaus have great significance: their existence allows provision of loans to borrowers who were not clients of a certaing bank, but whose credit history is available.

Now scoring is used not only for various kinds of credits, but also in marketing (for estimation if a certain group of clients will use a certain kind of product), for work with debtors (for decisions on the efficiency of various methods to be applied in case of next payment delay), for revealing credit card frauds, in case of suspicions that a client uses the competitor's service.

Prospects of the credit scoring in Russia

Russian citizens tend to take more and more credits for consumer needs — purchase of domestic appliances and long-term goods, purchase of automobiles and apartments, opening credit cards, etc. The volume of this market is growing very fast.

It is no wonder that the competition in this market has become more fierce — banks fighting for their market share offer various products, reduce interest rates, offer attractive crediting conditions. As a result of competition, it's not easy to ensure success in retail crediting business. The main question is: who should be given the credit and who should be not? Whereas formerly the non-return risk was covered by the size of interest rate, now the situation is different. Introducing severe restrictions for receiving the credit means losing potential profit that could be gained at more flexible conditions.

In order to ensure the profitability of retail crediting market, it is necessary to have the effective system for risk estimation, which would allow to cut unreliable borrowers at the early stages without rejecting the reliable ones. It is also necessary to correctly determine the payment size in the consumer credit or the limit on a credit card. It is such systems that create the bank's reliability, which enables it to introduce on the market the products attractive for borrowers.

In Russia, the popularity of credit cards is still insufficiently high. Aspiring to provide credit cards only to "the proven" clients, banks do not offer this service to a broad number of consumers, which could eventually increase the demand for it. The increasing demand from the less "proven" segment of potential borrowers will require designing of an effective system of risk estimation, which the banks do not possess at present.

So far, the volume of credits provided by banks to legal persons has been significantly higher in absolute expression than that offered to the population. But consumer crediting becomes one of the most dynamical directions of the bank sector development, which is primarily associated with the banks' need of new profitable credit products. Even now there is demand for automated systems of evaluation of individual borrowers.

Due to their sort of activity, large banks and insurance companies were the first in Russia to accumulate a great volume of statistical data. This was accompanied by the insurance companies offering their insurance services to the banks. The pioneer in this business was the insurance company "ROSNO." Banks can have the access to scoring system of the insurer, instantly estimating the borrower's capability. The "approved" credits are insured, and in case of payment non-return, the insurer compensates it. The settlement of conflicts with the debtor is also the responsibility of the insurer. The cost of this service for the bank amounts to 2–8% of the deposit.

Insurance cost may actually be included in the cost of the consumer credit. But, on the other hand, banks can economize on the estimation of the client's credit status and decrease the risk margin and also reduce the size of loan-ensuring reserve capital. The first client of "ROSNO" was "Delta-bank." As "ROSNO" believes though this service may prove to be unattractive to large banks that share about 80% of all the consumer crediting market, the remaining 20% may get interested sooner or later. Small banks won't have to invest non-return insurance scoring system, whereas market leaders have already got the expensive systems for express-estimation of borrowers' solvency.

The problem of the designing "your own scoring" can't be solved due to the absence of sufficient statistics of crediting histories. The establishment of the National Credit Bureau that would carry out borrowers' data exchange between banks might be a solution to the problem. Then there will be an opportunity to collect statistics for training banks' own scoring models. This bureau can independently perform customer's scoring estimation on the demand of concrete banks.

Russia has its special features associated with the instability of the economy as a whole to "skew" in the development of its branches and interbranch connections, big share of shadow incomes, etc. It distorts the parameters of potential individual borrowers. For example, one of the most significant parameters in the Western scoring systems is the age of a potential borrower (for Great Britain, France, and Germany). The older the person is, the higher is his estimation (he is considered to be a reliable borrower). The logic of such system in the West is obvious: a person who has worked all his life has had time to save up money, and has had a reliable credit history. In Russia, most likely, this logic will be inverted: the older is the borrower, the lower is his credit status estimation. Therefore it is impossible to merely transfer the model from one country into another, from one credit organization in another. It is impossible to create the uniform algorithm working for all countries equally well.

Moreover, risk estimation systems will be different for various regions of the Russian Federation because of different conditions of social and economic development. Each concrete model should correspond with a certain country, its economic and financial conditions, traditions, and to a certain credit organization.

Methods of credit scoring

In order to construct the credit classification model, it is necessary to do sampling of clients from the credit organization whose clients are already known. Such sampling can amount to several thousand clients. This does not present a problem in the Western countries where the company credit portfolio can contain tens of millions of clients. The sampling contains information on two groups of credits, recorded in the bank's activity: good and bad ones (the bad credit is a credit that is not returned).

There is a variety of credit classification methods based on linear multifactorial regress, logistical regress, the classification tree, the neural network, and data mining technology (DM). These methods will be briefly characterized.

Linear multifactorial regress is given by the expression

$$P = z_0 + x_1 z_1 + x_2 z_2 + \ldots + x_n z_n, \tag{7.1}$$

where P is default probability, x is a "weight" factor, z is client's characteristic. The disadvantage of this model is the following: in the left part of the equation there is probability that takes values from 0 to 1, whereas the variables in the right part can take any values.

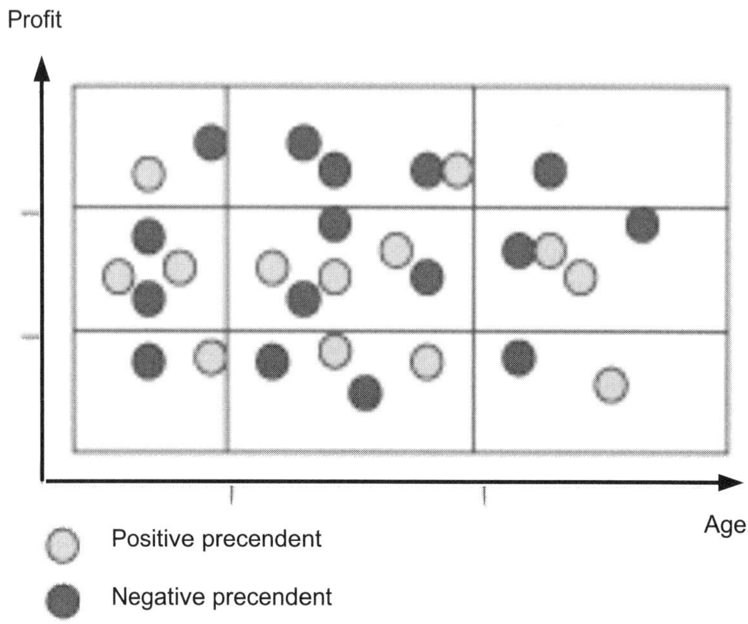

Fig. 7.1. Scoring-card on the method of logistical regress

Logistical regress performs the segmentation of precedents on the basis of splitting the factorial space by n-measured net, where n is the number of significant factors (Fig. 7.1). The initial assumption is: each cell of the net (n-measured rectangle) unites precedents from the training sample, which are characterized by the identical outcome probability. The unit coordinates of this net are computed on the basis of statistical criteria, proceeding from the principle of maximal distinction between the probabilities of credits outcomes for the adjacent precedents segments.

The ratio of positive and negative precedents in every segment is used for the calculation of scoring-points in the scoring card. The coordinates of net units in the factorial space determine the intervals of the attribute values. The technique describes the credits as good or bad ones, using a set of parameters, but we do not receive the answer as far as the credit is good.

The tree of classifications is a more general algorithm of segmentations of sample training by precedents than the logistical regress. Unlike the method of logistical regress in the tree classifications method, the segmentation of precedents is not performed by means of n-measured net, but by the consecutive dividing the factorial space into enclosed rectangular areas (Fig. 7.2).

In doing so, the following sequence of steps (Fig. 7.3) is performed. In the first step the precedent sample is divided into segments according to the most significant factor. In the second and subsequent steps, the procedure

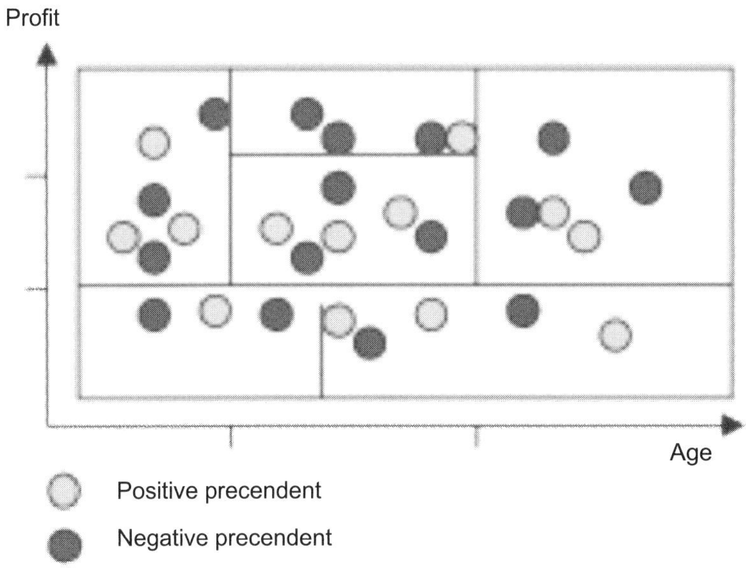

Fig. 7.2. Classification on good and bad objects

is repeated for each of the previously obtained segments until no variant of the further division results in significant difference between the proportion of positive and negative precedents in new segments. Each received before segment we divide again into subsegments by the next most significant factor and so forth. The number of branchings (segments) is selected automatically.

The above considered technique does not provide the answer whether the credit is good or bad. It does not allow us to obtain the exact quantitative estimation of the risk and to establish the tolerable risk.

Neural networks (NN) are used for establishing the credit status of legal persons where the analyzed samples are smaller than in the consumer credit. The most successful area of their application is the detection of credit

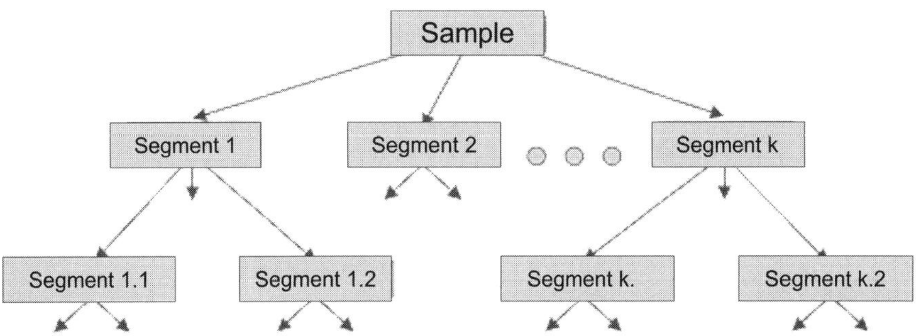

Fig. 7.3. Separation of good and bad objects on the method of classification tree

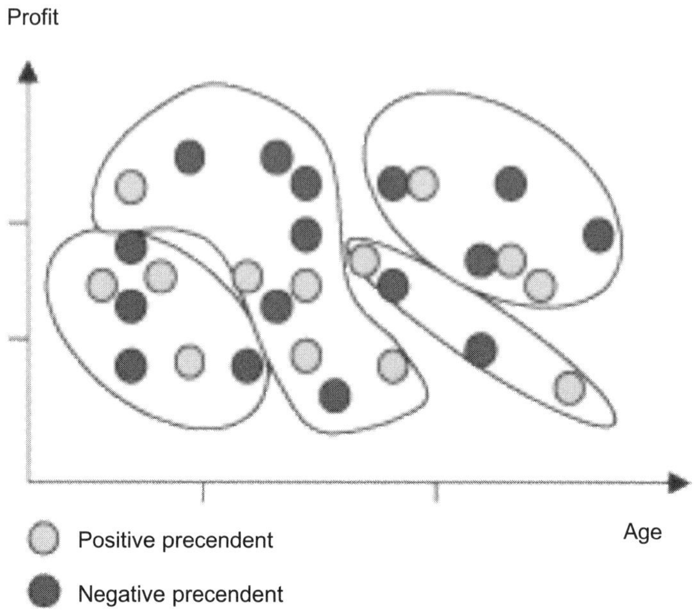

Fig. 7.4. Segments of separation of good and bad objects in NN

card fraud. The neural networks reveal the non-linear connections between the variables that can lead to a mistake in linear models. NN allow us to process the precedents of the training sample with more complex (than the rectangles) segments (Fig. 7.4). The form of segments depends on the NN internal structure. Formulae and factors of the risk models on the NN basis do not have physical and logic meaning.

The neural network is "a black box," whose internal contents (so-called weights of neurons) do not have any sense in terms of risk estimation. Such techniques do not allow one to explain why the credit is given or not given to a certain borrower. The NN-models of classification have low stability.

Technology of data mining (DM). The technology of *data mining* uses search algorithms for the legitimacies between various factors in great data volumes. Thus the DM-models analyze the dependence between all factors, but, as even with a small number of factors the number of possible combinations grows exponentially, the *data mining* algorithms use priori cutting off of weak dependence. Speaking in terms of the credit status analysis, *data mining*, application of statistical data on the given credits reveals those factors that essentially affect the borrower's credit status and also calculates the importance of this influence. Accordingly, the more strongly a certain factor affects the credit status, the greater point it is ascribed in the scoring technique. The more similar are the data of the credit card to the "the solvent citizen" data, the greater credit he can receive, the better conditions can be given to him.

Techniques based *on* data mining can work for small samplings. With big samples their accuracy, robustness, and transparency are insufficient. These models do not give the answer whether the credit is good or bad. The method does not allow one to receive the quantitative estimation of the risk and determine the tolerable risk and the risk price, and to reveal the contributions of factors and their grades to the risk.

The methods based on the group argument accounts (MGAA), the neural networks (NN), and the data mining are the methods for structural identification of models by statistical data. The risk estimation formula is not set beforehand. Both the formula itself and its coefficients are determined. The trained model is tested on the control sampling of statistical data.

There are a lot of various models based on statistics, as well as a lot of risk models for each method that are equally identical and have almost identical values of the target function. This results in the problem of the best model selection for the classification of objects on the basis of the major parameter — stability (robustness). As the studies have revealed, the risk estimation models based on MGAA, NN, and DM have low robustness. These models do not solve the problem of the risk analysis as they are devoid of physical and logical sense, and we cannot talk of their transparency.

As a result of the analysis of scoring techniques transparency for credits classifications, the following conclusions are arrived at:

1. The applied classification techniques are not sufficiently transparent.
2. At present commercial banks have problems acquiring (development) of accurate, robust, and transparent techniques and corresponding software for the estimation of the credit risks of physical and legal persons.
3. Western scoring techniques and corresponding software for the credit classification of physical and legal persons and decision-making concerning reservation amount offered by the market have low accuracy, robustness, and transparency.
4. It is necessary to develop more effective models and software for classification and estimation of credit risks of physical and legal persons that have sufficient accuracy, robustness, transparency, and the capability for analysis, estimation, and risk management automation.

7.2 Risk estimation method requirements

Problems of quantitative estimation and analysis of credit risks and borrowers' ratings and default reserve are urgent both for Western and Russian banks dealing with crediting of natural and legal persons.

The important characteristics for such methods are: (1) clarity of estimation results and credit risk analysis for banks employees; (2) transparency of methods for regulative authorities and clients; (3) accuracy and robustness. Transparency and validity of methods and results is achieved by the possibility to calculate the contributions of initiating events to the credit risk value.

Below are given some definitions of LP-approach to credit risks estimation [4, 92–94].

Credit risk of the borrower is estimated by the probability of credit non-return and possible loss of money. The determination of the probability (the risk) is based on statistical data of the success or failure of bank credits.

Attributes of credit risk. In a general case, credit risk is characterized by the following parameters: risk as probability of credit failure (non-returning), admitted risk, average risk, possible losses from the credit failure, average value of losses, maximal admitted losses, the number of credits in the bank, the possible number of different credits, the number of dangerous credits, entropy of dangerous credits.

Credit rating is the estimation of the borrower's quality and his classification as belonging to one of the categories from the viewpoint of his potential solvency.

Credit is described by signs (parameters), each having grades. There may be up to 40 signs and up to 10 grades in one sign. Credit signs and their grades are considered random variables and events: sign-events and grade-events, which with certain probability lead to credit failure.

Set of credits, any credit, any sign of the object, any grade of the sign have their own risk attributes.

Dependence of logic variables. Logical variables correspond with signs and grades. Logic variables (events) are dependent not a priori, but due to their expression by a logic formula that determines the relation between them.

Scenario of the credit non-success risk is formulated for a complete set of different credits in the following way: non-success credit occurs if any one, any two ... or all events initiating non-success occur.

Models of the non-success risk for complete and limited set of credits. It is necessary to be able to design the scenario of the credit non-success risk, the credit risk logic, and probabilistic models both for a complete set of different credits and for a limited set. The credit non-success risk LP-model can describe all possible different credits and be the most complete and accurate. Nevertheless in a number of cases, there is no need to take into account all states of credits. For example, it is known that credit non-success took place due to occurrence of one and no more than two events, though there are twenty initiating events. Then, for the sake of simplification, the LP-model should be written exactly for the limited number of credit states.

The logic model of credit risk is written on the basis of credit non-success in the form of a logic function in the disjunctive normal form.

The probabilistic model of credit risk (probabilistic risk polynomial or the arithmetical description of the risk model) is constructed after the orthogonalization of the non-success risk L-model. The value of non-success risk is within limits [0,1] for any values of probabilities of initiating sign-events and grade-events.

The admitted credit risk divides all credits into good and bad (or into several classes)according to their risk.

Credit risk analysis consists in calculating contributions of sign-events and grade-events in the credit risk, the average risk of credits in the bank, and in the accuracy of the credit risk model.

Credit risk management concerns the estimation of credit risk and decision making on credit provision, determination of credit cost, determination of the amount of risk reservation, changing of number of signs for the credit description, and numbers of grades in signs.

Risk costs of the credit is determined by the function of the average credit risk and the difference of a certain credit risk and the admitted risk for the bank credits.

Capital reservation for the credit depend on the value of the credit risk.

Transparency of the credit risk technique is the property of technique to reveal not only the phenomenon as a whole but its details, too. Transparency of the risk model involves: accuracy and robustness of recognizing, management of asymmetry recognizing, possibility to compute contributions of sign-events and grade-events to credit risk and bank's average credit risk.

Accuracy of credit risk technique is estimated by mistakes of recognizing bad and good credits (clients). Usually there is a requirement that bad credits should be recognized better. If credits must be classified into several classes, the problem of recognition accuracy is formulated similarly. The comparison of different techniques on the same data have revealed that different techniques of the credit risk estimation have almost twofold difference in accuracy [4].

Recognition asymmetry is the ratio of numbers of the incorrectly recognized good and bad credits. We introduce the recognition asymmetry factor due to the non-equivalence of losses caused by the wrong classification of good and bad ones. The value of the recognition asymmetry factor is from 1 to 10.

Robustness of the credit risk technique characterizes the stability of techniques in the credit risk estimation. Different techniques of the credit risk estimation classify credits into good and bad in different ways. The same credit can be recognized as bad by one technique and as good by another. Such instability amounts to 20% of the total credit number. The comparison of different techniques on the same data revealed [3, 4] that they can have sevenfold difference in robustness. Instability of estimation techniques of credit risks is pointed out in [97].

Basel II defines the concepts on capital reservation in case of default, and also the quality and accounting standards in the international banking system [4, 95]. The introduction of the agreement "Basel II" requires that banks should have accurate, robust, and transparent methods for the estimation of credit risks and reservations in case of default.

As a new approach to ensuring efficiency and transparency of techniques for the estimation of credit risks, we suggest the risk LP-theory with groups of incompatible events (GIE) [4, 30, 96]. The attractiveness of the risk LP-theory is its exclusive clarity and unambiguity of quantitative estimation of risk, in uniform approach to the risk problems in economics and engineering, in great opportunities for the analysis of the impact of any element including personnel, to the reliability and safety of the whole system. The risk LP-model may include the logic connections OR, AND, NOT between the elements of the system and cycles. The system elements can have several state levels. The credit risk dynamics can be taken into account by the changes of the state probabilities depending on time. The estimation of credit risks in the risk LP-model demonstrated two-fold more accuracy and seven-fold more robustness, as well as absolute transparency, in comparison with the well-known methods.

In the general case, the credit risk involved in crediting natural and juridical persons by commercial banks can be characterized by the following quantitative parameters:

- The risk as the probability of the non-success of each credit;
- The admitted risk;
- The average risk;
- The possible losses caused by any credit default;
- The average value of losses;
- The maximum admitted losses;
- The general number of credits in the bank;
- The possible number of different credits;
- The number of dangerous credits;
- The entropy of dangerous credits.

Using these parameters, it is possible to calculate the risk of the default and the possible losses of the bank.

Credit risk is defined as risk of emerging losses in the bank as the result of non-fulfillment, untimely, or insufficient fulfillment of debtors of their financial obligations according to the conditions of the loan contract. In bank balance credit risk is included in the assets in majority of positions: loans, remainders on correspondent accounts, short-term credits to commercial banks, securities acquired for resale or investing. Besides, credit risk arises in connection with a wide spectrum of banking activities including selection of investment portfolios, counteragent on dealership with derivative tools, and foreign currency.

Credit risk can arise in connection with business risk in a certain country as well as in performing guarantee functions. Credit risk directly affects banking capital state. As a result, high risk decreases market-value of the bank stocks and restricts the bank's ability to use the bonded debt for fund attraction.

There exist special requirements to the quantitative estimation methods and analysis of the credit risk: accuracy, robustness, and absolute transparency. Let us consider the definitions of these characteristics.

7.3 Transparency of estimation methods of credit risks

Transparency is the possibility to see not only the phenomenon itself, but also its details. Transparency becomes a significant characteristic of the methods for the credit risk estimation, due to the necessity of the adequate identification of both credit risk and the credit risk model. Transparency includes: strict mathematical methods, reduced subjectivity of expert estimations, clarity of risk estimation and analysis, complete understanding of the all processes by banking employees, availability of methods for regulators and debtors. Transparency of the methods and results is achieved also by the calculation of contributions of the initiating events to the credit risk [3, 4, 96].

In the proposed method, every credit is described by a set of 20 signs (parameters); each parameter having up to 10 levels of values (grades), whereas in reality a bank may have from some hundreds to some thousands of credits [4].

Transparency of credit risk can be determined by quantitative risk attributes of every grade, sign, credit and a set of all credits of the bank.

The quantitative attributes of the grade risk include:

- non-success credit probability,
- relative non-success probability among the sign grades,
- probability-frequency in the credits set,
- contribution to the risk model accuracy.

The quantitative attributes of the sign risk are:

- average non-success probability for the credit,
- structural weight and significance in the risk model,
- contribution to the credit risk,
- contribution to the average risk of the credit set.

The quantitative attributes of the credit risk include:

- non-success risk,
- possible losses,
- risk price,
- contribution to the average risk of the credit set.

The quantitative attributes of the risk of the credit set include:

- admitted risk,
- average risk,
- average losses,
- admitted losses,
- the number of credits,
- the number of dangerous credits,
- entropy of the risk of dangerous credits.

For every bank it is necessary to know how to determine the quantitative values of the above listed attributes, how to analyze the risk and to perform permanent monitoring of the credit risk attributes.

It is possible to optimize the credit risk model in order to increase its accuracy and robustness and to determinate of the optimal number of signs and grades in every sign.

7.4 Accuracy and robustness of credit risk estimation

Accuracy of recognition affects the following decisions: to give or not the credit, to determine the price (interest) for the credit risk, to compute the reservation level in case of the credit default. Accuracy is estimated by the number of relative errors in the recognition of bad and good credits (clients). The comparison of different methods on the same data showed that the risk estimation methods have a two-fold difference [3, 4].

Robustness characterizes the stability of estimation methods of credit risks. Different methods of risk estimation (or one method with the different training algorithms on statistical data) classify the credits as good and bad ones in different ways. A credit can be recognized as bad by one method and as good by another. Such instability in classification can amount to 20% of all credits. The comparison of methods on the same data showed (see [2]) that methods might have a seven-fold difference in robustness [3, 4, 30]. Several works indicated non-stability of the risk estimation methods [3, 97].

Recognition asymmetry is the ratio of the incorrectly recognized good and bad credits. The coefficient of the recognition asymmetry is introduced due to non-equivalence of losses caused by the wrong classification of good and bad credits. The value of the recognition asymmetry factor is preset from 1 to 10. Recognition asymmetry is achieved, for example, by the fact that the number of good credits for the model is not equal to the number of the good credits on statistics. Thus it is achieved what objects (good or bad) should be better recognized.

7.5 Specialization of banks and their risk models

Crediting juridical and natural persons is one of the main kinds of activity of national and commercial big, medium, and small banks. Every bank is special, as it works with different technologies, provides services to the specific segment of the market, and has its own strategic tasks. Competition also promotes the individuality of banks.

Credit business is connected with risk. The conditions of the credit activities are changing, the admitted risk on the market is changing as well. Credit activities are adapted to the conditions of the country's developing economics and to the population's living standards.

The methods of quantitative estimation and analysis of the credit risk have great importance for the maintenance of banks' stable functioning. The risk price should take into account the risk value of every credit. In addition to the mean value of the risk, which can be found by statistics of previous activities, the bank should also know the quantitative estimation of the risk for every credit.

Every bank develops its own model for the quantitative estimation and analysis of the credit risk with taking into account the general recommendations of Basel Committee on banking supervision. The higher accuracy of the credit risk estimation, the lower bank losses, banking interest, and higher bank's competitive ability. The society as a whole will benefit from higher accuracy, robustness, and transparency of risk estimation methods. The construction of the efficient risk model and the optimal credit risk management is possible only on the basis of constant quantitative analysis of statistical information on the credit success.

7.6 Axioms and models of credit risks

Axioms. There are objective foundations for the designing of the credit risk management system in the bank. They reflect the inevitability of losses and justify the necessity to use the admitted risk concept. These foundations include the following:

- Profit earning is inseparably accompanied by risk.
- Every banking employee can commit a fraud under pressure of life conditions, if it is possible to hide the fraud for some time and when adequate internal control of his or her activity is absent.
- Every commercial bank or company is able to commit fraud in the lack of transparency in its business and appropriate control of its activities.
- It is impossible to manage risk without quantitative methods of risk estimation.
- Behind the lack of transparency in the estimation methods of credit risk and ratings may hide frauds.
- Bank's chief management must think and act not only from positions of a bank's normal functioning maintenance but also to take into account possible actions of swindlers, business rivals, and government structures.
- One ought not to bring all possible innovations to bank management, because it is not practically possible to adjust the control system in case of such massive restructuring.
- Approved solutions from experience of other banks and companies should be adopted.
- It is necessary to take reasonable number of crediting process decisions, because it is not possible to provide high reliability for a big number of decisions.

It not possible to construct suitable technologies of the risk management system in the bank without taking into account the objective nature of the above listed regulations.

Models. Usually, the following models are to be used for estimation, analysis, and management of risk:

- Scenarios of non-success or default.
- Structural or graph models of the risk.
- Logical risk models.
- Probabilistic risk models.
- Models of trouble forecasting.
- Management models of the bank state and development of the bank.
- Models of restrictions imposed by regulators.
- Models of costs of decision-making and damage elimination in the absence of decisions.
- Models of the organizational management.
- Models of bank's state, in the form of table "State and parameters."

Information technology of bank management system should provide convenient implementation of these models and their connection through a database.

Admitted parameters values play a significant role in models of credit risk management, with admitted values parameters being considered as random variables. These variables are characterized by probabilities distribution and risk to exceed the admitted parameter values.

There are other alternative approaches when the only credit risk model is a postulated arithmetic expression with the expert weight for model parameters. That is, skipping the development of the credit risk scenario, its logic and probabilistic risk models, without taking into account the bank's individuality, users are offered some non-transparent program system, which in addition to that, is written in confidential program codes.

7.7 Bank management by risk criterion

The main disadvantage of conventional approaches to risk management is incapability to construct a unified (global) model of risk management for a lending agency as a whole, as well as corresponding technology and automated risk management system based on particular risk models on the bank's activity directions.

Control scheme. The problem of bank state and development management by risk criterion should be considered as a problem of complex object management. This task implies maintenance of the bank's state or changing its state from the initial to target one as a result of several steps. The bank's state at every stage will be assessed by non-success probability and possible financial losses values. [3, 4, 30].

Development can be considered as a process of bank state management (Fig. 7.5) with movement from the initial state (A) to the given final one (B) following the selected program path and with correction of the bank state

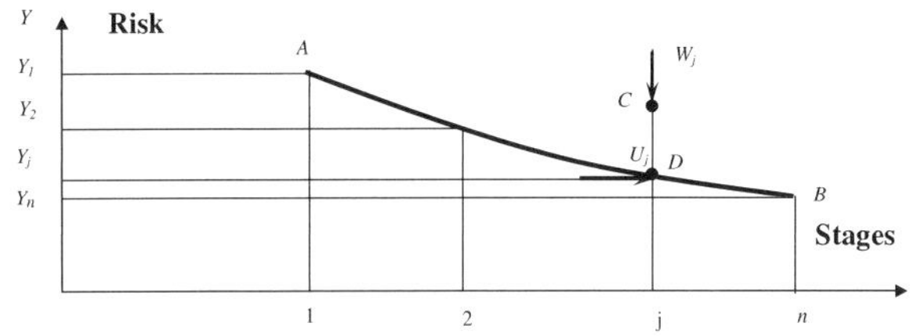

Fig. 7.5. The scheme of the bank management as the complex object management: Y are controlled parameters; U, W are managing and corrective actions

in case of deviation from the path. The program path (the line slope and curvature) are chosen by the bank's top management.

The program path passes inside the corridor of permissible values of controlled parameter Y. Such interpretation of the management problem of the bank's state and development uses the following concepts:

- $Y(P, T)$ are controlled parameters (risks and losses);
- $P(P_1, \ldots, P_n)$ are probabilities of failure on n directions of activities;
- $T(T_1, \ldots, T_n)$ are financial losses on n directions;
- $U(U_1, U_2, \ldots)$ are control actions (resources, assets, rates, volumes on n directions) that determine the states at stages;
- $W(W_1, W_2, \ldots)$ are corrective actions (resources, volumes on n directions) for the state retrieving to the program path in case of its deviation.
- $H(H_1, H_2, \ldots)$ are stages of bank progress.

For the development that follows the chosen path, risk parameter values Y can be obtained for different sets of controls U. It is also possible to return the bank from a state C to the given path $A - B$ with various sets of corrections W.

Information technology of bank management system by risk criterion should include the following procedures:

- non-success forecasting owing to parameters Y, i.e., when the parameters go beyond the corridor of their permissible values.
- Modeling or resource allocation to control the parameters Y, to control U and corrections W.
- Creation of development program with the determination of state parameters Y, controls U, and corrections W values at stages H.
- Processing current information and correction selection W.
- State and control models improvement.

For information technology of the bank's state and development management by risk criterion system, it is necessary to create a database that is to include the following structured information according to normative documents and the bank standards:

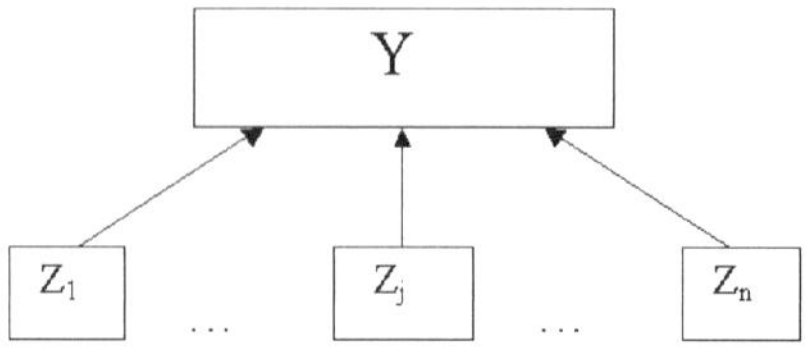

Fig. 7.6. Structural model of default (non-success) of bank

- Controlled parameters on n directions of the bank's activities.
- Credits' characteristics and gradations for natural and juridical persons; descriptions of other documents for the bank's activities directions according to Western standards.
- Resources on n directions of bank's activity that may be lost.
- Admissible probabilities of non-success and risk.
- Control actions (resources, assets, rates).
- Corrective actions (resources, assets, rates).

Total model of non-success risk. Total model of non-success risk can be constructed on the basis of particular risk models for the directions of the bank's significant activities (credits default, losses caused by poor investment decisions). Methods of risk models training (adjustment) can be designed by application of statistical data about successful banking operations.

The proposed method of logical-probabilistic modeling and the risk analysis of the bank's activities directions allows one to perform logical addition of all risks and to calculate the bank's total risk. Levels of the bank's default risk can be introduced.

Structural risk model of bank non-success is presented in Fig. 7.6. Here $Z_1, Z_2, \ldots, Z_n$ are independent binary variables for events of unsuccessful bank functioning on directions of the bank's activities that equal 1 (failure) or 0 (success) with the probabilities:

$$p\{Z_1 = 1\} = p_1, \ldots, p\{Z_n = 1\} = p_n;$$
$$p\{Z_1 = 0\} = 1 - p_1 = q_1, \ldots, p\{Z_n = 0\} = 1 - p_n = q_n. \tag{7.2}$$

Logical risk model of bank default (non-success) is as follows:

$$Y = Z_1 \vee Z_2 \vee \ldots \vee Z_n. \tag{7.3}$$

This model of the bank non-success (default) risk means that non-success occurs if any one, two ... or all directions of the bank's activities fail.

Let us write down L-function of credit non-success in an equivalent form after its orthogonalization:

$$Y = Z_1 \vee Z_2\overline{Z}_1 \vee Z_3\overline{Z}_2\overline{Z}_1 \vee \ldots. \tag{7.4}$$

Now we proceed from the logical description of non-success risk to its arithmetical description. P-model (P-polynomial) of the credit non-success risk is the following:

$$P = p_1 + p_2 q_1 + p_3 q_1 q_2 + \dots. \tag{7.5}$$

"Arithmetics" of P-model of risk implies that the total risk belongs to the interval [0,1] for any values of initial events probabilities. Possible bank losses as default consequences are equal to (see [2, 4]):

$$T = p_1 \cdot E_1 + p_2 \cdot E_2 + \ldots + p_n \cdot E_n, \tag{7.6}$$

where $E_1, E_2, \ldots, E_n$ are assets for the bank's activities directions; $p_1, p_2, \ldots, p_n$ are probabilities of default (non-success) on the bank's activities directions.

7.8 Conclusions

As a result of the fulfilled analysis, the following conclusions are made:

1. The scoring methods for credit classification currently applied by rating agencies and used by banks are not transparent enough;
2. Commercial banks are interested in using accurate, robust, and transparent methods and corresponding software for the estimation of credit risks of natural and juridical persons;
3. The requirements to the quality of credit risk estimation, namely, its accuracy, robustness, and transparency, are formulated. The transparency of estimation methods of credit risks has major significance for the analysis and management of credit risks;
4. The individuality of banks and their credit risk models is proved;
5. The axioms and models for credit risk estimation and analysis methods have been considered;
6. The scheme and procedures for the bank management by the risk criterion are proposed. The scheme of the risk summation for the bank's activities directions is considered;
7. The proposed approach to the logic and probabilistic modeling and analysis of the bank non-success risk has advantages over other methods.

8

Bases of Logic and Probabilistic Calculus

The majority of engineers is unfamiliar with mathematical logic and its symbolics. When it is difficult to explain with words, start to use symbols.

Sjamjuel H. Kolduell

In this chapter, using materials of works by I. A. Ryabinin [1, 2, 23, 24], we state the basic rules of the algebra of logic, necessary for constructed and investment of the LP-function of risk and safety of structure complex systems. The Boolean algebra or the algebra of logic is a part of the mathematical logic, studding logical operations on over the propositional. George Boulle is the founder of the algebra of logic because he first used the algebraic methods for decision of the logical problems.

8.1 Some information from Boolean algebra

Proposition in mathematical logic is understood as any statement such that there is sense in saying that it is true or false, so that the proposition may be either true or false. Propositions are commonly denoted by the capital letters A, B, C, etc.

A variable that takes only two values (1 or 0) is called a binary variable, and preassigned binary variables are named the arguments. There are propositions whose values are determined by the values of other propositions, i.e., they are their functions. The function that takes only two values (1 or 0) and determined by various set-ups of binary arguments is referred to as binary function or function of the algebra of logic

Mathematical logic studies the problems of presentation and transformation of binary functions of binary arguments by means of certain logical operations known as logical connections. Logical connections can be used to

E.D. Solojentsev, *Scenario Logic and Probabilistic Management of Risk in Business and Engineering*, DOI 10.1007/978-0-387-77946-1_8,

Table 8.1. Logic table for basic operations

x	y	$x \wedge y$	$x \vee y$	$\overline{x}$	$x \sim y$	$x \rightarrow y$
0	0	0	0	1	1	1
0	1	0	1	1	0	1
1	0	0	1	0	0	0
1	1	1	1	0	1	1

make from simple propositions compound ones taking the values "true" (1) or "false" (0) depending on the values of the incorporated simple propositions. Logical connections between propositions can be presented as operations with respect to binary variables. Let us now determine the basic logical operations.

Logic operations, such as conjunction (logic multiplication), disjunction (logic addition), negation, equivalence, implication, are usually denoted by the marks $\wedge, \vee, -, \sim, \rightarrow$, respectively and are represented by means of the truth Table 8.1.

It is the so-called tabular way of defining functions of algebraic logic (FAL). Along with the latter way, one can represent the functions with the help of the formulas in language containing variables x, y, z (possibly, with indexes) and symbols of some definite functions.

With the help of the equations of the algebra of logic, it is possible to describe conditions of efficiency or danger of systems. The equations show which elements (initiating conditions) and connections should be taken to guarantee performance of given system (or getting to a dangerous condition).

Let us consider now basic logical operations in more detail.

8.2 Basic logical operations

Conjunction. Conjunction or logical multiplication of propositions A and B is denoted by $A \wedge B$ (read: A and B). Sometimes, logical multiplication is denoted by symbols "&," "$*$," "x," or no sign at all is placed between the letters of the positions being multiplied:

$$A \wedge B = A \cdot B = A\&B = AB.$$

The value of truth of logic product $A\&B$ is determined depending on values of the truth of propositions A and B according to the following formulas:

$$0 \wedge 0 = 0; \quad 0 \wedge 1 = 0; \quad 1 \wedge 0 = 0; \quad 1 \wedge 1 = 1.$$

Conjunction $A\&B$ of two propositions is a compound proposition that is true if and only if its components A and B are true.

Disjunction. Disjunction or logical addition of two propositions A and B is denoted by the formula $A \vee B$ (to read: A or B). The value of the logical sum $A \vee B$ depending on the values of the components of propositions $A \vee B$ can be found from the following formulas:

$$0 \vee 0 = 0; \quad 0 \vee 1 = 1; \quad 1 \vee 0 = 1; \quad 1 \vee 1 = 1.$$

In the subsequent presentation, we shall also use (with the purpose of simplification of formulas) the matrix representation:

$$A \vee B = \begin{vmatrix} A \\ B \end{vmatrix}$$

Disjunction of two propositions A and B is a compound proposition that is false if and only if both addends A and B are false.

Negation. The negation of proposition A is denoted by A' (often $\overline{A}$) (to read: not A). The value of the proposition $\overline{A}$ can be found from the following expressions:

$$1' = 0; \quad 0' = 1.$$

Thus, the negation of proposition A is false when A is true, and true when A is false.

The above logical operations are not independent and can be expressed through each other. Logic expressions are transformed according to definite rules that will be considered below.

Rules for One Variable

1.	$A \wedge 1 = A;$	6.	$A \vee 0 = A;$
2.	$A \wedge 0 = 0;$	7.	$A \vee A = A;$
3.	$A \wedge A = A;$	8.	$A \vee A' = 1;$
4.	$A \wedge A' = 0;$	9.	$A'' = A;$
5.	$A \vee 1 = 1;$	10.	$A''' = A'.$

Rules 1–10 can easily be proved by replacing A by unity and zero. As a corollary, from rules 3 and 7 we have

$$A \wedge A \wedge \cdots \wedge A = A;$$
$$A \vee A \vee \cdots \vee A = A.$$

Unlike numeric algebra, the "multiplication of the variable by itself" or "canceling the similar terms" is carried out in the algebra of logic in conformity with the reduced identities without any exponents or coefficients.

Rules for Two or Three Variables

The conjunction and disjunction operations possess a number of properties analogous to the properties of ordinary operations of multiplication and addition. It is easy to see that *the associative law* holds for these functions:

11. $A \wedge (B \wedge C) = (A \wedge B) \wedge C = A \wedge B \wedge C;$
12. $A \vee (B \vee C) = (A \vee B) \vee C = A \vee B \vee C,$

along with the commutative law:

$$13.\quad A \wedge B = B \wedge A;$$
$$14.\quad A \vee B = B \vee A.$$

The rules 11–14 express the properties of conjunctions and disjunctions separately.

Because the associative and commutative laws are valid for logical multiplication and logical addition, the expression incorporating conjunctions and disjunctions may be written without brackets. In this case, it is agreed to consider the connection with the sign $\wedge$ to have higher priority than one with the sign $\vee$. This allows us to write the expression in the algebra of logic in the same way as in ordinary algebra (in the calculations the "primary" operations are performed before the "secondary" ones). This agreement allows us to write simply A $\wedge$ B $\vee$ C instead of (A $\wedge$ B)$\vee$ C.

Let us now consider the rules describing the connection between the operations of logic multiplication and addition taken together. It can be proved that *the distributive law* of conjunction with respect to disjunction holds for these functions:

$$15.\quad A \wedge (B \vee C) = (A \wedge B) \vee (A \wedge C), \tag{8.1}$$

along with the distributive law of disjunction with respect to conjunction:

$$16.\quad A \vee (B \wedge C) = (A \vee B) \wedge (A \vee C.$$

The latter law is not applied in ordinary algebra. Indeed,

$$a + bc \neq (a + b) \cdot (a + c).$$

It should be pointed out that all the three laws have a "symmetry" in the sense that a corresponding law for conjunction (disjunction) can be obtained from any law for disjunction (conjunction) by replacing the signs of disjunction by the signs of conjunction and vice versa. Indeed, if we take, for example, expression for (8.1)

$$A \wedge (B \vee C) = (A \wedge B) \vee (A \wedge C).$$

and change the signs we shall get

$$A \vee (B \wedge C) = (A \vee B) \wedge (A \vee C).$$

The following law is known in the literature under the name *of duality or inversion law*, and makes it possible to replace negation of conjunction by disjunction of negations and negation of disjunction by conjunction of negations:

$$\left.\begin{aligned} &17.\quad (A \wedge B)' = A' \vee B'; \\ &18.\quad (A \vee B)' = A' \wedge B'. \end{aligned}\right\} \tag{8.2}$$

If the rule 9 is applied to the expressions (8.2), we shall have:

$$\left.\begin{array}{ll} 19. & (A \wedge B) = (A' \vee B')'; \\ 20. & (A \vee B) = (A' \wedge B')'. \end{array}\right\} \qquad (8.3)$$

The latter two rules (8.3) are called de Morgan's formulas in honor of one of the founders of mathematical logic; the rules allow logical multiplication to be expressed through the negation of the logical sum of inverse propositions, and the logical sum through the negation of the logical product of inverse propositions. Formulas (8.3) can easily be generalized for any arbitrary number of logical variables, namely:

$$\bigwedge_{i=1}^{n} x_i = \left(\bigvee_{i=1}^{n} x_i'\right)';$$

$$\bigvee_{i=1}^{n} x_i = \left(\bigwedge_{i=1}^{n} x_i'\right)',$$

where the logical variables are denoted by one letter x with the index $i (i = 1, 2, \ldots, n)$, and the signs of conjunctions and disjunctions are used similarly to the signs of the product $\prod$ and the sum $\sum$ employed in ordinary algebra.

The above-given basic laws can be used to establish a number of other useful relations that make it possible to simplify composite logical expressions.

Let us first introduce operations of absorption and joining. The operation of absorption is determined by the relations

$$\begin{array}{ll} 21. & (A \wedge B) \vee A = A; \\ 22. & A \wedge (B \vee A) = A. \end{array}$$

The operation of joining is determined by the relations

$$\begin{array}{ll} 23. & (A \wedge B) \vee (A \wedge B') = AB \vee AB' = A(B \vee B') = A \cdot 1 = A; \\ 24. & (A \wedge B) \vee (A' \wedge B) = AB \vee A'B = B(A \vee A') = B \cdot 1 = B. \end{array}$$

where the operation of logical multiplication is written without the sign of conjunction.

Let us now simplify the expression $A \wedge (A' \vee B)$. On the basis of the distributive law of conjunction with respect to disjunction on rule 15, we have

$$A \wedge (A' \vee B) = (A \wedge A') \vee (A \wedge B).$$

In conformity with rule 4, $A \wedge A' = 0$ and therefore

$$A \wedge (A' \vee B) = 0 \vee (A \wedge B).$$

Utilizing rule 6, we finally obtain

$$25. \quad A \wedge (A' \vee B) = A \wedge B.$$

On the basis of the distributive law of disjunction, using rule 16, we have

$$A \vee (A' \wedge B) = (A \vee A') \wedge (A \vee B).$$

By rule 8, $A \vee A' = 1$ we have

$$A \vee (A' \wedge B) = 1 \wedge (A \vee B).$$

It can also be shown that

$$26. \quad A \wedge (A' \wedge B) = A \wedge B.$$

The operation of the generalized joining is determined by the relations

$$\left.\begin{array}{ll} 27. & AB \vee B'C = AC \vee AB \vee B'C; \\ 28. & \left|\begin{array}{c|c} A & B' \\ B & C \end{array}\right| = \left|\begin{array}{c|c|c} A & A & B' \\ C & B & C \end{array}\right| \end{array}\right\} \qquad (8.4)$$

The proof for the first formula of (8.4) is carried out by logic multiplication of the first term by $1 \vee C$ and of the second term by $1 \vee A$ and subsequent application of rules 15 and 23. The proof for the second formula of (8.4) is carried out by addition to the first cofactor the term $0 \wedge A$ and to the second cofactor the term $0 \wedge C$ and by application of rules 16 and 24.

Let us illustrate this proof for rule 27 in the matrix form:

$$\left|\begin{array}{c} AB \\ B'C \\ \\ \\ \end{array}\right| = \left|\begin{array}{c} AB(1 \vee C) \\ B'C(1 \vee A) \\ \\ \\ \end{array}\right| = \left|\begin{array}{c} AB \\ ABC \\ B'C \\ B'CA \end{array}\right| = \left|\begin{array}{c} AB \\ B'C \\ ABC \\ B'CA \end{array}\right| = \left|\begin{array}{c} AB \\ B'C \\ AC\left|\begin{array}{c} B \\ B' \end{array}\right. \end{array}\right| = \left|\begin{array}{c} AB \\ B'C \\ AC \\ \\ \end{array}\right|$$

8.3 Basic definitions and accepted notations

Let us introduce the "exponent" of argument x, which will be denoted by $x_i^{\alpha_i}$ where α_i is the binary variable. We shall assume that

$$x_i^{\alpha_i} = \left| \begin{array}{l} x_i, \ if\ \alpha_i = 1, \\ x_i', \ if\ \alpha_i = 0. \end{array}\right.$$

Definition 1. Expression of the type

$$K = x_1^{\alpha_1} x_2^{\alpha_2} \ldots x_r^{\alpha_r} \qquad (8.5)$$

we shall name elementary conjunction (K) of rank r. Because $x_i x_i' = 0$ and $x_i x_i \ldots x_i = x_i$, then all letters are different in the elementary conjunction.

There are exactly 2^n of different binary set $(\alpha_1, \alpha_2, \ldots \alpha_r)$ and, hence, 2^n of different type conjunctions.

Definition 2. Expression of the type

$$K_1 \vee \ldots \vee K_j \vee \ldots \vee K_s,$$

where K_j are elementary conjunctions of various ranks, we shall name *the disjunctive normal form (DNF)*. For example, the function

$$f(x_1, \ldots x_4) = x_1 x_2 \vee x_1 x_2 x_3' \vee x_1' x_3 x_4$$

is written down in DNF, as it has three terms being elementary conjunctions.

Definition 3. If the function $f(x_1, x_2, \ldots, x_n)$ is written down in DNF, and the rank of each elementary conjunction is equal to n, then such DNF is named *the perfect disjunctive normal form (PDNF)*, and conjunctions are members of $PDNF$.

Definition 4. Expression of the type

$$x_1^{\alpha_1} \vee x_2^{\alpha_2} \vee \ldots \vee x_r^{\alpha_r}$$

is named the elementary disjunction (D) of the rank r.

Definition 5. We shall say that two elementary conjunctions are *orthogonal* ones, if their product is equal to zero. For example, the product of elementary conjunctions $x_1 x_2'$ and $x_1 x_2 x_3 x_4$ is equal to zero, as one of them contains x_2', and another of them contains x_2 and, hence, they are orthogonal.

Definition 6. A DNF is called orthogonal disjunctive normal form $(ODNF)$, if all its members are orthogonal in pairs. According to this definition, $PDNF$ is $ODNF$, as all its members are orthogonal in pairs. But $PDNF$ is the most uneconomical of all forms of $ODNF$, as it contains the maximum quantity of the letters.

Definition 7. Iteration-free DNF is such DNF in which all letters have different indexes. The letters x_i and x_i' have the same index, therefore they cannot simultaneously enter in iteration-free disjunctive normal form $(IFDNF)$.

Definition 8. Iteration-free form of FAL is the form in which all letters have different indexes. The special case of the iteration-free form of FAL is $IFDNF$. For example, the function

$$f(x_1, \ldots, x_8) = \left| \begin{array}{c|c|c|c|c} x_1 & x_2 & \vee & x_5 & x_6 \\ & x_3 & & & x_7 x_8' \\ & x_4' & & & \end{array} \right|$$

is written down in the iteration-free form, as all letters have different indexes.

Definition 9. The probabilistic function (PF) is the probability that FAL is true

$$P\{f(x_1, \ldots, x_n) = 1\}.$$

Definition 10. Functions of the algebra of logic, admitting direct transition to the PF by replacement of logic variables by probabilities and logic operations by the appropriate arithmetic operations, are named the forms of transition to replacement (FTR).

Definition 11. The probability function mixed form ($MFFP$) is the form of the function obtained as the result of partial replacement in FAL of the logic variables by the probabilities and containing simultaneously two types of variables (logic variables and probabilities) and two systems of operations (logic and arithmetic ones).

The feature of $MFFP$ is as follows: all the relations to arguments are determined in the explicit form through used elementary operations (logic and arithmetic ones). It cannot contain the operators of the type $P\{f = 1\}$, if the explicit expression of such functions as PF or $MDNF$ is not known. The mixed form has a simple probabilistic sense. If the function after the replacement of some logic variables has yet no replaced variables of the vector X, then $P\{f = 1\} = P(X)$. This expression has a sense of the conditional probability that $f = 1$. Moreover, the conditions are written down with the help of not replaced logic variables. On assigning value of the vector X, the probability $P(X)$ turns into the conditional probability, written down in the form usual for the probability theory.

Definition 12. The form of FAL, admitting the transition from $MFFP$ by replacement of a part of logic variables by the appropriate probabilities and the logic operations by arithmetic ones, and by moving not replaced logic variables to parameters of a degree of probabilities, is named the form of transition to the partial replacement ($FTPR$). $FTPR$ is special case of FTR with the form of transition to complete replacement ($FTCR$), in which the replacement of all logic variable is simultaneously made.

Definition 13. Operation of equivalence of propositions A and B is identified by a symbol " $\sim''$. The value of the expression $A \sim B$ is determined from the following relations:

$$0 \sim 0 = 1; \quad 0 \sim 1 = 0; \quad 1 \sim 0 = 0; \quad 1 \sim 1 = 1. \tag{8.6}$$

Definition 14. The *negation* of the equivalence of the propositions A and B (this operation names as logic summation modulo two) is denoted by the symbol $\dot{\vee}$ or $\oplus$. The value of *negation* of equivalence of propositions A and B is determined from the following relations:

$$0 \oplus 0 = 0; \quad 0 \oplus 1 = 1; \quad 1 \oplus 0 = 1; \quad 1 \oplus 1 = 0.$$

Sometimes this operation is named the strict disjunction and denoted by "$\vee\vee$" (or $\dot{\vee}$). In this case, the sign, connecting the propositions A and B, is understood not in sense "or" but in sense "or–or." From the relation (8.6) it is obvious that the strictly dividing proposition $A \dot{\vee} B$ is true only when A is false, B is true and when A is true, B is false.

For the logic summation modulo two the commutative and associative laws take places, along with the distributive law over the operation of the conjunction:

$$29.\quad A \oplus B = B \oplus A;$$
$$30.\quad A \oplus (B \oplus C) = (A \oplus B) \oplus C;$$
$$31.\quad A \wedge (B \oplus C) = (A \wedge B) \oplus (A \wedge C).$$

The obvious relations can be derived:

$$32.\quad A \oplus 1 = A'; \qquad 34.\quad A \oplus A = 0;$$
$$33.\quad A \oplus 0 = A; \qquad 35.\quad A \oplus A' = 1.$$

The above mentioned basic logic operations are connected with the logical summation modulo two by the following formulas:

$$\left.\begin{array}{r} 36.\quad A \vee B = A \oplus B \oplus AB; \\ 37.\quad A \wedge B = A \oplus AB'; \\ 38.\quad A \oplus B = AB' \vee A'B. \end{array}\right\} \qquad (8.7)$$

Definition 15. The Boolean difference (or the logical difference) of the function $f(x_1, \ldots, x_n)$ with respect to argument x_i is the result of the logic summation modulo two of the initial function and the function received from the initial one by the replacement of argument x_i by its *negation*:

$$\Delta_{x_i} f(x_1, \ldots, x_n) = f(x_1, \ldots, x_i, \ldots, x_n) \oplus f(x_1, \ldots, x_i', \ldots, x_n). \qquad (8.8)$$

Definition 16. We shall say that a function $f(x_1, x_2, \ldots, x_n)$ is symmetric with respect to x_i, if

$$f_{x_i'}(x_1, \ldots, x_n) = f(x_1, \ldots, x_i', \ldots, x_n). \qquad (8.9)$$

Definition 17. Functions, obtained by replacement in the initial FAL of the argument x_i by 1 and 0, is named the unit and zero functions with respect to argument x_i and depicted accordingly:

$$f_1^i(x_1, \ldots, x_n) = f(x_1, \ldots, 1, \ldots, x_n); \qquad (8.10)$$

$$f_0^i(x_1, \ldots, x_n) = f(x_1, \ldots, 0, \ldots, x_n). \qquad (8.11)$$

Definition 18. The function $f(x_1, \ldots, x_n)$ is the monotonous one, if for any sets $(a_1, \ldots, a_n)$ and $(b_1, \ldots, b_n)$, such that $a_i \leq b_i$, the following relation holds:

$$f(\alpha_1, \ldots, \alpha_n) \leq f(\beta_1, \ldots, \beta_n)$$

Definition 19. The function, written down in the form of a matrix, in which conjunctions are the logic symbols in a line, and disjunctions are in a column, is named *the logic matrix.*

To the logic matrixes all the known transformations of the algebra of logic can be applied. So, the commutative law for conjunction allows rearrangement of symbols in a line, and the commutative law for disjunction allows rearrangement of lines of the logic matrix.

Let FAL have the form:

$$\begin{aligned} f(x_1,\ldots,x_8) = \{\{x_1 \wedge x_3 \wedge [x_5 \vee (x_4 \wedge x_6 \wedge x_8)]\}\vee \\ \vee\{x_2 \wedge x_4 \wedge [x_6 \vee (x_3 \wedge x_5 \wedge x_8)]\}\} \wedge x_7. \end{aligned} \tag{8.12}$$

In the matrix form the equation (8.12) can be written as follows:

$$f(x_1,\ldots,x_8) = \left|\begin{array}{l|c|} x_1x_3 & x_5 \\ & x_4x_6x_8 \\ x_2x_4 & x_6 \\ & x_3x_5x_8 \end{array}\right| x_7 = \left|\begin{array}{c} x_1x_3x_5x_7 \\ x_1x_3x_4x_6x_8x_7 \\ x_2x_4x_6x_7 \\ x_2x_4x_3x_5x_8x_7 \end{array}\right| \tag{8.13}$$

The second matrix of the equation (8.13) is written down in DNF.

The inversions law (8.2) for logic matrixes is applied by replacement of conjunctive connections of logic symbols in a line to disjunctive connections of negations of these symbols, placed in a column, and disjunctive connections between lines to conjunctive connections between columns, formed from these lines. Applying the inversions law to the logic matrix (8.13), we obtain

$$f(x_1,\ldots,x_8)' = \left|\begin{array}{c} x_1x_3x_5x_7 \\ x_1x_3x_4x_6x_8x_7 \\ x_2x_4x_6x_7 \\ x_2x_4x_3x_5x_8x_7 \end{array}\right|' = \left|\begin{array}{c|c|c|c} x_1' & x_1' & x_2' & x_2' \\ x_3' & x_3' & x_4' & x'4 \\ x_5' & x_4' & x_6' & x_3' \\ x_7' & x_6' & x_7' & x_5' \\ & x_8' & & x_8' \\ & x_7' & & x_7' \end{array}\right|$$

8.4 Theorems of Boolean algebra and probabilistic logic

The close connection between event probability theory and mathematical logic was observed a long time ago. Now mathematical logic and probability theory are united on the new basis of the logic and probabilistic calculus.

The probability theory quantitatively estimates the reliability or safety of systems, which structure is described by means of the mathematical logic.

In practical application of the logic and probabilistic methods of research of reliability and safety of structural-complex systems, the basic difficulty is transformation of arbitrary FAL to the forms of transition to complete replacement ($FTCR$). In order to make this transformation standard and mathematically strict, it was necessary to construct a novel “bridge” between the algebra of logic and the probability theory. The history of creation of logic and probabilistic methods (LPM) and contribution of the individual scientists to its creation and development are described in work [25]. Omitting the strict proofs of the special theorems, properties, and algorithms that form the mathematical basis of LPM, here we formulate only their essence needed for the subsequent practical application.

Theorem 1. Arbitrary FAL, depending on n arguments $n \geq 1$, can be given in the form:

$$f(x_1, \ldots, x_n) = x_i f_1^{(i)}(x_1, \ldots, x_n) \vee x_i' f_0^{(i)}(x_1, \ldots, x_n). \tag{8.14}$$

The expression (8.14) is known as Shannon's formula of expansion. It is correct also for the algebra of modulo two. Applying rule 36 to the right part (8.14), we obtain

$$\begin{aligned} f(x_1, \ldots, x_n) &= x_i f_1^{(i)}(x_1, \ldots, x_n) \vee x_i' f_0^{(i)}(x_1, \ldots, x_n) \\ &= x_i f_1^{(i)}(x_1, \ldots, x_n) \oplus x_i' f_0^{(i)}(x_1, \ldots, x_n) \\ &\quad \oplus x_i x_i' f_1^{(i)}(x_1, \ldots, x_n) f_0^{(i)}(x_1, \ldots, x_n) \\ &= x_i f_1^{(i)}(x_1, \ldots, x_n) \oplus x_i' f_0^{(i)}(x_1, \ldots, x_n). \end{aligned}$$

Corollary 1. The Boolean difference of arbitrary FAL with respect to argument x_i can be written as follows:

$$\Delta_{x_i} f(x_1, \ldots, x_n) = f_1^{(i)}(x_1, \ldots, x_n) \oplus f_0^{(i)}(x_1, \ldots, x_n). \tag{8.15}$$

For the proof of the equivalence of expressions (8.8) and (8.15), the formula of decomposition and rules 29–38 are used. According to (8.8) and (8.9), we have

$$\begin{aligned} \Delta_{x_i} f(x_1, \ldots, x_n) &= f(x_1, \ldots, x_n) \oplus f_{x_i'}(x_1, \ldots, x_n) = \\ &= [x_i f_1^{(i)}(x_1, \ldots, x_n) \oplus x_i' f_0^{(i)}(x_1, \ldots, x_n)] \\ &\quad \oplus\ [x_i' f_1^{(i)}(x_1, \ldots, x_n) \oplus x_i f_0^{(i)}(x_1, \ldots, x_n)] \\ &= (x_i \oplus x_i') f_1^{(i)}(x_1, \ldots, x_n) \oplus (x_i' \oplus x_i) f_0^{(i)}(x_1, \ldots, x_n) \\ &= f_1^{(i)}(x_1, \ldots, x_n) \oplus f_0^{(i)}(x_1, \ldots, x_n). \end{aligned}$$

Corollary 2. The Boolean difference of arbitrary FAL with respect to argument x_i can be presented using basic operations of conjunction, disjunction, and negation in the following form:

$$\Delta_{xi} f_1(x_1, \ldots, x_n) = \left| \begin{matrix} f_1^{(i)}(x_1, \ldots, x_n) * f\prime_0^{(i)}(x_1, \ldots, x_n) \\ f\prime_1^{(i)}(x_1, \ldots, x_n) * f_0^{(i)}(x_1, \ldots, x_n) \end{matrix} \right|$$

The latter corollary follows from formula (8.15) and rule 38.

Theorem 2. It is possible to present any FAL, depending on n arguments ($n \geq 1$), in the form:

$$f(x_1, \ldots, x_i,\ x_{i+1}, \ldots, x_n) =$$
$$\bigvee x_1^{\alpha_1} x_2^{\alpha_2} \ldots x_i^{\alpha_i} f(\alpha_1, \alpha_2, \ldots, \alpha_i,\ x_{i+1}, \ldots, x_n). \tag{8.16}$$

This theorem is named the theorem of decomposition of any FAL with respect to any number of arguments $x_1, x_2, \ldots, x_i$. It is also correct to name the expression (8.16) the decomposition formula by D. A. Pospelov, who proved Theorem 2 in 1964.

After decomposition of FAL in all n arguments, we obtain the perfect disjunctive normal form (PDNF) of initial function, which can be written down in the form

$$f(x_1, \ldots, x_n) = \bigvee_1 x_1^{\alpha_1} x_2^{\alpha_2} \ldots x_n^{\alpha_n},$$

where the symbol $\bigvee_1$ means that the disjunction is taken only on sets $\langle \alpha_1, \alpha_2, \ldots, \alpha_n \rangle$, satisfying the equality

$$f(\alpha_1, \alpha_2, \ldots, \alpha_n) = 1.$$

Theorem 3. For all monotonous FAL, the set of sets on which the zero function with respect to argument x_i accepts unit value is the subset of the set of sets on which the unit function with respect to the argument x_i is equal to unit

$$\{(x_1, \ldots, x_n) : f_0^{(i)}(x_1, \ldots, x_n) = 1\}$$
$$\subset \{(x_1, \ldots, x_n) : f_1^{(i)}(x_1, \ldots, x_n) = 1\}.$$

The proof of the theorem is given in [24]. From Theorem 3 follows five consequences. The formulas essentially facilitate logic transformations for monotonous FAL.

$$\begin{aligned}
&1) \{(x_1, \ldots, x_n) : f_0^{(i)}(x_1, \ldots, x_n) = 1\} \subset \\
&\quad \subset \{(x_1, \ldots, x_n) : f(x_1, \ldots, x_n) = 1\} \subset \\
&\quad \subset \{(x_1, \ldots, x_n) : f_1^{(i)}(x_1, \ldots, x_n) = 1\} \\
&2) f_1^{(i)}(x_1, \ldots, x_n) \vee f_0^{(i)}(x_1, \ldots, x_n) \equiv f_1^{(i)}(x_1, \ldots, x_n); \\
&3) f_1^{(i)}(x_1, \ldots, x_n) \wedge f_0^{(i)}(x_1, \ldots, x_n) \equiv f_0^{(i)}(x_1, \ldots, x_n); \\
&4) f_1^{(i)}(x_1, \ldots, x_n) \wedge f\prime_0^{(i)}(x_1, \ldots, x_n) \equiv \Delta_{x_i} f(x_1, \ldots, x_n); \qquad (8.17) \\
&5) f\prime_1^{(i)}(x_1, \ldots, x_n) \wedge f_0^{(i)}(x_1, \ldots, x_n) \equiv 0.
\end{aligned}$$

Theorem 4. The partial derivative from the probability of the truth of monotonous FAL $f(x_1, \ldots, x_n)$ in the probability of the truth of the argument

x_i is equal to probability of the truth of the Boolean difference of this function with respect to the argument x_i:

$$\frac{\partial P\left\{f(x_1,\ldots,x_n)=1\right\}}{\partial P\left\{x_i=1\right\}}=P\left\{\Delta_{x_i}f\left(x_1,\ldots,x_n\right)=1\right\}.$$

Theorem 5. The probability of the truth of any FAL, presented in $ODNF$, is equal to the sum of probabilities of the truth of all orthogonal members in this FAL:

$$P\left\{f(x_1,\ldots x_n)=\bigvee_{i=1}^{s}O_i=1\right\}=\sum_{i=1}^{s}P\left\{O_i=1\right\},$$

where O_i can be not only elementary orthogonal conjunctions of $ODNF$ but also arbitrary FAL, orthogonal in pairs.

Theorem 6. The disjunction of orthogonal iteration-free forms in the basis of conjunction-negation is the transition form to the complete replacement ($FTCT$).

This assertion follows from Theorem 5 and the fact that each term in the initial disjunctive form is $FTCT$.

Now some forms of transition to the complete replacement are known: $PDNF$, $ODNF$, iteration-free FAL in the basis of conjunction-negation.

If FAL is presented in $FTCT$, then the transition to the probabilistic function is carried out by the following rules:

(1) Each letter in $FTCT$ is replaced by the probability of its equality to unit

$$p\left\{x_i=1\right\}=r_i,\quad p\left\{x_i=0\right\}=p\left\{x_i'=1\right\}=q_i=1-r_i;$$

(2) The function negation is replaced by the difference between unit and the probability of equality of this function to unit, for example,

$$\begin{aligned}&p\{f(x_1,\ldots,x_7)=[(x_1x_2)'(x_3x_4)'(x_5(x_6'x_7')')']'=1\}\\&\quad=1-(1-r_1r_2)(1-r_3r_4)[1-r_5(1-q_6q_7)];\end{aligned}$$

(3) The operations of the logic multiplication and additions are replaced with operations of the arithmetic multiplication and addition.

PF for FAL, written in any iteration-free form, can be found from its expression in the basis of conjunction-negation, which is obtained by repeated application of de Morgan's rules (8.3).

Let, for example,

$$f(x_1,\ldots,x_8)=x_1(x_2\vee x_3\vee x_4')\vee x_5(x_6\vee x_7x_8')$$

and one should find $p\{(x_1,\ldots,x_8)=1\}$. As this function is iteration-free FAL (though it is not DNF), we have

$$f(x_1, \ldots, x_8) = \{[x_1(x_2'x_3'x_4)']'\{x_5[x_6'(x_7x_8')]'\}'\}';$$
$$P\{f(x_1, \ldots, x_8) = 1\} = 1 - \{1 - r_1[q_2q_3r_4]\}$$
$$\times\{1 - r_5[1 - q_6(1 - r_7q_8)]\}.$$

In the conclusion, we emphasize again that it is only the bases of the logic and probabilistic calculus. Specific logic and probabilistic methods will be described later. Besides, it is necessary to keep in mind that in the fundamental mathematical encyclopedias and directories, *LPM* are not mentioned yet, and, hence, are not studied by pure mathematicians at universities. Practically all creators of modern understanding of the probabilistic logic, as well as J. Boole, did not have special mathematical education, being high quality engineers and applied mathematicians.

9

LP-Modeling and Analysis of Risk in Engineering

The script of serviceability (danger) of the system, formalized on language of algebra of logic, is the main attribute of logic and probabilistic calculus of safety of structural-complex system.
I. A. Ryabinin

The logic and probabilistic method (LP-method), created by I. A. Ryabinin, is presented below mainly on the basis of his works [2, 98, 99].

9.1 Basic concepts and definitions of the theory of safety

The scientific approach to the problem of safety requires carrying out the integrated analysis, classification of failures and accidents, basic influencing factors, behavior of environment, and actions of the personnel. For answering these questions, appropriate methods of mathematical modeling, physical and economic models of origin, and development of technical and economic accidents are necessary.

We do not have pretensions to create the general theory of safety and risk of any complex system, and below we shall only consider approaches to development of the risk and safety LP-theory of the structurally complex systems. However, this LP-theory can be the basis for other concepts of safety and risk that take into account not only logic connections but also other connections: physical, functional, economical, organizational, financial, etc.

First of all, by structurally complex systems (SCS) we understand the systems that at their mathematical description cannot be reduced to consecutive, parallel, or treelike structures. The structurally complex systems are described by a scenario of the network type with cycles and recurrence of arguments at their formalization. Secondly, by structurally complex systems we understand also the systems with a large number of states both for system elements and the system itself.

E.D. Solojentsev, *Scenario Logic and Probabilistic Management of Risk in Business and Engineering*, DOI 10.1007/978-0-387-77946-1_9,

The structure of connections inside society, business, finance, etc., is not simple. These systems have large number of interconnections and states of elements and the fact is not always taken into account in mathematical models of risk. In this chapter and later, the original risk LP-theory for complex technical, organizational, and banking systems is stated; the theory takes into account the above mentioned features of SCS.

Fundamental concepts of the risk LP-theory are the concept of the dangerous state of the system, characterized by the large-scale damage, and the concept of the danger, that is, the ability of the system to get to a dangerous state [2, 98, 100, 101].

In each particular case, it is necessary to give the analytical description of that dangerous state of SCS that can result in the accident. In the risk LP-theory, such description begins with drawing up *the scenario of the dangerous state*, which is done with the help of conjunctions AND and disjunctions OR of *initiating events and conditions.* They are various external and internal influences, failures, violating the service and storage conditions, mistakes of personnel, etc.

In the reliability theory, the formalization of the concept of efficiency is carried out with the help of the structural diagrams of functioning. In the risk theory, the formalization of the concept of *the dangerous state* is done with the help of *the scenario of the dangerous state.*

9.2 The basic principles of the LP-method

To each logically connected element of system corresponds the logic variable Z_j, $j = 1, 2, \ldots, n$ that describes the state of the element j:

$$Z_j = \begin{cases} 1, & \text{if the element } j \text{ is serviceable,} \\ 0, & \text{if the element } j \text{ is out of order.} \end{cases}$$

Then all the set of the possible states of the system can be presented by the set of vectors $\{Zn\}$ consisting 2^n various vectors.

In some tasks on the graphs, the routes as sequence of graph edges are sought. The route, which does not contain repeating arches in the oriented graph, is named a path. The presence of the path in the graph of the structural system corresponds with its serviceable condition. In LP-method, the following definition is used. The minimum path of system functioning ($MPSF$) represents such a conjunction of its elements in which none of its components may be withdrawn without disturbing the functioning of the system. Such a conjunction may be written in the form of the following FAL:

$$W_l = \bigwedge_{j \in Kw_l} Z_j,$$

where Kw_l is the set of the numbers corresponding with the given path.

In other words, $MPSF$ describes one of the possible independent variants of fulfilling the task by the system with the minimum set of efficient elements that are absolutely necessary to perform the given work.

In some tasks on the graphs, it is required to find the set of edges, named the cut or section, such that in moving away from the cut, the graph loses its integrity and breaks up into two or more components. From the point of view of reliability, it would indicate loss of serviceability. LP-method defines the minimum section of system failures ($MSSF$) to be such a conjunction of the negations of its elements, in which none of its components may be withdrawn without disturbing the condition of inefficiency of the system. Such conjunction may be written in the form of the following FAL:

$$S_i = \bigwedge_{i \in K_{S_j}} x_i',$$

where K_{S_j} means the set of the numbers corresponding with the given cut.

In other words, $MSSF$ describes one of the possible versions of disturbing the capability of the system to work with the help of minimum set of failed elements.

$MPSF$ and $MSSF$ are dual in the relation to each other. Each real system has a finite number of minimum paths (l=1,2, ..., d) and minimum sections (t=1,2, ...,n). Utilizing these concepts, the conditions of capability of a system to work may be written differently:

(a) either in the form of $MPSF$

$$y(x_1, \ldots, x_m) = \bigvee_{l=1}^{d} W_l = \bigvee_{l=1}^{d} \left[\bigwedge_{i \in K_l} x_i \right]; \tag{9.1}$$

(b) or as the conjunction of negations of $MSSF$)

$$y(x_1, \ldots, x_m) = \bigvee_{j=1}^{n} S_j' = \bigvee_{j=1}^{n} \left[\bigvee_{i \in K_{S_j}} x_i \right]. \tag{9.2}$$

Thus, the conditions of capability of a real system to work may be represented in the form of conditions of capability of a certain equivalent system (in the sense of reliability) whose structure is taken to be parallel connection of $MPSF$, or any other equivalent system the structure of which is a serial connection of negations of $MSSF$.

$MPSF$ or $MSSF$ form the complete probabilistic space of events. Formula of the type (9.2) as disjunctions of conjunctions and without brackets is named the disjunctive normal form (DNF). The number of conjunctions in DNF defines the dimension of DNF.

Example. As an example, we write down the logic expressions for $MPSF$ and $MSSF$ for "bridge" (Fig. 9.1), which we consider as the electrical circuit:

$$y = y_3 \vee y_4 = z_1 z_3 \vee z_2 z_4 \vee z_1 z_4 z_5 \vee z_2 z_3 z_5; \tag{9.3}$$

$$\overline{y} = \overline{y_3 y_4} = \overline{z_1 z_2} \vee \overline{z_2 z_3 z_5} \vee \overline{z_1 z_4 z_5} \vee \overline{z_3 z_4}. \tag{9.4}$$

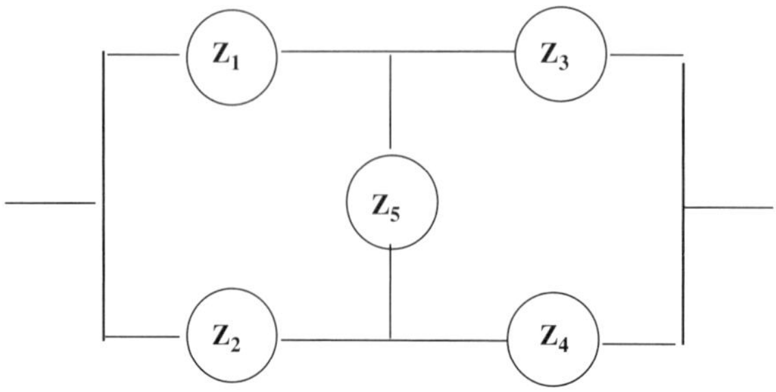

Fig. 9.1. Structural model of risk of "bridge" type

9.3 Transformation of L-function to P-polynomial

The algorithm of orthogonalization is based on transformation of $FALs$ to $ODNF$. The transition from the risk logic function to the risk probabilistic function (the polynomial) is not simple. It is connected with orthogonalization of the risk L-function that is written down in DNF. Only for the orthogonal DNF it is possible to replace the variables z_j and $\overline{z}_j$ by the probabilities p_j and q_j. We replace the mark of disjunction $\vee$ by the mark of addition $+$, and the mark of conjunction $\wedge$ by the mark of multiplication $\cdot$. Let us describe some methods of orthogonalization of L-functions.

Method of the direct orthogonalization. Let DNF be written down in the tabular form [1, 2]:

$$Y(Z) = \begin{vmatrix} z_1 \wedge z_3 \wedge z_5 \\ z_2 \wedge z_4 \wedge z_6 \\ z_1 \wedge z_3 \wedge z_4 \wedge z_6 \wedge z_8 \end{vmatrix} = \begin{vmatrix} K_1 \\ K_2 \\ K_3 \end{vmatrix},$$

where conjunctions stand in lines and disjunctions stand between lines.

Procedure of the direct orthogonalization is performed following the scheme:

$$Y(Z) = \begin{vmatrix} k_1 \\ \overline{k}_1 \wedge k_2 \\ \overline{k}_1 \wedge \overline{k}_2 \wedge k_3 \end{vmatrix};$$

$$\overline{k}_1 = \begin{vmatrix} \overline{z}_1 \\ z_1 \wedge \overline{z}_3 \\ z_1 \wedge z_3 \wedge \overline{z}_5 \end{vmatrix}.$$

The dimension of the function $Y(Z)$, as it is easy to see, may essentially increase in result of the orthogonalization. Therefore other methods of orthogonalization of logic functions were suggested and developed. They are adapted for calculation on the computer and not so demanding to resources of memory. Amidst the methods we should name the algebra of cortege and the algebra of mixed forms and recurrence sequences [1, 2].

Orthogonalization of logic function by the method of conditional probabilities. As an example, let us consider the orthogonalization of the logic function by the method of conditional probabilities for a "bridge" (Fig. 9.1). We shall denote conditions by the symbol "|".

$$y = y_3 \vee y_4 = z_1 z_3 \vee z_2 z_4 \vee z_1 z_4 z_5 \vee z_2 z_3 z_5;$$
$$Y = K_1 \vee K_2 \vee K_3 \vee K_4.$$

For the probability of the first logic term we have

$$P\{K_1\} = p_1 p_3 = A_1.$$

The probability of the sum of two logic terms is as follows

$$P\{K_1 \vee K_2\} = P\{K_1\} + P\{K_2\} - P\{K_2\} \cdot P\{K_1 \mid K_2 = 1\} =$$
$$= p_1 p_3 + p_2 p_4 - p_2 p_4 \cdot P\{z_1 z_3 \mid K_2 = 1\} = p_1 p_3 + p_2 p_4 - p_1 p_2 p_3 p_4 = A_{12}.$$

Probability of the sum of three logic terms is equal to:

$$P\{K_1 \vee K_2 \vee K_3\} = P\{K_1 \vee K_2\} + P\{K_3\} - P\{K_3\} P\{K_1 \vee K_2 \mid K_3 = 1\} =$$
$$= A_{12} + p_1 p_4 p_5 - p_1 p_4 p_5 \cdot P\left\{ \begin{vmatrix} z_1 & z_3 \\ z_2 & z_4 \end{vmatrix} \mid K_3 = 1 \right\} =$$
$$= A_{12} + p_1 p_4 p_5 - p_1 p_4 p_5 \cdot P\left\{ \begin{vmatrix} 1 & z_3 \\ z_2 & 1 \end{vmatrix} \right\}$$
$$= A_{12} + p_1 p_4 p_5 - p_1 p_4 p_5 \cdot P\{\overline{\bar{z}_2 \cdot \bar{z}_3}\} =$$
$$= A_{12} + p_1 p_4 p_5 - p_1 p_4 p_5 (1 - q_2 q_3) = A_{12} - p_1 p_4 p_5 q_2 q_3 = A_{123}.$$

Above we used the rule of variable replacement taking into account conditional probabilities and de Morgan's theorem of replacement of disjunction by negation of conjunction.

For probability of the sum of all four logic terms, we get

$$P\{K_1 \vee K_2 \vee K_3 \vee K_4\} = P\{K_1 \vee K_2 \vee K_3\} + P\{K_4\} -$$
$$- P\{K_4\} \cdot P\{K_1 \vee K_2 \vee K_3 \mid K_4 = 1\} =$$
$$= A_{123} + p_2 p_3 p_5 - p_2 p_3 p_5 \cdot P\left\{ \begin{vmatrix} z_1 & z_3 & \\ z_2 & z_4 & \\ z_1 & z_4 & z_5 \end{vmatrix} \mid K_4 = 1 \right\} =$$
$$= A_{123} + p_2 p_3 p_5 - p_2 p_3 p_5 \cdot P\left\{ \begin{vmatrix} z_1 & 1 & \\ 1 & z_4 & \\ z_1 & z_4 & 1 \end{vmatrix} \right\} =$$
$$= A_{123} + p_2 p_3 p_5 - p_2 p_3 p_5 \cdot P\left\{ \begin{vmatrix} z_1 \\ z_4 \end{vmatrix} \right\} =$$
$$= A_{123} + p_2 p_3 p_5 - p_2 p_3 p_5 (1 - q_1 q_4) = A_{123} + p_2 p_3 p_5 q_1 q_4 = A_{1234}.$$

Here we also used the absorption law.

The final expression for P-polynomial can be obtained by making substitutions of P_i, Q_i instead of A_1, A_{12}, A_{123}:

$$P\{Y=1\} = A_{123} + p_2p_3p_5q_1q_4 = A_{12} - p_1p_4p_5q_2q_3 + p_2p_3p_5q_1q_4 = \\ p_1p_3 + p_2p_4 - p_1p_2p_3p_4 - p_1p_4p_5q_2q_3 + p_2p_3p_5q_1q_4. \quad (9.5)$$

9.4 "Weight" of the argument in the L-function

Calculation of Boolean difference. The Boolean difference for FAL with respect to x_i can be calculated using the expression (8.15):

$$\Delta f(X_m) = f(x_1,\ldots,x_i,\ldots,x_m) \oplus f\left(x_1,\ldots,\overline{x}_i,\ldots,x_m\right), \quad (9.6)$$

where $\oplus$ means summation in modulo 2. We shall need the following notations:

$A = f(x_1,\ldots,x_i,\ldots,x_m)$ is the initial function of FAL;

$B = f(x_1,\ldots,\overline{x}_i,\ldots,x_m)$ is the function with is symmetric in x_i to the initial one.

Using the relationships between operation $\oplus$ of summation in modulo 2 and basic logic operations (see rule 38)

$$A \oplus B = AB' \vee A'B,$$

we write down the expression (9.6) in the following form:

$$\Delta_{xi} f(X) = f(x_1,\ldots,x_i,\ldots,x_m) \oplus f\left(x_1,\ldots,\overline{x}_i,\ldots,x_m\right) = A \oplus B \\ = A\overline{B} \vee \overline{A}B.$$

The Boolean difference for FAL with respect to x_i may also be calculated from expression (8.15):

$$\Delta_{xi} f(x_1,\ldots,x_m) \quad = \quad f_1^{(i)}(x_1,\ldots,x_m) \oplus f_0^{(i)}\left(x_1,\ldots,x_m\right),$$

where

$f_1^{(i)}(x_1,\ldots,x_m) = f\left(x_1,\ldots,1,\ldots,x_m\right)$ is the unit function with respect to argument x_i (8.10);

$f_0^{(i)}\left(x_1,\ldots,x_m\right) = f(x_1,\ldots,0,\ldots,x_m)$ is the zero function with respect to argument x_i (8.11).

Calculation of element's weight in L-functions. The "weight" of the Boolean difference of argument x_i is the number of sets on which $\triangle_{x_i} f(x_1, x_2,\ldots,x_m)$ takes value equaled to 1:

$$G\{\triangle_{x_i} f(x_1,\ldots,x_m) = 1\}\ . \quad (9.7)$$

The "weight" of the Boolean difference (9.7) characterizes the role of an element x_i in structural reliability of the system.

It is convenient to use relative value for measurements of "weight" of an element. The weight of the element x_i in the system, having m elements, is the the ratio the Boolean difference of argument x_i to the number of all sets of the m-dimensional logic space

$$g_{x_i} = \frac{G\{\triangle_{x_i} f(x_1, \ldots, x_m) = 1\}}{2^m}. \tag{9.8}$$

If the L-function is given in $ODNF$, the "weight" of the logic function by definition can be written down in the following form:

$$g_{Y(x_1,\ldots,x_m)} = \frac{\sum\limits_{f=1}^{k} 2^{m-r_f}}{2^m} = \sum_{f=1}^{k} 2^{-r_f},$$

where k is the number of orthogonal conjunctions in the logic function; m is the number of arguments of function; r_f is the rank of elementary orthogonal conjunction.

For monotonic FAL, according to Theorem 3 and formula (9.17) we have

$$\Delta_{xi} Y(x_1, \ldots, x_m) = Y_1^{(i)}(x_1, \ldots, x_m) Y\prime_{0)}^{(i)}(x_1, \ldots, x_m) \tag{9.9}$$

Substituting Boolean difference (9.9) in the form of difference of sets

$$\Delta_{xi} Y(x_1, \ldots, x_m) = Y_1^{(i)}(x_{m-1}) / Y\prime_0^{(i)}(x_{m-1}),$$

where the functions $Y_1^{(i)}, Y_0^{(i)}$ are written down in $ODNF$, we find

$$G\{\Delta_{xi} Y(x_m) = G\{Y_1^{(i)}(x_{m-1})\} - G\{Y_0^{(i)}(x_{m-1})\} =$$
$$\sum_{f=1}^{k} 2^{m-(r_f-1)} - \sum_{j=1}^{l} 2^{m-(r_j-1)}, \tag{9.10}$$

where

k, r_f are the number and the rank of orthogonal conjunctions containing argument x_i;

l, r_j are the number and the rank of orthogonal conjunction, containing negation of the argument $\overline{x}_i$.

Dividing expression (9.10) by 2^m, we obtain the formula for calculation of the "weight" of an element x_i in the system

$$g_{xi} = \sum_{f=1}^{k} 2^{-(r_f-1)} - \sum_{j=1}^{l} 2^{-(r_j-1)}. \tag{9.11}$$

Example 1. Let us determine the "weight" of elements x_1 and x_5 in DNF (9.3) for the system shown in Fig. 9.1. By applying algorithm of orthogonalization to FAL (9.3) written down in DNF, we obtain

$$Y(x_1,\ldots,x_5)=\left|\begin{array}{l} x_1\ x_3 \\ x_2\ \overline{x}_3\ x_4 \\ \overline{x}_1\ x_2\ x_3\ x_4 \\ \overline{x}_1\ x_2\ x_3\ \overline{x}_4\ x_5 \\ x_1\ \overline{x}_2\ \overline{x}_3\ x_4\ x_5 \end{array}\right| . \tag{9.12}$$

The above formula differs from (9.5), because we used another algorithm of orthogonalization and the terms in the initial L-function were written down in another sequence. Such variation is a usual phenomenon in orthogonalization.

To make clear the formula (9.10), we write it for calculation of "weight" of Boolean difference of argument x_1 for (9.12) in detail:

$$\begin{aligned} &G\{\Delta_{x1}Y(x_5)\} \\ &= G\left\{\left|\begin{array}{l} x_1 \quad\quad x_3 \\ x_1\ \overline{x}_2\ \overline{x}_3\ x_4\ x_5 \end{array}\right|\right\} - G\left\{\left|\begin{array}{l} \overline{x_1}\ x_2\ x_3\ x_4 \\ \overline{x}_1\ x_2\ x_3\ \overline{x}_4\ x_5 \end{array}\right|\right\} \\ &= [\ 2^{5-(2-1)} + 2^{5-(5-1)}\ ] - [\ 2^{5-(4-1)} + 2^{5-(5-1)}\ ] \\ &= [\ 2^4 + 2^1\ ] - [\ 2^2 + 2^1\ ] = 18 - 6 = 12. \end{aligned} \tag{9.13}$$

From (9.8) we have:

$$G_{x1} = 12/32 = 0.375.$$

And by (9.11) we obtain

$$\begin{aligned} g_{x1} &= [2^{-(2-1)} + 2^{-(5-1)}\ ] - [\ 2^{-(4-1)} + 2^{-(5-1)}] \\ &= [2^{-1} + 2^{-4}] - [2^{-3} + 2^{-4}] \\ &= \ (0.5 + 0.0625) - (0.125 + 0.0625) = 0.05625 - 0.1875 = 0.375. \end{aligned}$$

For argument x_5, we write down

$$\begin{aligned} G\{\Delta_{x5}Y(x_5)\} &= G\{\left|\begin{array}{l} \overline{x}_1\ x_2\ x_3\ \overline{x}_4\ x_5 \\ x_1\ \overline{x}_2\ \overline{x}_3\ x_4\ x_5 \end{array}\right|\} - G\{\Theta\} \\ &= \ (2^{5-(5-1)} + 2^{5-(5-1)}) - 0 = 2 * 2^1 = 4, \end{aligned}$$

$$g_{x5} = 4/32 = 0.125.$$

Thus, not using probabilities, we have managed to estimate the structural reliability of elements x_1, x_5 by determination of their "weight":

$$g_{x1} = 3g_{x5}.$$

Example 2. Let us find the weight of the first element z_1 in the following L-function

$$y_1\,(z_1, z_2, z_3) = z_1 \wedge z_2 \wedge z_3 = z_1\, z_2\, z_3\,, \tag{9.14}$$

It is determined from the expression for the logic difference of argument z_1:

$$\Delta_{z1} Y_1 = z_1 z_2 z_3 \oplus \overline{z}_1 z_2 z_3 =$$

$$z_1 z_2 z_3 \begin{vmatrix} z_1 \\ \overline{z}_2 \\ \overline{z}_3 \end{vmatrix} \vee \begin{vmatrix} \overline{z}_1 \\ \overline{z}_2 \\ \overline{z}_3 \end{vmatrix} \overline{z}_1 z_2 z_3 = \begin{vmatrix} z_1 & z_2 & z_3 \\ \overline{z}_1 & z_2 & z_3 \end{vmatrix} = \begin{vmatrix} z_1 \\ \overline{z}_1 \end{vmatrix} z_2 z_3 = z_2 z_3.$$

As it was shown in [2], weights should be calculated for $R_1 = R_2 = R_3 = 0.5$:

$$g_{z_1} = P\{z_2 z_3 = 1\} = r \cdot r = 0.5 \cdot 0.5 = 025.$$

Thus, in consecutive structure y_1, a dangerous situation appears with probability 1/4 because of the initiating condition z_1.

Example 3. Let us find the weight of the first element z_1 in the following L-function

$$y_2 = z_1 \vee z_2 \vee z_3. \tag{9.15}$$

Let us write down in $ODNF$

$$y_2 = z_1 \vee z_2 \overline{z}_1 \vee z_3 \overline{z}_1 \overline{z}_2. \tag{9.16}$$

Let us inscribe the expression for the weight of Boolean difference of the first element z_1

$$G(z_1) - G\left\{\begin{vmatrix} \overline{z}_1 & z_2 & \\ \overline{z}_1 & \overline{z}_2 & z_3 \end{vmatrix}\right\} = 2^{3-(1-1)} - [2^{3-(2-1)} + 2^{3-(3-1)}] = 2.$$

Because $2^3 = 8$, we have: $g_{z_1} = 2/8 = 0.25$, thus, in the parallel structure y_2 a dangerous situation appears with probability 1/4 because of the initiating condition z_1.

9.5 Importance of elements in a system

Importance of elements in the system, in contrast with the "weight," is determined not by the logic model but by the probabilistic model. The following expression is used (see [2]):

$$\xi = \partial P_c / \partial P_i = P\{\Delta_{xi} Y(X_m) = 1\}. \tag{9.17}$$

For the L-function (9.12), the probabilistic polynomial can be written as follows:

$$Pc = p_1p_3 + p_2q_3p_4 + q_1p_2p_3p_4 + q_1p_2p_3q_4p_5 + p_1q_2q_3p_4p_5. \tag{9.18}$$

Let us calculate the "importance" of the elements x_1 and x_5:

$$\partial P_c/\partial p_1 = p_3 - p_2p_3p_4 - p_2p_3q_4p_5 + q_2q_3p_4p_5; \tag{9.19}$$

$$\partial P_c/\partial p_5 = q_1p_2p_3q_4 + p_1q_2q_3p_4. \tag{9.20}$$

In work [2], it was shown that by substituting values 0.5 for all $i = 1, 2, \ldots, m$ into the formula (9.17) instead of probabilities of arguments x_i and $\overline{x}_i$, we obtain one more expression for the "weight" of elements:

$$g_{xi} = P\{\Delta_{xi}Y(X_m) = 1\} \mid r_1 = \ldots = r_m = 0.5. \tag{9.21}$$

Let us check this rule using expressions (9.19) and (9.20):

$$g_{x1} = \partial P_c/\partial P_1 = 0.5 - 0.5 \cdot 0.5 \cdot 0.5 - 0.5 \cdot 0.5 \cdot 0.5 \cdot 0.5 + \\ +0.5 \cdot 0.5 \cdot 0.5 \cdot 0.5 = 0.375,$$
$$g_{x5} = \partial P_c/\partial P_5 = 0.5 \cdot 0.5 \cdot 0.5 \cdot 0.5 + 0.5 \cdot 0.5 \cdot 0.5 \cdot 0.5 = 0.125.$$

It is easy to see that the analytical expressions for "weight" and "importance" may be cumbersome and very labor-intensive in derivation, but usually it is possible to avoid them, if there is a computer program (see Chapter 14). Then the "importance" of an element is easily determined:

$$\xi_i = P_c\,|_{p_i=1} - P_c|_{p_i=0}.$$

9.6 Example of construction of the L-function of danger

Let us estimate a submarine sinking danger. For the submarine sinking, it is enough to fill with water one of the compartments. Filling the compartment can occur through a hole or other violation of water tightness. To struggle for survivability of a boat in each compartment, there is the pump that can take out certain quantity of water. Besides, between pumps H_1 and H_2 there is dam CR, enabling under certain conditions removal of water with the help of the adjacent module pump (Fig. 9.2).

It is required to evaluate submarine sinking risk during a certain period. It is natural to accept that the dangerous state is the fact of submarine loss. The scenario of dangerous states, resulting in destruction of the submarine, after accounting all possible states z_i is represented schematically in Fig. 9.3. Here initiating events are z_1 and z_2, the holes in compartments No 1, No 2, and the events z_3 and z_4 — failures of pumps H_1, H_2. The event z_5 means that access to the dam valve of an emergency compartment is impossible.

Let us construct the function of dangerous condition with the help of $MPDF$:

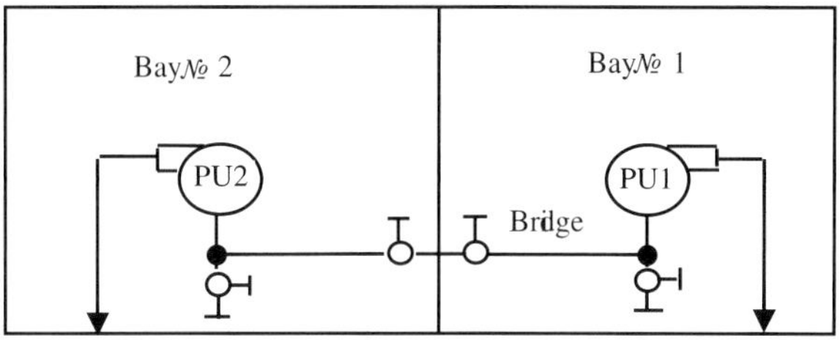

Fig. 9.2. Object of accident

$$y(z_1, \ldots, z_5) = z_1 z_3 z_4 \vee z_1 z_3 z_5 \vee z_2 z_4 z_3 \vee z_2 z_4 z_5, \tag{9.22}$$

$$y(z_1, \ldots, z_5) = \left| \begin{array}{l} z_1\ z_3 \left| \begin{array}{l} z_4 \\ z_5 \end{array} \right| \\ \\ z_2\ z_4 \left| \begin{array}{l} z_3 \\ z_5 \end{array} \right| \end{array} \right|$$

The function of the algebra of logic (9.22), written down in DNF, is monotonic and repeated one. Inverting FAL (9.22), we obtain the function of system safety (FSS):

$$y'\{z'_1, \ldots, z'_5\} = \left| \begin{array}{l} z'_1\ z'_2 \\ z'_1\ z'_4 \\ z'_2\ z'_3 \\ z'_3\ z'_4 \\ z'_3\ z'_5 \\ z'_4\ z'_5 \end{array} \right| .$$

Six $MSDS$ in the latter formula

$$S_1 = z'_1 z'_2, \ S_2 = z'_1 z'_4, \ S_3 = z'_2 z'_3, \ S_4 = z'_3 z'_4, \ S_5 = z'_3 z'_5, \ S_6 = z'_4 z'_5$$

indicate those conjunctions, which completely "protect" the system from danger (in this case, from submarine sinking).

9.7 Explosion in a submarine: scenario and risk LP-model

In this chapter we shall consider some examples of modeling, analysis, and management of risk in structural-complex systems (SCS) in various areas of engineering. We consider that examples are not less instructive than the theory.

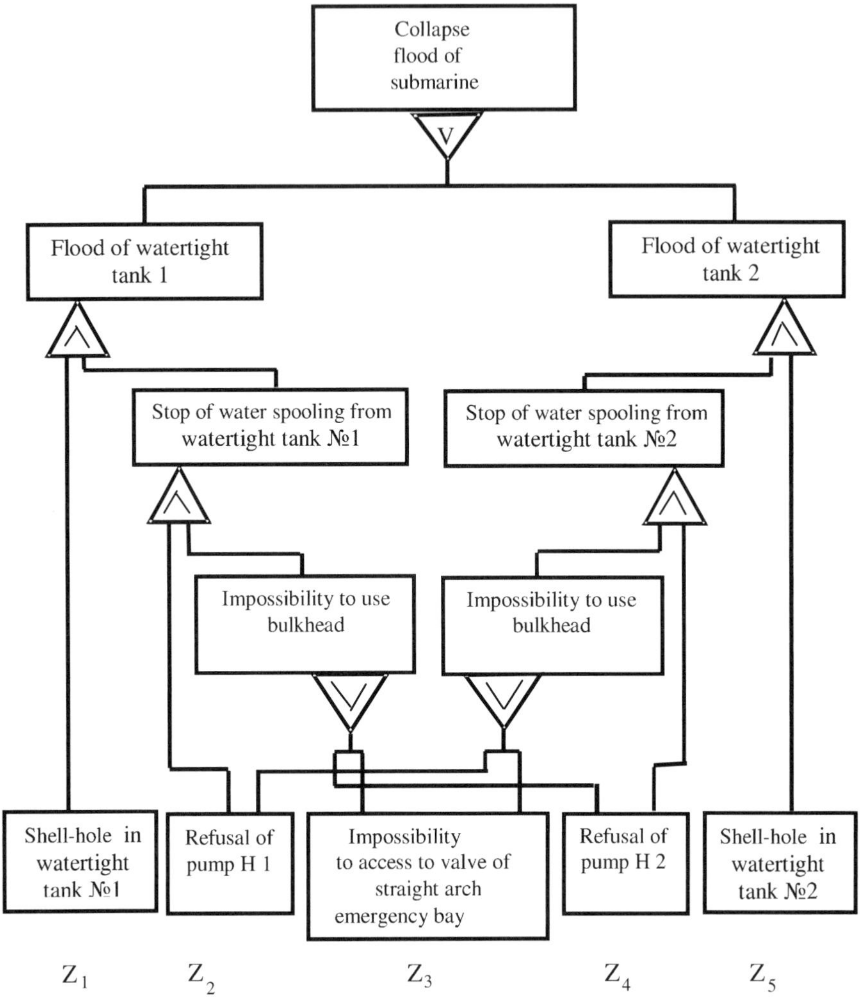

Fig. 9.3. Scenario of dangerous state

A complex system can consist of equipment, sensors, computers, programs, instructions, and actions of personnel, including management, testing, repair and service. We consider examples of construction of risk LP-models of *SCS* in which risk elements are actions of personnel, too. In this section, we study modeling, estimation, and analysis of explosion risk in the storage-battery tank of the submarine [2].

It is known that for prevention of the explosion of the mixture of air and hydrogen, evaporating from the batteries, a number of special precautions are taken. The intensity of gasing from the battery depends on a mode of its usage, service life, temperature of environment, etc. Hydrogen is eliminated by the

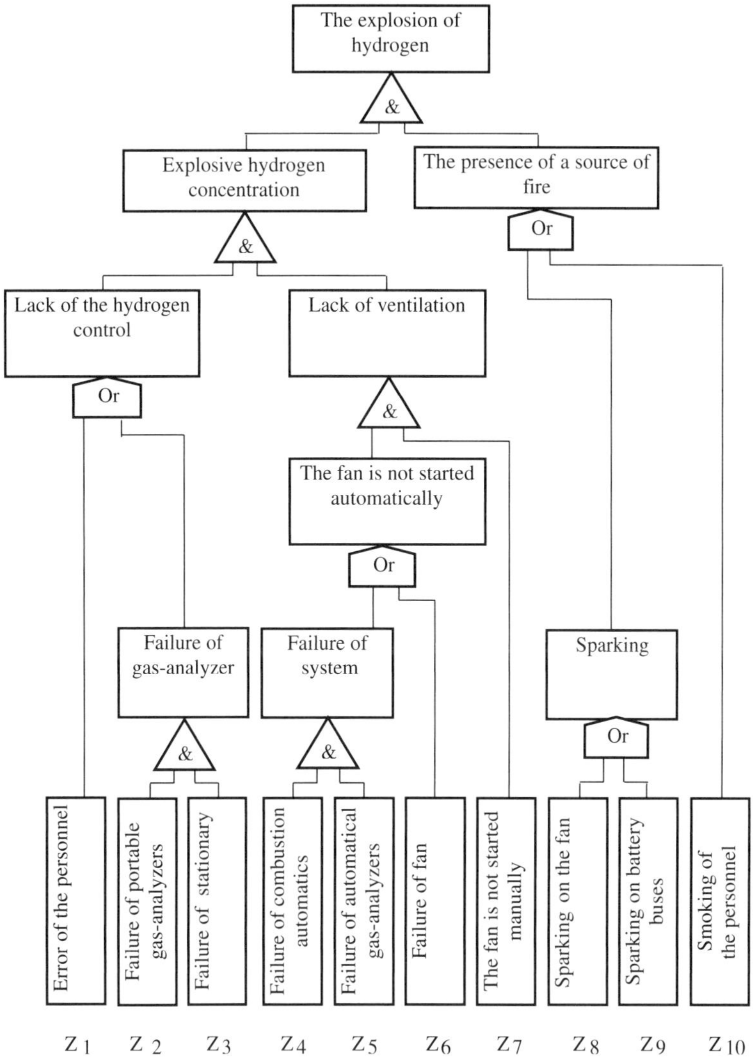

Fig. 9.4. The scenario of the dangerous state

system of ventilation or by burning in special devices. The concentration of hydrogen in the atmosphere of the room is constantly supervised by automatic and portable gas analyzers.

Explosion will necessarily occur (Fig. 9.4) if the explosive concentration of hydrogen is achieved due to failure of ventilation (initiating conditions Z_4, Z_5, Z_6, Z_7) and control of hydrogen concentration (initiating conditions Z_1, Z_3, Z_3), and due to the presence of source of fire (initiating conditions Z_8, Z_9, Z_{10}).

The scenario of the dangerous condition is shown in Fig. 9.4. Drawing up such scenario is a creative part of safety analysis, the most difficult and non-formalizable one. In the given concrete case, as the dangerous condition we understand the explosion of hydrogen in the room where the storage battery is placed. Such explosion can result (and resulted repeatedly in practice) in destruction of personnel and objects, that is, in a damage of big scale.

The philosophical problem of uniqueness and completeness of safety research raise two questions:

(1) Will the specialists give the unique interpretation of ways the system gets to the dangerous condition?

(2) Will all circumstances, leading to explosion, be taken into account?

In our opinion, positive answers to these questions can be given with organizing role of mathematics of the LP-methods and pragmatic assignment of scales of the system under consideration (i.e., the account of all circumstances only within the limited volume and limited resources).

When it is wished to obtain as many as possible concrete recommendations on active protection of system to avoid getting in a dangerous condition, it should not be thought that the purpose is reached only by account of as many as possible initiating conditions. Movement from small to big, i.e., from minimal number of taken into account conditions ("core" of system) to consideration of the additional circumstances added to the "core" is more correct. In our example, it would be possible to attribute to the "core" of system only conditions Z_4, Z_5, Z_6, Z_7, and then to recollect other Z_i. There is opportunity of generalized interpretation of both mistakes of people Z_1 and ways of infringement of instructions Z_7. However, in the process it is necessary to be able to stop — "to see forest behind trees."

The final event is the explosion of hydrogen Z_{19} in the battery-storage tank; it occurs at achievement of explosive concentration Z_{18} and simultaneous action of a source of ignition Z_{16}. The explosive concentration of hydrogen is formed if ventilation Z_{17} and checking Z_{14} of concentration of hydrogen are absent. The checking is absent due to a mistake of staff or failure of gas analyzer Z_{11}. Failure of a gas analyzer means failure of portable Z_2 or stationary gas analyzers Z_3. The absence of ventilation is by failure of manual Z_7 and automatic Z_{15} start. The latter occurs because of failure of fan Z_6 or because of switching-off the system of automatics Z_{12}. The switching-off of the system of automatics occurs because of simultaneous failure of automatics of after-burning Z_4 and gas analyzer Z_5. The presence of a source of ignition is caused by possible smoking staff Z_{10} or presence of any sparking Z_{13}. The sparks occur in the fan Z_8 or on the contacts of battery.

The scenario of explosion of hydrogen in the battery-storage tank of the submarine can be written down as follows (events and relations are shown in capital letters):

Rule 1. **Explosion of hydrogen** occurs,
if there is **explosive concentration**
and there is a **source of ignition**;

Rule 2. **Explosive concentration** occurs,
if **checking of hydrogen**
and **ventilation** fail;
Rule 3. **Checking of hydrogen** is absent
if **mistake of personnel**
or failure of **gas analyzer:**
both **stationary** and **portable** take place);
Rule 4. **Source of ignition** is present
if **personnel smoking**
or **sparking in the fan**
or **sparking on battery buses** take place;
Rule 5. **Ventilation** is absent
if **the fan** is not started up **manually**
and **the fan** is not started up **automatically**;
Rule 6. **Fan** is not started up **automatically**
if **failure of the fan**
or **failure of system of automatics**:
and **afterburning of hydrogen**
and **the gas analyzer** take place.

The above stated scenario fixes only the events leading to explosion, and does not fix event when explosion is impossible though some elements fail.

The reasons of explosion at the lowermost level of the tree of events Z_1–Z_{10} are named initiating conditions and considered as independent random events. We note that the phenomenon is investigated "top–down": first we formulate the dangerous condition (explosion), and then define its possible reasons. For each dangerous condition, system failures of its elements or chains of failures are analyzed until the primary failure (of a single unit or a mistake of the person) is found.

Organizing role of mathematics can be seen when forming L-function of a dangerous condition. If the creative part of research is finished with the scenario of the dangerous condition (Fig. 9.4), then L-function of dangerous condition can be written down as a logic matrix of events Z_i:

$$Y(Z_1, \ldots, Z_{10}) = \begin{vmatrix} Z_1 & Z_4 Z_5 & Z_7 & Z_8 \\ Z_2 Z_3 & Z_6 & & Z_9 \\ & & & Z_{10} \end{vmatrix} . \tag{9.23}$$

After removal of brackets (logic multiplication), we get the L-function of dangerous condition as the disjunction of twelve minimal path of dangerous condition (functioning):

$$Y(Z_1,\ldots,Z_{10}) = \begin{vmatrix} Z_1Z_6Z_7Z_8 \\ Z_1Z_6Z_7Z_9 \\ Z_1Z_6Z_7Z_{10} \\ Z_1Z_4Z_5Z_7Z_8 \\ Z_1Z_4Z_5Z_7Z_9 \\ Z_1Z_4Z_5Z_7Z_{10} \\ Z_2Z_3Z_6Z_7Z_8 \\ Z_2Z_3Z_6Z_7Z_9 \\ Z_2Z_3Z_6Z_7Z_{10} \\ Z_2Z_3Z_4Z_5Z_7Z_8 \\ Z_2Z_3Z_4Z_5Z_7Z_9 \\ Z_2Z_3Z_4Z_5Z_7Z_{10} \end{vmatrix} . \qquad (9.24)$$

Thus, it is necessary to understand that in this case, explosion of a mixture of air and hydrogen can develop only in 12 different ways, and not one more.

Inverting (9.24), we obtain the function of dangerous condition as the disjunction of six minimal cross-sections of prevention of danger:

$$\overline{Y}(\overline{Z}_1,\ldots,\overline{Z}_{10}) = \begin{vmatrix} \overline{Z}_7 \\ \overline{Z}_1\overline{Z}_2 \\ \overline{Z}_1\overline{Z}_3 \\ \overline{Z}_4\overline{Z}_6 \\ \overline{Z}_5\overline{Z}_6 \\ \overline{Z}_8\overline{Z}_9\overline{Z}_{10} \end{vmatrix} . \qquad (9.25)$$

Uniqueness in the given context is understood as the possibility of prevention of explosion by six minimal sets $\overline{Z}_i$, only, and not one more.

From (9.24) and (9.25) it is visible that the event Z_7 (the fan is not started up manually) is included into all 12 minimal paths of dangerous functioning, and at the same time, it is the most "profitable" in minimal sections of prevention of danger, i.e., development of explosive conditions is impossible without Z_7, and for prevention of explosion it is enough to start the fan Z_7 manually. In Table 9.1, weights of arguments Z_j are given.

Weighing initiating events Z_j one at a time, two at a time, etc., allows us to estimate their role in creation of the dangerous condition of system only by their place, i.e., by taking into account only the logic of development of possible events. However, it is already quite precious. Specialists of the system are to come to the unambiguous and explained result. At the same time, it is necessary not to forget about enormous influence of probabilities of the events P_j.

Table 9.1. "Weights" of arguments Z_j of dangerous function

Z_j	Z_1	Z_2	Z_3	Z_4	Z_5	Z_6	Z_7	Z_8	Z_9	Z_{10}
Gz_j	0.205	0.068	0.068	0.068	0.068	0.205	0.342	0.049	0.049	0.049

Table 9.2. Initial probabilities of danger of arguments

Z_j	Z_1	Z_2	Z_3	Z_4	Z_5	Z_6	Z_7	Z_8	Z_9	Z_{10}
P_j	0.01	0.001	0.001	0.001	0.001	0.001	0.0001	0.01	0.01	0.01

Table 9.3. Contributions of initiating events to the danger of the system

Z_j	Z_1	Z_2	Z_3	Z_4	Z_5	Z_6	Z_7	Z_8	Z_9	Z_{10}
B_{zj}	0.99	0.0001	0.0001	0.0009	0.0009	0.99	1.0	0.83	0.83	0.83

The efforts of specialists, directed to more objective estimation of initial probabilities of initiating conditions Z_j, are rather useful and productive. In case of successful overcoming the specified information barrier, the further development of the analysis of safety of SCS should be continued in the direction of specification of the real contribution of events to development of dangerous conditions (or their prevention).

We set some simple initial data reflecting our idea of possible values of probabilities (Table 9.2).

In this simple example, in which the function of dangerous condition does not contain repeated arguments, we shall solve the problem omitting orthogonalization, that is, we shall search explosion risk from expression (9.24):

$$P_y = P\left(\left|\begin{matrix} Z_1 \\ Z_2Z_3 \\ \end{matrix}\right|\begin{matrix} Z_4Z_5 \\ Z_6 \\ \end{matrix}\left|\begin{matrix} Z_7 \\ \\ \end{matrix}\right|\begin{matrix} Z_8 \\ Z_9 \\ Z_{10} \end{matrix}\right| = 1\right)$$
$$= p_7[1 - q_1(1 - q_2q_3)] \cdot [1 - q_6(1 - p_4p_5)] \cdot [1 - q_8q_9q_{10}]. \quad (9.26)$$

On substituting initial data from Table 9.2 in (9.26), we get: $P_y = 0.1180398 \cdot 10^{-9}$. Individual contributions B_{zj} as parts of the system risk P_y are presented in Table 9.3.

The considered risk LP-model of explosion and fire in dangerous places (on ships, in apartments, at gas and oil transfer stations) is an example of not only constructions of the risk LP-model, but it is also a demonstration of the risk LP-model for the risk estimation in insurance. Indeed, the risk of insurance of the system is equivalent to non-success risk of the system.

9.8 Risk LP-model of the structural-complex system

LP-models can be rather complex and have some hundreds of elements and some cycles. Before we build a scenario and a risk LP-model, we need to become acquainted with some examples of risk LP-models in the field of engineering. Below we shall consider an example of the risk LP-model of the

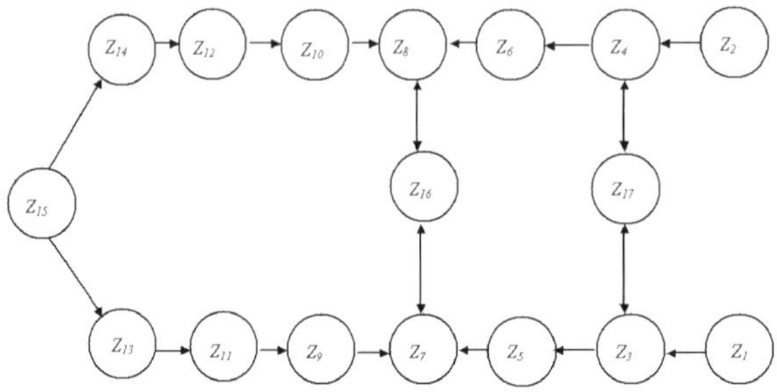

Fig. 9.5. The structural risk model of the complex object: electric power plant

complex technical system and show the possibility of its training on statistical data by using the methods proposed above.

Consider the risk LP-model of a ship power plant [2], which diagram is given in Fig. 9.5. Here we have the following elements:

z_1, z_2 are main generators of alternating current;
z_3, z_4 are main switchboards of disconnected load;
z_5, z_6 are automatic switches;
z_7, z_8 are sections of fixed load;
z_9, z_{10} are automatic section;
z_{11}, z_{12} are reversible converters of direct-to-alternating current;
z_{13}, z_{14} are silicon shutoff devices;
z_{15} is direct current source;
z_{16}, z_{17} are boards jumpers with automatic switches.

The capacity of any of main generator is sufficient for supplying the electric energy consumers connected by network with switchboards z_3, z_4, z_7, z_8. The capacity of reserve source z_{15} is enough for supplying with electric energy all consumers with fixed load (from switchboards z_7, z_8), or only with load from one of switchboard $z_7 \oplus z_8$. Carrying capacity of the jumper z_{17} corresponds with the power of one main generator; carrying capacity of the jumper z_{16} corresponds with the power of the reserve source. We could analyze 16 conditions of the electrical system. We write down the risk LP-model of the system:

$$Y = z_3 z_4 \vee z_7 z_8. \tag{9.27}$$

The LP-model was trained by the statistical information, which we used for training the credit risk LP-model. Each sign-element had some grades. The functioning LP-model of the power plant is taken from work [2].

The logic risk function is orthogonalized by using a program working on the basis of algebra of corteges [102, 103]. In orthogonal disjunctive normal form, the risk L-function contains 139 conjunctions, logically connected by the sign of disjunction. This orthogonalized LP-model cannot be given here (it would take many pages), neither the corresponding risk P-model. However, the risk P-model was brought in the program for identification and was trained by method of random search with taking into account the GIE. The results of investigation of this LP-model are given in detail in [2, 3, 30].

10

Automated Structural and Logical Modeling

Any sufficiently advanced technology is undistinguishable from magic.
Any advanced technology is indistinguishable from a miracle.
Arthur Clark

A. S. Mozhaev's methodology and technology of automated structural and logical modeling suggested and developed by him in [5, 6, 104, 105] is described. His proposals have much larger capabilities in comparison with the methods based on trees of events and failures, graphs of arcwise connectedness, structural and logical methods of risk modeling, and calculation of parameters of reliability, safety and risk of structurally complex systems of objects and processes. By a simple example, we shall describe the technique for construction of schemes of the functional integrity (SFI) and algorithmic methods for automatic construction of LP-models. These tools are realized in the software complexes for automated modeling reliability, safety, and risk of systems. Essentially, SFI are the graphical scenarios of successful operation or emergency condition of the complex system.

10.1 Problems of LP-modeling

In the past years, some classes of methods of structural analysis were used for calculation of probabilistic parameters of reliability, safety, and risk of complex systems. The most frequently used of these are the methods of event trees, failure trees, logical-probabilistic, topological, logical-graphic ones, GO-technology, etc. The mentioned methods have common methodological basis, which can be characterized by the following positions:

- All the above mentioned methods for description of elements in models of reliability, safety, and risk (we will name them as risk models) of systems

E.D. Solojentsev, *Scenario Logic and Probabilistic Management of Risk in Business and Engineering*, DOI 10.1007/978-0-387-77946-1_10,

use simple (binary) random events with two incompatible outcomes, such as success or failure, an operation is done or it fails, a device is turned on or off, etc.

- The main way of statement of problems is the construction of a structural model (scheme) of operation of the investigated system or a scenario of emergency appearance.
- The algebra of logic is the mathematical base for modeling in these methods.
- The main form of description of the risk models determined model of the system is the logical function.
- Different types of computational probabilistic (analytical, statistical, Markov) or other models for the quantitative estimation of various risk models properties of the investigated system are determined on the basis of the logical model.
- The values of parameters characterizing reliability, safety, and risk of the system are determined on the basis of logical and computational probabilistic models. The same models are used for various techniques of development and substantiation of research, design, operational and other management solutions.
- All the most cumbersome and labor-consuming stages of structural and logical modeling (construction of logical functions, computational probabilistic and other models, carrying out calculations and result application) are aimed as much as possible to be automatized and realized on a computer.

In the current work, we generalize the experience of development and application of one of the new trends in the structural and logical modeling reliability, safety, and risk of operation of systems on the basis of generalization and development of the logical and probabilistic method and of the special tools for construction of structural models called by the schemes of functional integrity.

An important part in successful development of the theory and practice of the logical-probabilistic modeling has been played by the fact that by their nature, these methods are well adapted to full automation of the most complex, laborious, and cumbersome processes of construction of computational mathematical models of various properties of investigated systems. The realization of such new information technology of the automated structural-logical modeling (ASLM) has allowed the general audience to access the software for operating and multivariate analysis of various system objects and processes, models of which, in view of the structural complexity and high dimension, cannot be constructed by traditional manual ways of modeling. The development of such software is founded on development of a set of special algorithmic methods for modeling, which permits us to realize by a computer all main stages of construction of logical and probabilistic models of systems of arbitrary given structure.

10.2 Risk scenario of a railway accident

The SFI tool and basic algorithmic methods of modelings intended for realization by a computer of all stages of ASLM technology are considered. The SFI tool and the algorithmic methods of ASLM technology are illustrated by the simple example of probabilistic safety analysis of a hypothetical railway segment; its scheme is shown in Fig. 10.1. In this example, we take into account two initial causes of possible emergency: the fracture of the rails and/or appearance of an object on the way. In case of the fracture of the rails and failure-free operation of the indicator, the green traffic light is replaced by the red signal. If the engine driver sees the object on the rails and/or the red signal of the traffic light, he will switch on the train braking system. Then, under condition of failure-free operation of the brake system, the railway accident is prevented. It is required to construct models and to calculate the probabilistic characteristics of safe operation of a segment of the railway and appearance of emergency, and also to determine the significance and contributions of each indicated element to safety of the considered system as a whole.

Basically (with some accuracy), the considered problem can be solved by any of the above listed methods of structural modeling that allows the interested reader to test propriety and to estimate efficiency of the considered below algorithmic methods of ASLM technology.

10.3 Idea of development of LP-modeling

This idea is very simple and founded on the following two rules:

1. All elements $i = 1, 2, \ldots, n$ of the modeled object or process are represented by simple, binary events, which, during operation of the system, can be only in two positions. Such binary models of elements are denoted by simple logical variables $\tilde{x}_i = \{x_i, \overline{x}_i\}$. The direct notation of the logical variable

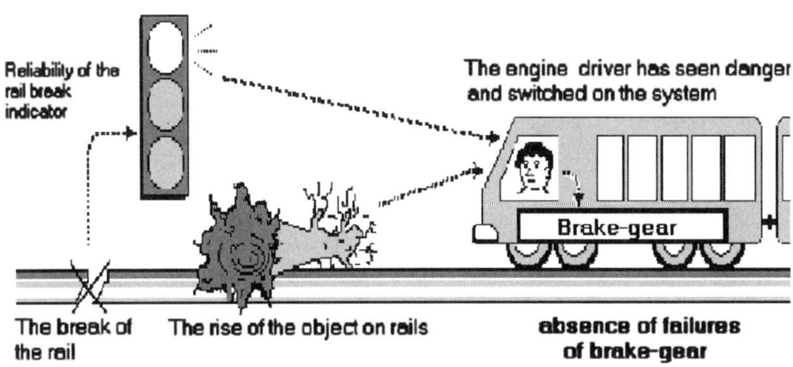

Fig. 10.1. The segment of railway traffic

is attributed to one, and the inverse one to another possible outcome of the appropriate binary event. The own probabilistic parameters $p_i, q_i = 1 - p_i$ of realization, respectively, direct x_i and inverse $\overline{x}_i$ outcomes of each binary event are considered as directly given or calculated with the help of known techniques. In the considered example (Fig. 10.1) for the probabilistic analysis of safety of a segment of the railway, five binary events are fixed. In Fig. 10.2 these events are indicated by the numbered circles and attributed to parts of the investigated system process.

2. In the system, each element i can execute (or not execute) some one (or many) output system function. The conditions of realization of this function by the element i are denoted by the integrative function y_i. The conditions of non-realization are denoted by the integrative function $\overline{y}_i$. In Fig. 10.3, the plots and substantive descriptions of output functions of each of five binary elements of the considered segment of the railway are given.

The reasonable and purposeful division of the investigated system into logically connected sets of simple binary $\tilde{x}_i \in \{x_i, \overline{x}_i\}$ and composite functional $\tilde{y}_i \in \{y_i, \overline{y}_i\}$ events is called the event-logical approach to the analysis of systems [5, 6]. The possibility of the event-logical description of the system is the necessary and sufficient condition of applicability of the logical and probabilistic methods for model construction and analysis of different properties of the considered system object.

With the help of one or several output functions $\tilde{y}_i \in \{y_i, \overline{y}_i\}$ that composite event is represented, which on intention of the model author characterizes the investigated property of the system (reliability, safety, risk, etc.). The simple logical variables $\tilde{x}_i$ of outcomes of the indicated binary events are used as parameters of the formed logical models, and own probabilistic characteristics of the elements p_i, q_i are used as parameters of the formed probabilistic models of the investigated system.

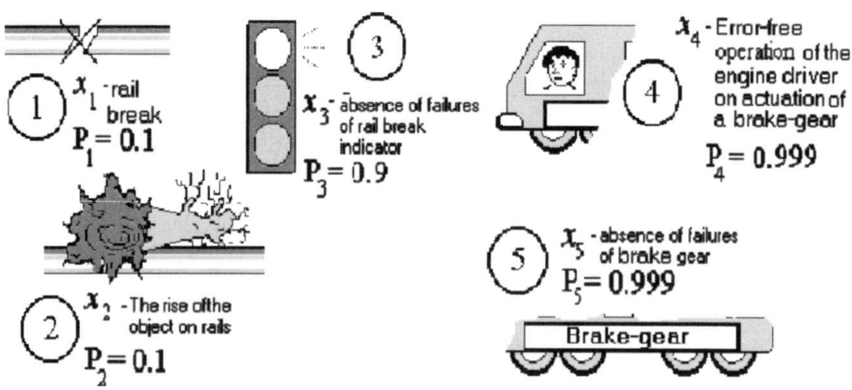

Fig. 10.2. The examples of binary models of elements

It is necessary to note that the principle difference of the event-logical approach from the classic logical-probabilistic [2] and all other structural methods is the possibility of using in structural models not only y_i but also the inverse $\overline{y}_i$ of the output functions. This is the main reason that makes it possible to realize in ASLM technology all possibilities of the basic apparatus of modeling the algebra of logic on the functionally complete set of operations AND, OR, and NOT. On this basis, it is also possible to automatize completely the processes of construction of all types of known monotonic models, as well as of the essentially new class of non-monotone models of reliability, safety, and risk of operation of complex system objects and processes.

10.4 Basic stages of LP-modeling

Usually in logical-probabilistic methods, four sequentially implemented stages are considered. In the automated structural-logical modeling, the stages have the following contents:

1. Primary structural and logical modeling. At this stage, the full formulation of the logical-probabilistic modeling problem is made. The formulation consists of three interdependent parts:

1.1. The scheme of functional integrity $G(X, Y)$ of the investigated system is designed on the basis of the fixed set of simple binary $\tilde{x}_i$ and composite functional $\tilde{y}_i$ events. Here X means the set of nodes (binary events $\tilde{x}_i$), and Y means the set of connecting arcs, (output and ensuring functions $\tilde{y}_i$). SFI should be the analytically precise and rigorously formalized mapping all the knowledge of conditions, which make it possible (impossible) for each element of the considered system to realize its output functions.

1.2. The L-criterion of operation of the investigated system $Y_c = Y_c(\tilde{y}_i)$, $i = 1, 2, \ldots, n$ is set with the help of one or several output functions. This criterion determines (in a generalized form) that mode of operation or usages of the system for which mathematical model should be constructed for quantitative estimation of the investigated property of the system as a whole. The complex and multifunction system objects can be characterized by several logical criterions, for each of which we should construct a mathematical model of the system.

1.3. The probabilistic p_i, q_i and other parameters of all binary events $\tilde{x}_i$, $i = 1, 2, \ldots, n$ representing elements of the modeled system are directly determined or set.

2. Definition of the logical function of the system efficiency. At this second stage of construction of the determine logical model of the process of operation of the investigated system $Y_c = Y_c(x_i, \overline{x}_i), i = 1, 2, \ldots, n$ is realized. The model represents the so-called logical function of efficiency of the system (FES) or the logical function of transitions. Arguments of this logical function are the simple binary events $\tilde{x}_i$; their own probabilistic parameters p_i, q_i are known. In this function, with the help of the logical sums, products, and supplements (inverses) of simple random events, we determine, precisely and uniquely, the composite random event of realization by the system of given criterion of its operation, i.e., modeled property of reliability, safety, or risk of the investigated system as a whole. It is necessary to note that (unlike the known structural methods) SFI of L-criteria operation and FES are not limited by the condition of monotonicity [2] and can be both the shortest paths of successful operation and the minimum cross-sections of failures and their arbitrary non-monotonic combinations.

3. Determination of the computational probabilistic model of the system. At the third stage, the transformation of logical FES (and/or the logical function of transitions) is made into one of the forms that permits direct calculation of probabilistic and other parameters of reliability, safety and risk of the investigated system $P_c = P_c(p_i, q_i, t)$, $i = 1, 2, \ldots, n$. At the moment there exist methods for obtaining four forms of computational P-models: polynomials of probabilistic functions, logical-statistical models, and some sorts of Markov and network (combinatorial and series) system models [5, 104–108].

4. Calculation of the system characteristics. At this final stage, the parameters that are needed for the solution of different problems of systems analysis risks are calculated with the help of the obtained computational models. The outcomes of the calculations can be used, for example, for the normative estimation of properties of the system, comparison and selection of variants of its structure, optimization and synthesis of systems during scientific researches, designing, exploitation, development, and substantiation of the management solutions in the field of reliability, safety, and risk of operation of investigated objects.

The stage of primary structural and logical modeling in its creative part, certainly, cannot be automated. Here only the service components can be automated. However, all the subsequent and most cumbersome stages of the logical, probabilistic modeling and calculations are completely automated [5, 6]. It became possible only after development of the algorithmic methods of modeling, that is maintenance of a very high, machine level of the formalism not only of the ways of representation of the initial, intermediate, and final data, but, mainly, of the procedures of construction of logical and probabilistic mathematical models of systems. The most important setting and procedural ways of formalizing logic and probabilistic modeling are stated below.

10.5 Algorithmic methods of primary structural and logical modeling

The central place in the primary modeling, unconditionally, takes the process of construction of SFI. On the one hand, the tools of description of SFI should correspond with the abilities of the logical and probabilistic modeling (algebra

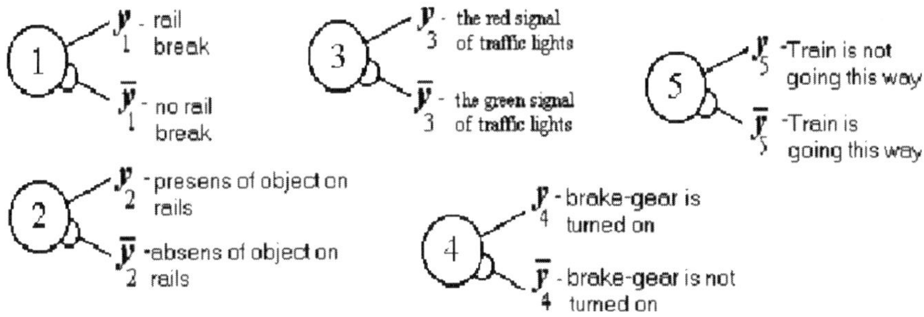

Fig. 10.3. Description of output (integrative) functions

of logic and probability theory), on the other hand, they should have the level of formalism sufficient for full automation of all subsequent stages.

Partly the SFI apparatus has already been used (Fig. 10.3) at description of output functions of elements of the railway segment in the above considered example. In full the representational tools of the SFI are shown in Fig. 10.4. First of all, we emphasize that in SFI, all ways of description of system structures, which were traditionally used in event trees, failure trees, and graphs of cohesion (consecutive and parallel connection of elements and cycles), are preserved. At the same time, in SFI some new tools of the graphic description

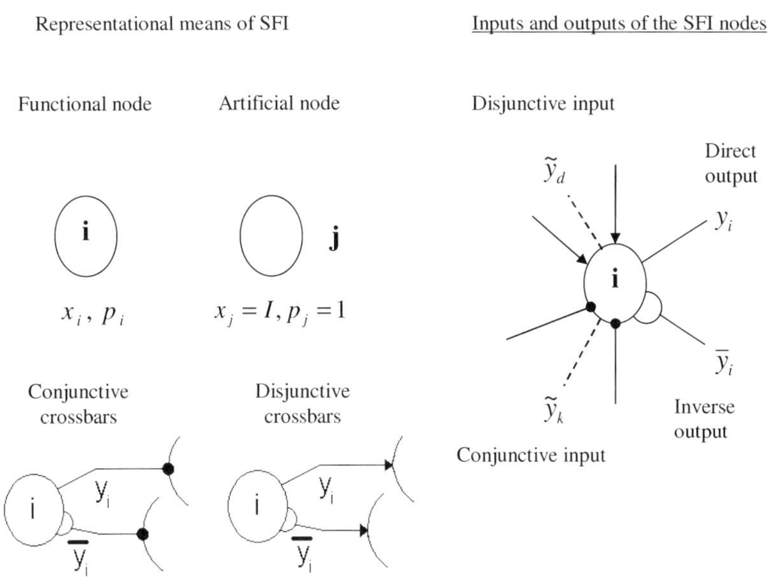

Fig. 10.4. The apparatus of functional integrity schemes

of systems ensuring full realization of all possibilities of the logic algebra are introduced.

The functional nodes of SFI are intended for description of elements i of the modeled system with the help of direct outcome x_i of the appropriate binary event. The fictitious nodes do not represent any elements, but allow us to depict on the graph the composite logical conditions, connections, and relations between elements in the investigated system. On the output of each SFI node (both functional and fictitious) can be two types of output branches: lines y_i (they mean realization of the output function of the element) and inverse $\overline{y}_i$ (they mean failure in realization of the output function).

The realization or not of the output function $\tilde{y}_i \in \{y_i, \overline{y}_i\}$ is determined both by the state of the corresponding element $\tilde{x}_i \in \{x_i, \overline{x}_i\}$, and, generally, by certain sets (combinations) of output functions $\tilde{y}_j \in \{y_j, \overline{y}_j\}$ of other providing elements j systems. In the SFI graph, the logical conditions of maintenance are represented by branches, which come to the given SFI node. The orientation of arcs is denoted by the point symbol (the conjunctive arc), or by the arrow mark (the disjunctive arc). The conjunctive arcs determine not reserved groups of maintenance functions, and disjunctive arcs correspond with the reserved groups. In the SFI apparatus between conjunctive and disjunctive groups of maintenance, the conjunctive connection is established. It means that the direct output function y_i of the node i of SFI will be realized only in the case if the own event x_i occurs *AND* realized all without exception the output functions $\tilde{y_k}$ of the conjunctive group *AND* at least one output function $\tilde{y}_d$ of the disjunctive group is realized.

The realization of the inverse output function of the element is determined by the full logical inverse of all graphically mapped conditions of realization of the direct output function. The letter can be represented by the generalized structural fragment of SFI. Thus, by construction the SFI tool realizes all possibilities of the algebra of logic, and the apparatus is the analytically rigorous graphical form of the system of logical equations. Such system can always be restored from the graph SFI.

The process of construction of SFI includes some informal (creative) procedures, for example, choice of binary events (models of system elements), calculus of their probabilistic characteristics, determination of the contents of output functions of elements, and graphic mapping logical conditions of their realization. In practice, it is convenient to define conditions of realization of output functions of elements at first fragmentary (separately for each node), and then to combine them in the SFI of the system as a whole (see Fig. 10.5):

$$\begin{aligned} Y_l &= X_i \cdot \left(\bigvee_{d \in D} \tilde{Y}_d \right) \cdot \left(\bigwedge_{k \in K} \tilde{Y}_k \right), \\ \overline{Y}_l &= \overline{X}_i \cdot \left(\bigvee_{d \in D} \overline{\tilde{Y}_d} \right) \cdot \left(\bigwedge_{k \in K} \overline{\tilde{Y}_k} \right), \end{aligned} \tag{10.1}$$

The graphic fragments of realization conditions of output functions by each element of the considered railway segment are shown in Fig. 10.6.

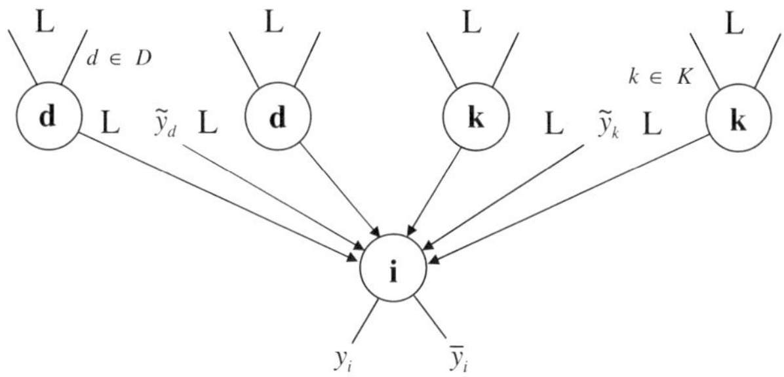

Fig. 10.5. Generalized fragment and base logic equations

These conditions are necessary for the analysis of safety. The formulation of these conditions is founded on the following factors: knowledge of the investigated system as the whole (Fig. 10.1); definition of the semantic contents of fixed binary events (Fig. 10.2), direct and inverse output functions of each element (Fig. 10.3); usage of graphic tools of the functional integrity schemes (Fig. 10.4), and base logical equations (10.1). The semantic contents of all the SFI fragments, shown in Fig. 10.6, can be defined as follows.

The formed model does not take into account any concrete causes of appearance of the initial dangers $\tilde{y}_1$ and $\tilde{y_2}$. Therefore, the sufficient conditions for realization of these output functions are only own binary events $\tilde{x}_1$ and $\tilde{x}_2$ (the break of the rail, the presence of an obstacle on rails). Such SFI nodes are called prime nodes, as no arcs of functional maintenance come to them.

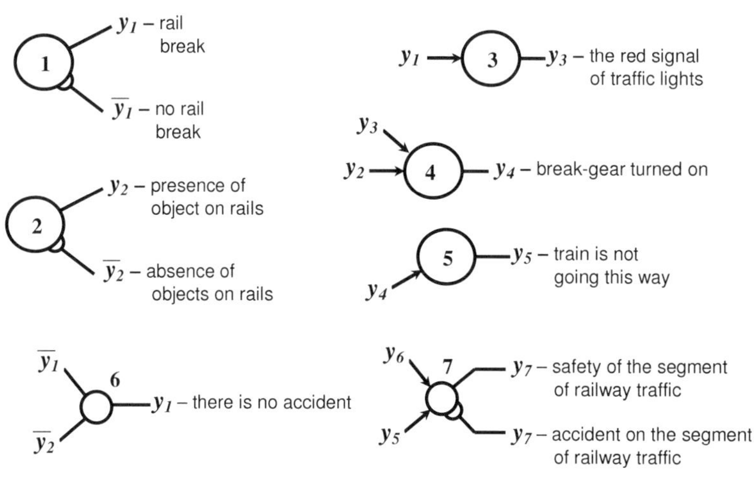

Fig. 10.6. Fragments of realization conditions of input functions

The necessary internal (own) condition of realization of the output function y_3 (the signal of the red traffic light) is the failure-free operation of the indicator x_3 AND realization of the function y_1 of the rail fracture.

The function y_4 (turning on the brake system of the train) is realized by error-free action of the engine driver x_4 AND provided by at least one of two conditions: the red signal is lit (y_3) OR there is an obstacle on rails (y_2).

The train does not get to the dangerous segment of the railway (y_5) if the brake system was timely switched on (y_4) AND the switched brake system functions (x_5).

For the well-directed safety analysis with the help of the fictitious nodes 6 and 7 in Fig. 10.6, two groups of output functions are chosen. The function $y_6 = \overline{y}_1 \cdot \overline{y}_2$ determines the total absence of the initial causes of appearance of the railway accident. The function $y_7 = y_6 \vee y_5$ determines two possible variants of safe operation of the considered segment of the railway: y_6 is realized for the total absence of possible causes of the accident, OR y_5 is realized for all possible variants of the correct (nominal) safety system operation in conditions of appearance of at least one initial cause of the accident. It is natural that the inverse of safety conditions ($\overline{y}_7$) should correspond with all possible variants of events that result in railway accident on the considered segment.

Now it is necessary only to combine all the fragments, shown in Fig. 10.6, and we shall obtain the complete scheme of the functional integrity of the structural safety model of the considered segment of the railway. This SFI is shown in Fig. 10.7. The corresponding system of logical equations for this SFI are

$$\begin{aligned} y_1 &= x_1; & \overline{y}_1 &= \overline{x}_1; \\ y_2 &= x_2; & \overline{y}_2 &= \overline{x}_2; \\ y_3 &= x_3 * y_1; & \overline{y}_3 &= \overline{x}_3 \vee \overline{y}_1; \end{aligned}$$

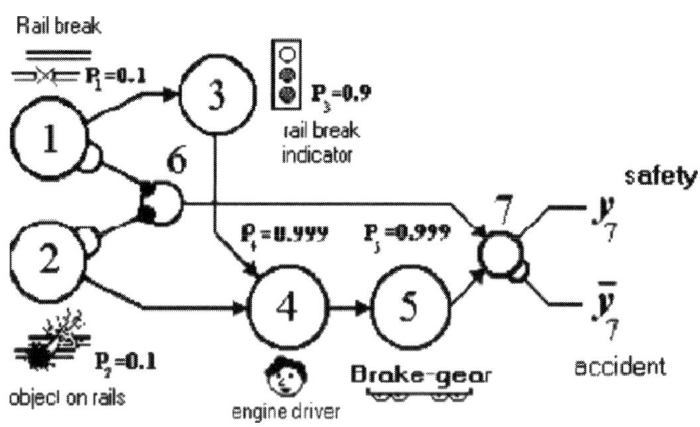

Fig. 10.7. Logical equations of model of safety of the railway segment

$$\begin{aligned} y_4 &= x_4 * (y_2 \vee y_3); & \overline{y}_4 &= \overline{x}_4 \vee \overline{y}_2; \\ y_5 &= x_5 * y_4; & \overline{y}_5 &= \overline{x}_5 \vee \overline{y}_4; \\ y_6 &= \overline{y}_1 * \overline{y}_2; & \overline{y}_6 &= y_1 \vee y_2; \\ y_7 &= y_6 \vee y_5; & \overline{y}_7 &= \overline{y}_6 * \overline{y}_5. \end{aligned} \tag{10.2}$$

At the final stage of primary structural and logical modeling, the probabilistic parameters (P_i, $i = 1, 2, \ldots, n$) of elements are determined (in the example they are indicated in Fig. 10.7) and the logical criteria of operation of the investigated system are fixed. In the considered example, the purpose of modeling is the probabilistic analysis of the safe operation or the accident of the railway segment. The corresponding models can be obtained on the basis of SFI, shown in Fig. 10.7, and any of the following two logical criteria of operation:

(1) the safety criteria

$$Y_{c_1} = y_7; \tag{10.3}$$

(2) the criteria of the accident appearance

$$Y_{c_2} = \overline{y}_7. \tag{10.4}$$

All the considered forms of representation of input data of the automated structural-logical modeling are easily represented in a computer [106–108] and can be the basis for the full automation of all the subsequent stages of construction of computational models.

10.6 Graphical-analytic method of determination of L-function

At the stage of logical modeling, the logical function of system efficiency is determined that should precisely correspond with SFI of the system and to the given logical criterion of its operation:

$$Y_c = Y_c\left(\{x_i, \overline{x}_i\},\ i = 1, 2, \ldots, n\right). \tag{10.5}$$

The universal graphical-analytic method of determination of all the types of monotonic and non-monotonic logical function from any SFI (the system of logical equations) and from any logical criterion of operation was designed for correct solution of the problem (10.5) by a computer. This method also takes into account groups of incompatible events and logical sequences of events [5, 6]. The main contents of the universal graphical-analytic method is determined by the following principles.

Logical function is determined with the help of the substitution method by consecutive deployment of all the output functions of the given L-criteria

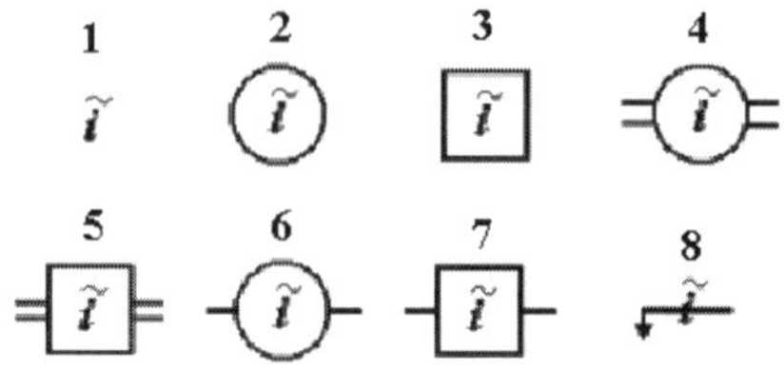

Fig. 10.8. Types of units of the tree of decisions

of operation following the functional integrity scheme of the system with the help of the base logical equations (10.1). The definition of function conjunctions is done by method of backward search in depth, realized in the form of construction of columns of the special decision-tree of the logical equations system, representing SFI of the system. For construction of the decision-tree, the special symbolic apparatus is used. The structure and description of its symbols are shown in Table 10.1 and in Fig. 10.8. Two decision-trees of the system of logical equations (SFI) by criterions (10.3) and (10.4) respectively are shown in Figs. 10.9 and 10.10.

The aggregative algorithm of the universal procedure of definition of logical FES includes the following steps:

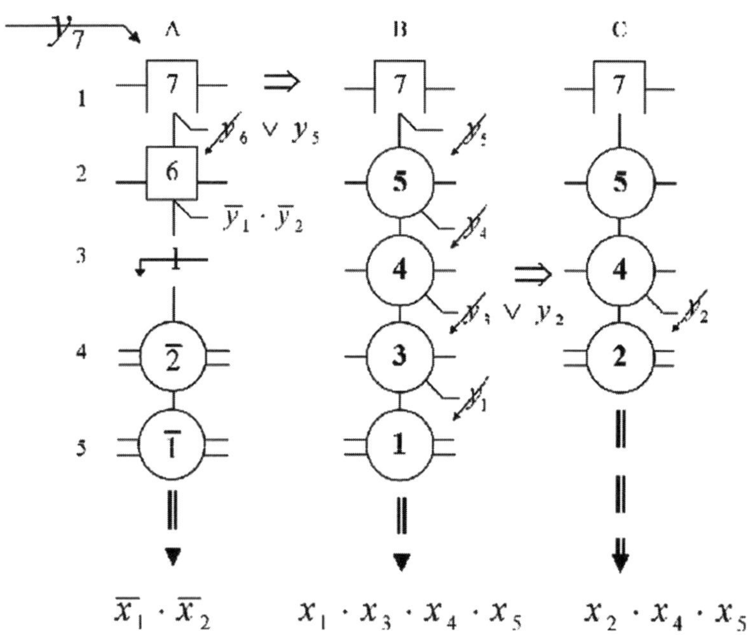

Fig. 10.9. Tree of decisions for finding logical function of safety

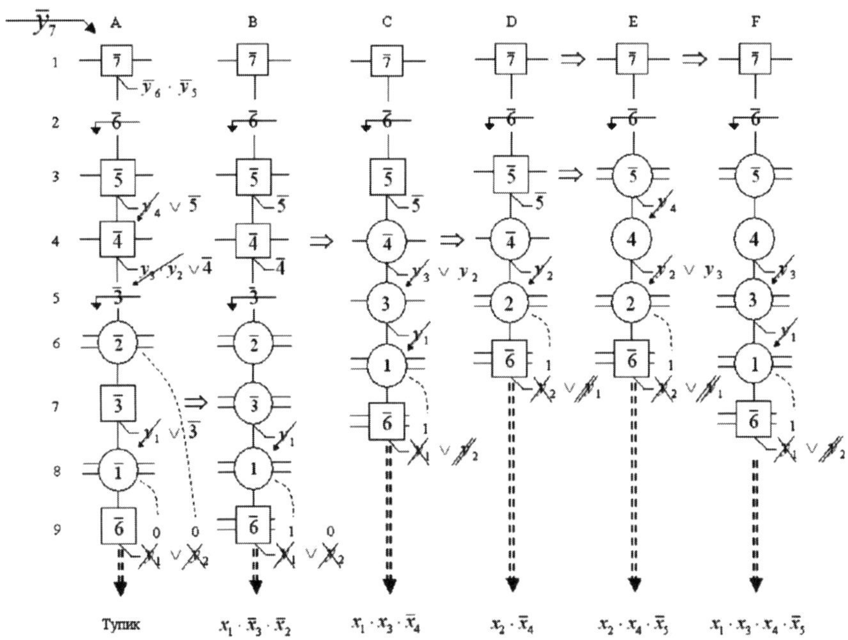

Fig. 10.10. Tree of decisions for logical function of accident

1. Processing L-criteria of operation. The next left conjunction of output functions $\overline{y}_i$ is rewritten from L-criteria of operation to the next new column of the decision-tree, as a sequences of initial nodes (Table 10.1, item 1), and then step 2 is done. If there are no more conjunctions in L-criteria of operation, then construction of the decision-tree is finished and the algorithm goes to the step 9.

2. Formation of the maintenance function. In a column of the decision-tree, the first from below initial node is taken to deployment and is marked by circle or square (Table 10.1, items 2, 3). On branch of the node, the function of maintenance is written. For the prime SFI nodes, the maintenance function is equal to I (logical one). For remaining nodes, the direct maintenance function includes the right-hand sides of logical equations (Fig. 10.7) of SFI without the simple variables $\overline{x}_i$ (Fig. 10.9, a1, a2, b3). For inverse output functions $\overline{y}_i$ of functional nodes of SFI, chosen for deployment, the sign of own inverse is written rightmost in maintenance function (Fig. 10.10, a4, a7).

3. Processing the maintenance function. All the maintenance function components are checked up on cyclicity, opening, and logical contradictions by the rules, which correspond with the laws of the algebra of logic to the methods of taking into account groups of incompatible events and logical sequences. In checking, the cyclical and contradictory output functions $\overline{y}_i$ in the maintenance function are replaced by the logic zero (Fig. 10.10, a9, b9),

Table 10.1. The node types of decision trees

Number and name of node	Description of node of decision tree of the logical equation system
1. The initial node	In the decision tree this node represents the integrated function, set in the formed column, but not accepted for opening
2. The opening real node	This node represents the integrated function $\widetilde{Y}_i$ of the functional node i of SFI, accepted for opening, and means the realization of corresponding own binary event
3. The opening conditional node	This unit represents the integrated function $\widetilde{Y}_i$ of the fictitious node i of SFI or the initial stage of opening of inverse output function of the functional node
4. The opened real node	This node represents either the prime functional node of SFI or the functional node, such that the ensuring condition has been realized in higher part of the formed column of the decision tree
5. The opened condition node	This unit represents either the prime fictitious node of SFI or the fictitious node, such that the ensuring condition has been realized in higher part of the formed column of the decision tree
6. The opened on guarantee real node	This node represents the functional node of SFI, such that the ensuring condition is realized in lower part of the formed column of the decision tree
7. The opened on guarantee condition node	This node represents the fictitious node of SFI, such that ensuring condition is realized in lower part of the formed column of the decision tree
8. The displaced node	This node represents the transfer of the initial node i to the lower position of the column of the decision tree for further opening

and the open ones are replaced by the logical unit (Fig. 10.10, b9, c7, d6, e6, i7). Then, the maintenance function is converted under the rules of the algebra of logic. If maintenance function became equal to the logical zero (Fig. 10.10, a9), then the formation of the column is finished (deadlock) and the algorithm gets to step 8. If the maintenance function became equal to the logical unit, then the unit of the decision-tree that is being opened is considered as the open one (Table 10.1, items 4, 5) and the step 5 is further executed. In all other cases, the algorithm proceeds to the step 4.

4. Displacement of conjunction of maintenance function. The leftmost conjunction of output functions $\overline{y}_i$ is selected from maintenance function and rewritten at the end of the given column of the decision-tree as a sequences of initial nodes (in Figs. 10.9 and 10.10 all displacement conjunctions are crossed by inclined arrows). The step 2 is further executed (main cycle).

5. Transformation of the column by the properly open node. All higher nodes of the column sequentially (bottom-up) are converted by the following rules. Properly open, open on guarantee, and the displaced units (Table 10.1) are omitted. The accepted for deployments real and conditional nodes are considered as open on guarantee (Table 10.1, items 6, 7). If all higher part of the column is converted, then the formation of the logic function of conjunction is finished and the algorithm comes to step 7. If during transformation an initial node was met, then step 6 is further executed.

6. Processing an initial node. If the lower properly open column node corresponds with prime SFI node, then the initial node is crossed out with an angular arrow (Table 10.1, item 8) and is rewritten into the lower part of the formed column (Fig. 10.9, a3). Then step 2 is executed. If the lower properly open node of the column is inverted and it does not correspond with a prime SFI node, then (if it is necessary to take into account the sequences) the direct maintenance function (Fig. 10.10, c4, e3) for this node is formed and we further get to step 3.

7. Reading conjunction. The next formed in the column conjunction of the required L-function contains logical variables; the numbers are indicated at real nodes (Table 10.1, items 4, 6). If these conjunctions are written from the column in the bottom-up order, then the actual sequence of events in them is saved. In Figs. 10.9 and 10.10, the ordered L-function of conjunctions of safety and accident are written under each column (down directed arrows). The step 8 is further executed.

8. Looking for branches. The units of the column are viewed sequentially, bottom-up. Those units that do not have the maintenance function are removed and, respectively, the previous levels of deployment of units of the higher part of the column are restored. If all the units are removed (branches are not found), the algorithm goes to step 1. If the first branches from below are found (marked by horizontal arrows in Figs. 10.9 and 10.10), then step 4 is run.

9. Transformation of L-function. The minimizing transformations of the obtained L-function and its reduction to the form convenient for further application (for example, taking into account availability of groups of incompatible events, logical sequences, different initial system conditions, etc.) are done.

The logical function of safety (Fig. 10.8) and accident (Fig. 10.9) obtained with the help of the graphical-analytic method are as follows:

$$Y_{c_1} = y_7 = \overline{x}_1 \cdot \overline{x}_2 \vee x_1 \cdot x_3 \cdot x_4 \cdot x_5 \vee x_2 \cdot x_4 \cdot x_5 \quad (10.6)$$

$$Y_{c_2} = \overline{y}_7 = x_1 \cdot \overline{x}_3 \cdot \overline{x}_2 \vee x_1 \cdot x_3 \cdot \overline{x}_4 \vee x_2 \cdot \overline{x}_4 \vee x_2 \cdot x_4 \cdot \overline{x}_5 \vee \\ \vee x_1 \cdot x_3 \cdot x_4 \cdot \overline{x}_5 \quad (10.7)$$

respectively. Once again we note that the considered graphical-analytic method of the solution of systems of logical equations is strictly formalized and analytically universal. First it allowed us to realize all the possibilities of the algebra of logic and to solve with the help of program complexes all, without exception, known problems of monotonic logical-probabilistic modeling, and also to build essentially new non-monotonic logical-probabilistic models of operation of complex objects and processes. Secondly, it is possible to overstep the bounds of the classic algebra of logic and to take into account two significant types of relations: groups of incompatible events and logical sequences of events in time. Then, in the obtained L-function (10.6), (10.7) all the possible combinations of initial dangers (events $\overline{x}_1, \overline{x}_2$), and also minimum safe (10.6) and emergency (10.7) sequences of events of operation of system components of safety management of motion ($\overline{x}_3, \overline{x}_4, \overline{x}_5$) are presented. These combinatorial sequences determine all (within the framework of adopted assumptions) variants of correct, i.e., safe operation of the system (10.6) and all variants of appearance of accidents, resulting in railway accidents. Taking into account sequences allows us not only to obtain computational probabilistic models but also to formulate and to solve problems of the determined management of system safety.

10.7 Combined method of construction of P-polynomials

At the stage of probabilistic modeling, the processes of construction of several types of computational models are automated. Here we will dwell only on original positions of the algorithmic method of construction of probabilistic function polynomials (P-function), which is called the combined method [5, 6, 106, 107].

The given problem of P-function determination consists in transformation of the logical function into the probabilistic function polynomial of the following type

$$Y_c = Y_c(\{x_i, \overline{x}_i\},\ i = 1, \ldots, n) \Rightarrow P_c(\{P_i, Q_i\},\ i = 1, \ldots, n) \tag{10.8}$$

For the correct solution of the problem (10.8) by the combined method, one uses, generally, two types of transformations of the initial logic function Y_c. At first quasi-orthogonalization on one logical variable is made, and then the symbol transition to the polynomial of the required P-function is executed. Under assumption of independence in aggregate of all elementary binary events $\tilde{x}_i, i = 1, 2, \ldots, n$, the rules of fulfillment of these two stages are given below.

Rules of quasi-orthogonalization on one variable. All pairs of non-orthogonal logic functions of conjunctions are checked for the possibility of orthogonalization using the following rule

$$\gamma \cdot \tilde{x}_i \vee \gamma \cdot \varphi = \gamma \cdot x_i \vee \gamma \cdot \varphi \cdot \overline{\tilde{x}_i} \tag{10.9}$$

Here γ and φ are those parts of checked conjunctions that do not include the variable $\tilde{x}_i$. The orthogonalizing transformation (10.1) does not increase the total number of conjunctions of the initial FSE. Executing the indicated transformations for the functions (10.6) and (10.7), we obtain:

$$\begin{aligned} Y_{c1} = y_7 &= \overline{x}_1 \cdot \overline{x}_2 \vee x_1 \cdot x_3 \cdot x_4 \cdot x_5 \vee x_2 \cdot x_4 \cdot x_5 = \\ &= \overline{x}_1 \cdot \overline{x}_2 \vee x_1 \cdot x_3 \cdot x_4 \cdot x_5 \cdot \overline{x}_2 \vee x_2 \cdot x_4 \cdot x_5; \end{aligned} \tag{10.10}$$

$$\begin{aligned} Y_{c2} = \overline{y}_7 = x_1 \cdot \overline{x}_3 \cdot \overline{x}_2 \vee x_1 \cdot x_3 \cdot \overline{x}_4 \vee x_2 \cdot \overline{x}_4 \vee x_2 \cdot x_4 \cdot \overline{x}_5 \vee \\ \vee x_1 \cdot x_3 \cdot x_4 \cdot \overline{x}_5 = x_1 \cdot \overline{x}_3 \cdot \overline{x}_2 \vee x_1 \cdot x_3 \cdot \overline{x}_4 \cdot \overline{x}_2 \vee \\ \vee x_2 \cdot \overline{x}_4 \vee x_2 \cdot x_4 \cdot \overline{x}_5 \vee x_1 \cdot x_3 \cdot x_4 \cdot \overline{x}_5 \cdot \overline{x}_2. \end{aligned} \tag{10.11}$$

Rules of symbol transition to the probabilistic polynomial.
Here the logical function is considered as the rigorous analytic form of determination of that composite random event, the probability of which should be correctly calculated with the help of the required P-function. In the stated sense, the conjunction is a product, the disjunction is a sum, and the inverse is the complement of simple random events, probabilistic parameters of which are known. Therefore, for obtaining the P-function polynomial, it is sufficient to make symbol transformations of notations of logical variables and operations to notations of probabilistic variables and arithmetic operations using the precise conformity with the laws of the probability theory. The full set of such rules of symbol transition from L-function to P-function is given below.

1. Transformation of simple logic variables. In the initial logic function, all single logic variables $\tilde{x}$ are substituted by notations of corresponding probabilistic parameters:

$$\begin{aligned} x_i &\to p_i \\ \overline{x}_i &\to \overline{p}_i \end{aligned} \tag{10.12}$$

2. Transformation of conjunctions. The rules of transformation of conjunctions in combined P-function are determined by the following formula:

$$\tilde{p}_j \cap \tilde{p}_k \to \left| \begin{array}{cl} \tilde{p}_j & if\ x_j = x_k;\ \tilde{x}_j = \tilde{x}_k \\ 0 & if\ x_j = x_k;\ \tilde{x} \neq \tilde{x}_k\ GIE \\ \tilde{p}_j * \tilde{p}_k & if\ \ x_j \neq x_k \end{array} \right. \tag{10.13}$$

Here the first rule intends for refraining from repetition, if they there are in logic products. The second rule is used for variables that belong to one GIE. The third rule is used, if multiplied events are different in totality.

3. Transformation of disjunctions. The rules of transformations of disjunctions are the following:

$$\tilde{p}_j \cup \tilde{p}_k \to \left| \begin{array}{cl} \tilde{p}_j & if\ x_j = x_k;\ \tilde{x}_j = \tilde{x}_k \\ 1 & if\ x_j = x_k;\ \tilde{x}_j \neq \tilde{x}_k\ GIE \\ \tilde{p}_j + \tilde{p}_k - \tilde{p}_j * \tilde{p}_k & if\ \ x_j \neq x_k \end{array} \right. \tag{10.14}$$

Here the first rule intends for refraining from repetition, if they there are in logic sums. The second rule is used for variables that belong to one GIE. The third rule is used, if adding different and independent events in totality.

4. Transformation of inversions. All inversions in combined P-function determine events opposite to given ones:

$$\overline{p}_i \to p_i = 1 - p_i$$
$$\overline{x}_j \to 1 - x_j \qquad (10.15)$$

Providing the total orthogonality L-function, the symbol transformations precisely coincide with the known rules of the direct replacement. Converting L-functions (10.3) and (10.4) into P-function polynomials, we obtain probability of safety

$$p_{c_1} = p\{y_7\} = q_1q_2 + p_1p_3p_4p_5q_2 + p_2p_4p_5; \qquad (10.16)$$

probability of emergency:

$$p_{c_2} = p\{\overline{y}_7\} = p_1q_3q_2 + p_1p_3q_4q_2 + p_2q_4 + p_2p_4q_5 + p_1p_3p_4q_5q_2. \qquad (10.17)$$

With the help of methods for taking into account groups of incompatible events, the possibility of using in models of reliability, safety, and risk of systems of elements with any number of proper conditions and stochastic dependence is realized. At this, some laws of the algebra of logic and rules of construction of probabilistic function polynomials change.

10.8 Calculation of standard P-characteristics of systems

The automatically formed P-function polynomials in themselves are algorithms of calculation of the general system probabilistic characteristics of the system. Then, substituting into (10.16) and (10.17) values of probabilistic parameters of elements (Fig. 10.7), we obtain
probability of safety of the segment railway

$$p_{c_1} = p\{y_7\} = 0.99064, \qquad (10.18)$$

probability of emergency on the segment railway

$$p_{c_2} = p\{\overline{y}_7\} = 0.00936. \qquad (10.19)$$

For development and substantiation of various management solutions, of great importance is the objective grading that role played by different elements in maintenance of the general system parameter of reliability, safety or risk of the system operation. The parameters of significance and contribution of separate elements are applied for these purposes [5, 106]. These parameters determine the change of the general system characteristic p_c when values

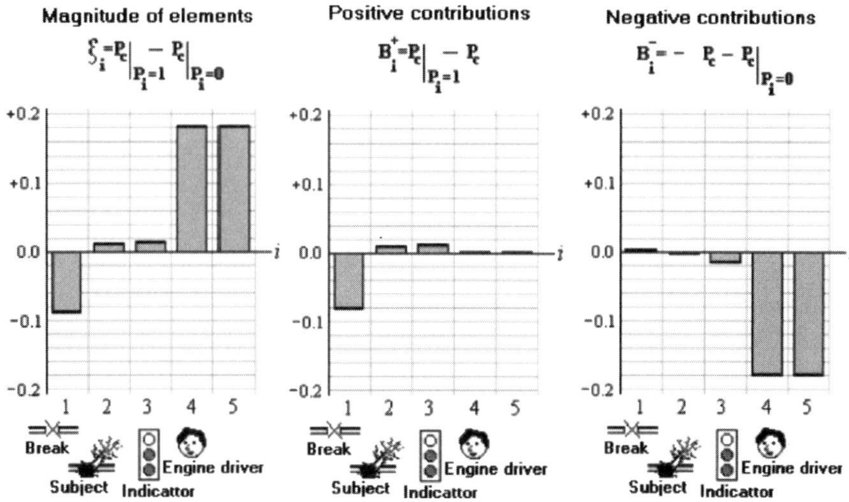

Fig. 10.11. The results of calculation of magnitudes and contributions of elements

of separate parameters of elements p_i vary. The outcomes of calculations of element parameters role of the considered example are given in Fig. 10.11.

For past years, certain scientific and practical experience of development and application of the theory, technology, and program complex for the automated structural and logical modeling and calculation of reliability, safety, and risk parameters of systems of different types, classes, and destination has been accumulated. The realization of all abilities of the main modeling tool of the algebra of logic made it possible to automate the construction processes not only for all models, represented by event trees, failures, and graphs, but also to formulate and solve many essentially new and actual problems of the structure analysis of reliability, safety, and risk of complex systems.

On the basis of further development of the considered algorithmic methods, it is possible to automate construction processes of some new classes of mathematical system models: statistical, Markov's, and network ones. In formed models of dependent events, the possibility to take into account availability in the system of elements with any number of conditions and different sequences of random events in the time is realized. Tens of investigations are made with usage of the ASLM theory and technology. For example, the successful calculations of reliability and failure-resistance of the automated management system of the technological process at the stage of design were made. Development of the commercial software has started. In the program complex, it is planned to realize the latest progress in ASLM technology. The basic ASLM characteristics can be better than those of the known program system *Risk Spectrum* of the company *Relcon AB* for the automated modeling and calculation of reliability and safety, working in "trees of failures" technology.

11

Logical and Probabilistic Theory of Risk with Groups of Incompatible Events

There is nothing more practical than the good theory.

Robert Kirchoff

We start to consider the logic and probabilistic non-success risk theory with groups of incompatible events (the non-success risk LP-theory with the GIE). It includes:

1. The construction of non-success risk LP-models with the GIE,
2. The identification of risk LP-models with the GIE on statistical data,
3. The estimation and analysis of the risk,
4. The management of risk.

Applications of the non-success risk LP-theory are the following:

1. The credit risks,
2. The security portfolio risk,
3. The risk in efficiency and quality,
4. The non-success risk of company management,
5. The risk models in problems of bribes and corruption, etc.

The risk LP-models with the GIE allow one to realize the active scenario management of the non-success risk in complex systems on the stages of designing, testing, and operating.

The classifier of scientific area - YDK 519.862.6, Econometrics.

The problem - Modeling and analysis of risk in complex systems.

The essence of techniques - Introduction in statistical database (DB) of the groups of incompatible events or finite sets that allows one to receive the LP-equation system (knowledgebase (KB)), to use the LP-calculus, and to solve new tasks: risk, efficiency, management.

The non-success risk is a generalized term that we will use for all risk problems and tasks in order to avoid such various concepts as a risk of accidents, a risk of crashes, a risk of security portfolios, a risk of swindles, a risk of bribes, etc.

E.D. Solojentsev, *Scenario Logic and Probabilistic Management of Risk in Business and Engineering*, DOI 10.1007/978-0-387-77946-1_11,

It is considered the modeling and analysis of the non-success risk in systems with GIE. The elements and output characteristics of these systems have several levels of values. Examples of these systems are credits in a bank, diagnostics and monitoring, quality of productions, precision of products, etc. In these areas, the non-success risk is the usual and mass phenomenon, and there are the statistical data on the risk objects or the states of the risk object [3, 4, 27–30].

Structurally complicated object of risk. To structurally complex risk objects we relate objects (systems) where elements are logically interconnected and have several levels of states (values). For such objects there are the huge computing difficulties at designing the risk LP-models on the statistical data, which can be overcome only by special techniques and using special logic software.

We classify the direction of such investigations as the risk analysis in systems with groups of incompatible events, or the analysis of reliability of systems with multiple states [109, 110].

The non-success risk LP-theory with GIE includes the construction of the risk model, the identification of the risk model on basis of the statistical data, and the risk analysis. In its turn, the construction of the risk model includes the development of the structural risk model, the logical risk model, and the probabilistic risk polynomial. We consider the risk homogeneous objects (for example, credits) or the conditions of the system in different moments of the time (for example, the security portfolio). The non-success risk LP-models can be any logical complexity but in business the risk LP-models are usually not complex.

The structural risk model in business most often is associative, and takes into account all or limited set of objects or states of one object. For example, the scenario of the credit risk is as follows: non-success occurs, if any one, any two, ... or all sign-events occur.

Using the structural risk models, we construct the corresponding L-functions of various complexity with connections AND, OR, NOT, cycles and repeated elements. Any risk L-function after orthogonalization can be written in the form of the probabilistic polynomial or the probabilistic model (P-model, P-function). Then we can calculate the risk of the object, if we know the probabilities of grade-events. The considered approach allows us to construct the well-organized risk P-polynomial. As examples show, the accuracy of the risk LP-model is almost two times higher and the robustness is almost seven times more than those of methods based on the discriminant analysis and the neuron networks.

11.1 Converting database to knowledgebase

Similar statistics tabular presentation (database) is common for problems of classification (securities portfolio), efficiency (social process and quality), management (company failure), bribery and corrupt practices [3].

Table 11.1. Objects and parameters

Objects i	Parameter z_1	Parameter z_2	Parameter z_j	Parameter z_n	Parameter y
1					
...	...	...	...	...	...
			z_{jr}		
...	...	...	...	...	...
N					

Database. Tabular database (Table 11.1) contains statistics on congeneric objects (credits) or states of one object at different points of time (securities portfolio). DB table can have up to several dozens of columns and up to several hundreds of lines. The values of parameters (characteristics) in DB can be integral or non-integral and are considered as statistics on objects or states of one object.

Table cells show values (quantitative or qualitative) characterizing the object or its state. To measure parameters, the following scales are used: logical, qualitative, numeric, etc. The last column is an efficiency characteristic of the object or object's state. We shall call characteristics describing the object with lowercase letters $z_1, \ldots, z_j, \ldots, z_n$, object's efficiency characteristic with a lowercase letter $y_i, i = 1, 2, \ldots, N$. Table 11.1 shows characteristics values z_{ij} and efficiency characteristic value y_i in the last column.

Database modification. We are changing initial statistics representation replacing values of characteristics by their grades (numbered intervals).

In scripts (scenarios) and LP risk models of classification, investment, efficiency, management, corrupt practices, and bribery problems, there exist a great variety of objects N (up to 1000 and more), parameters-events n (up to 20 and more), and grades-events within each parameter-event (from 2 to 40).

Thus, the modified tabular DB (Table 11.2) now has denumerable sets with finite number of elements (grades) for each characteristic. Unlike Table 11.1 where characteristics could have an indefinite number of different values, now each range for each parameter has a denumerable number of elements equal to the number of grades it is broken down into.

Table 11.2. Objects and grades

Objects i	Parameter Z_1	Parameter Z_2	Parameter Z_j	Parameter Z_n	Parameter Y
1					
...	...	...	...	...	...
			Z_{jr}		
...	...	...	...	...	...
N					

Knowledgebase. In the modified DB (Table 11.2), parameters describing the object shall be called events-parameters or logical variables and shall be denoted with uppercase letters $Z_1, \ldots, Z_j, \ldots, Z_n$, object efficiency parameter shall be called event-efficiency parameter and shall be denoted with uppercase letter Y. Cells of Table 11.2 hold events-grades $Z_{jr}, j = 1, 2, \ldots, n; r = 1, 2, \ldots N_j$ for parameters $Z_1, \ldots, Z_j, \ldots, Z_n$. The last column holds events-grades $Y_r, r = 1, 2, \ldots, N_y$ for efficiency parameter Y. In general, grades are not linearly ordered, and one cannot say if grade 3 is worse or better than grade 4.

As a result, we get a set of N logic equations with the left-hand side of the equation and the right-hand side of the equation. Let's correlate the probability of validity or invalidity of each logic variable of the right- and the left-hand side with it. This logic equations set shall be called a knowledgebase (KB). It is easy to calculate the frequency (probability) of events-grades for this KB. They are equal to the ratio of the number of objects or states with grade to the total number of objects or states in KB of N. This system shall be also considered a set of logical propositions and shall be used to acquire new knowledge.

In scenarios and LP risk model, events-parameters are linked with the help of logic operations AND, OR, NOT and can be arranged in cycles. Events-descriptors (triggering events) correspond with logical variables that can be dependent though not initially but just because they are within some definite logical formula determining connection between them. Events-grades for each groups of incompatible events are dependent. They form a group of incompatible events for each characteristic [3].

Introduction in statistical DB the groups of incompatible events or finite sets allows one to receive the LP-equation system or KB, to use the LP-calculus, to use the arbitrary distributions of parameters (not only the normal low), and to solve new tasks: risk, efficiency and management.

11.2 Structure risk models

In multicomponent systems with GIE in problems of classification, investment, efficiency, and management, we use the cascading structure for the non-success risk model. We bring this structure (Fig. 11.1):

(1) the set of objects or object states for the component of the system,

(2) the object or the state,

(3) the signs (parameters) for the description of the object (the state) and its efficiency,

(4) the grades of the signs.[1]

[1] In the most cases and in the subject index we use the term "the object," but definitions and equations are true for "the state" of the system or the object.

Variables and indexes. The random events correspond with grades and signs that lead to the non-success [3, 30]. In above named problems, we use the following designations for indexes of components, objects, signs, grades, and corresponding events and logical variables:

$k = 1, 2, \ldots, K$ are indexes of different components of the system;

$i = 1, 2, \ldots, N$ are indexes of different objects or states, or efficiency parameters;

$j = 1, 2, \ldots, n$ are indexes of different signs (parameters), influencing on the efficiency parameter;

$r = 1, 2, \ldots, N_j$ are indexes of grades of signs influencing on the efficiency parameter.

Here are:

K is the number of components in the system;

N is the number of objects or states in the statistics on the component;

n is the number of signs influencing the efficiency parameter;

N_j is the number of grades in the discrete distribution of the parameter;

N_y is the number of grades in the discrete distribution of the efficiency parameter.

In above named problems, we use the following designations of random events and corresponding logical variables:

$Y_1, Y_2, \ldots, Y_k, \ldots$ are components of systems;

$Y_{kr};\ k = 1, 2, \ldots, K;\ r = 1, 2, \ldots, N_k$ are grades of components of the system;

Y is the efficiency parameter of the system or the system component;

$Y_r,\ r = 1, 2, ..., N_y$ are grades of the efficiency parameter;

$Z_1, \ldots, Z_j, \ldots, Z_n$ are parameter-events;

Z_{jr} are grade-events.

In risk scenario, the parameter-events are connected by the logical operations OR, AND, NOT. The grade-events for each sign (parameter) form the group of incompatible events (GIE). The maximum number of combinations (different objects or states) is the following:

$$N_{max} = N_1 \cdot N_2 \cdot \ldots \cdot N_j \cdot \ldots \cdot, N_n, \tag{11.1}$$

where $N_1, N_2, \ldots, N_j, \ldots, N_n$ are numbers of grades in signs.

Let us bring examples of some systems.

The bank as the system. Components of the system are the directions of the bank activity (crediting natural and juridical persons, investment portfolio of securities, etc.), the set of credits, the credit, the parameters (signs) of the credit, the grades of the signs.

The rocket launcher as the system. Components of the system are the subsystems (refueling, etc.), the set of the states of the subsystem, the parameters of the state, the grades of parameters of the state.

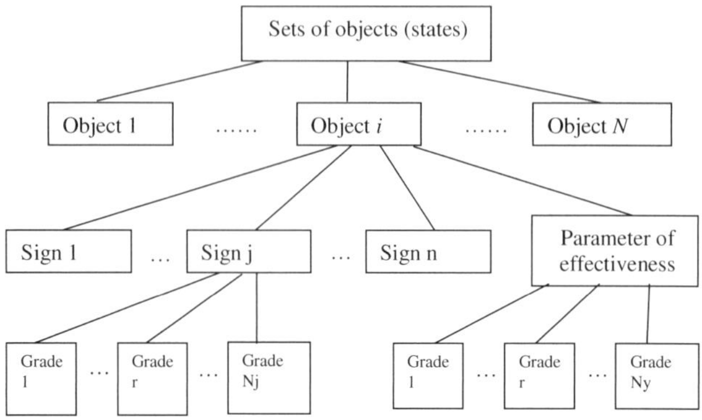

Fig. 11.1. Structure of models of many component system with GIE

Government of the country as the system. Components of the system are the criterion projects, the set of regional projects, the attributes of regional projects, the grads of the attributes.

11.3 Groups of incompatible events and the property of orthogonality

Groups of incompatible events. Representative table of statistical data (Table 11.1) can hold statistical data with any distribution function and arbitrary connections between parameter-events. The frequencies (probabilities) of grades are calculated from the table by the expression:

$$P2_{jr} = N_{jr}/N, \quad j = 1, 2, \ldots, n; \quad r = 1, 2, \ldots, N_j, \tag{11.2}$$

where N_{jr} is the number of objects (states) in the table for the parameter j with the grade r.

From (11.1) and (11.2) it follows, too, that it is desirable to have in Table 11.1 as many statistical data as possible and to define more exactly values of probabilities of grade-events P_{jr} as we get new information on objects or their conditions.

The research of actual "tables" confirms validity of passage to discrete non-parametric probability distributions, connected with grade-events. For example, each of 1000 credits was described by 20 parameters and each parameter had from 2 to 11 grades. As a result of the analysis, the simplified empirical discrete distributions of frequencies of grade-events were as follows: uniform, on a straight line, sloping up or down; in a triangle turned up or down; symmetric, displaced to the left or to the right. From such variety of the forms of distributions, it is necessary to set distributions of grade-events in GIE by probabilities (Fig. 11.1).

The risk in real systems depends on many parameters. For example, the investment portfolio can have tens of securities with different yield and risk values. More often different parameters have different dimension, distribution laws of parameters are different and are not normal.

Usually multidimensional normal distributions are applied. That is, each single parameter is distributed normally, and its mean value and dispersion are known. The dispersing matrix of all parameters is considered to be known. Any means for calculation of risk in realistic multidimensional systems, which influencing parameters have different non-normal distribution laws, are absent. The LP-theory of risk with GIE offers a solution to this problem on the basis of LP-calculus.

In the LP-theory of risk with GIE, a transition from continuous distribution of a random variable to a discrete distribution is used. The range of values of the continuous random variable is split into intervals, not necessarily having identical length (Fig. 11.2 a).

The probability of random variable Z_j to belong to a given interval is determined by statistical data (Fig. 11.2 b). Naturally, the sum of probabilities over all intervals for one parameter is equal to 1. Numbers of intervals are equal to the numbers of grades. Suppose that such transition for the LP-model of risk with the GIE, which has several parameter-events connected by logical connections AND, OR, NOT, is done. Then calculation of final event probability is fulfilled by the rules of LP-calculus.

For every GIE, the following logical equations hold [3]:

$$\begin{aligned} z_{jr} \wedge z_{jk} &= 0; \\ \overline{z}_{jr} \vee \overline{z}_{jk} &= 1; \\ \overline{z}_{jr} \wedge z_{jk} &= z_{jk}; \\ z_{jr} \vee \overline{z}_{jk} &= \overline{z}_{jk} \end{aligned} \tag{11.3}$$

along with the substitution rules of incompatible events by their probabilities:

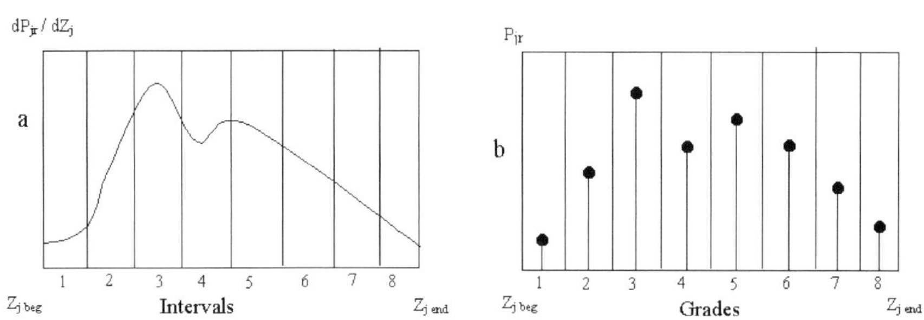

Fig. 11.2. Continuous (a) and discrete (b) distributions

$$P\{z_{jr} \wedge z_{jk}\} = 0;$$
$$P\{z_{jr} \vee z_{jk}\} = P\{z_{jr}\} + P\{z_{jk}\};$$
$$P\{\overline{z}_{jr} \vee \overline{z}_{jk}\} = 1;$$
$$P\{\overline{z}_{jr} \wedge \overline{z}_{jk}\} = 1 - (P\{z_{jr}\} + P\{z_{jk}\}). \quad (11.4)$$

The expressions (11.3) and (11.4) are not convenient for calculations because of unhandness of character transformations in the case when there are some GIE with a great number of grade-events. Because we would not like to idealize and simplify solutions of real problems, as the basic method for solving problems of risk we choose the algorithmic approach.

Orthogonality of L-functions for different objects. We write the logical function for possible objects or states of the table (Table 11.1) in the perfect disjunctive normal form (PDNF) [3, 30]:

$$Y = Y_1 \vee Y_2 \vee \ldots \vee Y_i \vee \ldots \vee Y_N, \quad (11.5)$$

where each object or state is determined by the logical function including all logical variables:

$$Y_i = Z_1 \wedge \ldots \wedge Z_j \ldots \wedge \ldots \wedge Z_n. \quad (11.6)$$

In engineering, each logic variable j accepts only two values, Z_j and $\overline{Z}_j$. Thus, the logic function (11.6) determines $N = 2^n$ conditions of the object. The logic functions of any two conditions, for example

$$Y_i = Z_1 \wedge \ldots \wedge Z_j \wedge \ldots \wedge Z_n;$$
$$Y_{i+1} = Z_1 \wedge \ldots \wedge \overline{Z}_j \wedge \ldots \wedge Z_n \quad (11.7)$$

are orthogonal in view of the following identity for the logical product

$$Y_i \wedge Y_{i+1} = 0, \quad (11.8)$$

which holds because $Z_j \wedge \overline{Z}_j = 0$.

Therefore, $PDNF$ (11.5) is orthogonal, because it can be written as the logic sum of all $Y_i, i = 1, 2, \ldots, N$, where any two logic items (conjunctions) are orthogonal.

In business, at description of objects or conditions of objects (Table 11.1), each logic variable in (11.6) accepts many values, equal to the number of grades or intervals on which the yield is split. The logic functions for two different conditions of the portfolio, for example

$$Y_i = Z_1 \wedge \ldots \wedge Z_{jr} \wedge \ldots \wedge Z_n; Y_{i+1} = Z_1 \wedge \ldots \wedge Z_{j\ r+1} \wedge \ldots \wedge Z_n \quad (11.9)$$

are orthogonal, too (11.8), because

$$Z_{jr} \wedge Z_{jr+1} = 0, \quad (11.10)$$

because Z_{jr} and Z_{jr+1} belong to the same GIE. The property of orthogonality of logical items of the risk L-function (11.5) allows us to proceed from logical functions to algebraic expressions.

Dependent and independent events. Sign-events (initiating events) correspond with logical variables, which may be dependent, but not initially, and only because they belong in certain logical formula, which defines the dependency between them. For every GIE, the grade-events are dependent in corresponding with expressions (11.3).

If parameter-events $Z_1, \ldots, Z_j, \ldots, Z_n$ are independent and given by their grade-events with arbitrary discrete distribution lows (it may happen that some of them are constructed on basis of normal distribution lows), then we have the right to consider all the variety of objects (11.1).

The probability for arbitrary object or its state i by (11.6) is equal:

$$P\{Y_i\} = P_{1r_1} \cdot P_{2r_2} \cdot \ldots \cdot P_{nr_n}, \tag{11.11}$$

where the index r_j for each parameter j is equal to one of the values $1, \ldots, r, \ldots, N_j$.

Based upon the orthogonal logical function (11.5), the condition is fulfilled

$$\sum_{i=1}^{N} P_i = 1. \tag{11.12}$$

Starting from the expression (11.1) for the number of possible objects (states) and from the expression (11.11) for calculation of the probability for the object, we conclude that computing complexity of algorithm is exponential:

$$N_c \approx a^n, \tag{11.13}$$

where n is the value of the determining factor of the problem; a is a constant. For example, for algorithms of the risk LP-theory of the security portfolio, the parameters are the following: n is the number of securities in the portfolio, a is the number of intervals on which the security yield is split.

However, the computing complexity should not shock the reader. Though values of the parameters a and n are rather great ($n = 10 \div 100$ and $a = 30 \div 100$), the engineering approach lets us reduce the complexity drastically. Under the approach, the problem does not need the complete evaluation of all possible conditions of the portfolio, which is replaced by construction of the total distribution of the portfolio yield. It is known that 13 points of statistical data are sufficient to restore distribution. Therefore, if the number of such points are 1300, or $13,000$, or $130,000$, then it is quite enough. Modern computers proceed with some millions of operations per second, so the above mentioned calculations can be carried out quickly and in a real-time scale. The limited number of random combinations we receive by Monte Carlo modeling.

We can recall the same situation exists in the theory of optimal planning the multiple-factor experiments, where the experiments are conducted with variables on the unit sphere at points where it meets orthogonal axes. Therefore, not all experiments are fulfilled, but only a limited number of experiments, corresponding with fractional replicas.

11.4 Logical and probabilistic risk models

The structural risk model can be equivalent to a realistic one (for example, electrical system), it can be associative and constructed for all or limited set of objects (states of objects).

Node. Let us consider the non-success risk LP-model of type "node" (Fig. 11.3 a). Here $Z_1, Z_2, \ldots, Zn$ are independent (events), binary variables, accepting value 1 (non-success) or 0 (success) with probabilities:

$$P\{Z_1 = 1\} = p_1, \ldots, P\{Z_n = 1\} = p_n;$$
$$P\{Z_1 = 0\} = 1 - p_1 = q_1, \ldots, P\{Z_n = 0\} = 1 - p_n = q_n. \tag{11.14}$$

The non-success risk L-model of the "node" (Fig. 11.3a) is

$$Y = Z_1 \vee Z_2 \vee \ldots \vee Z_j \ldots \vee Z_n. \tag{11.15}$$

In words, it mean that the non-success occurs if any one, or two, ..., or all initiating events occur.

The risk L-function (11.15) after orthogonalization is

$$Y = Z_1 \vee Z_2 \overline{Z_1} \vee Z_3 \overline{Z_1}\ \overline{Z_2} \vee \ldots. \tag{11.16}$$

The probabilistic risk model (P-model, P-polynomial) is

$$P = p_1 + p_2 \cdot q_1 + p_3 \cdot q_1 \cdot q_2 + \ldots \tag{11.17}$$

In engineering, where more often we estimate the reliability as success, the formula (11.17) usually is described as follows:

$$P = 1 - q_1 \cdot q_2 \cdot q_3 \cdot \ldots \cdot q_n.$$

In the risk LP-model, the "arithmetics" is such that for the final event, the risk value belongs to [0,1] for any values of probabilities of initiating events.

Bridge. The non-success risk L-model of the "bridge" type (Fig. 11.3 b) is represented in the normal disjunctive form as a logic sum of the shortest paths of successful operation:

$$Y = Z_1 Z_3 \vee Z_2 Z_4 \vee Z_1 Z_5 Z_4 \vee Z_2 Z_5 Z_3. \tag{11.18}$$

Orthogonalization of (11.18) provides the non-success risk P-model:

$$P_i = p_2\, p_4 + p_1\, p_3 + q_1\, p_2\, p_3\, q_4\, p_5 + p_1\, q_2\, q_3\, p_4\, p_5 - p_1\, p_2\, p_3\, p_4. \tag{11.19}$$

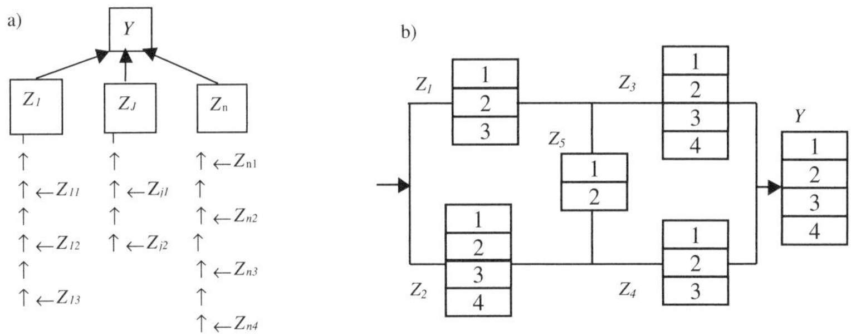

Fig. 11.3. Structural models of risk: (*a*) unit type; (*b*) bridge type

11.5 Risk parameters, measure of risk, and cost of risk

In considered problems of management of risk on statistical data, two concepts are basic:

Risk is a probability of non-success or losses,

Y_{ad} is an efficiency (minimum or maximum yield, success or non-success, etc.).

In problem of classification (credit risks, bribes, etc.), one minimize the risk at the given efficiency:

$$Risk \rightarrow min; Y_{ad} = const.$$

In problem of investment (security portfolio, management of social processes, etc.), one can maximize the efficiency at given the risk:

$$Y_{ad} \rightarrow max; Risk = const.$$

Risk parameters. In systems with GIE, we consider the follow parameters to characterize the risk (Fig. 11.4): Y_{ad} is the admitted value of the efficiency parameter; *Risk* is the probability to have the value of the efficiency parameter less than the admitted one; N_{ad} is the number of objects (states of the object) in "the tail" of the efficiency parameter distribution; H_{ad} is the entropy of probabilities of objects (states of the object) in "the tail" of the efficiency parameter distribution.

Calculation of the admitted value of the efficiency parameter Y_{ad} at a given value of the risk *Risk* is the complex algorithmic problem. We consider three different methods for its solution.

1. Method of interpolation. We construct the differential discrete distribution of the efficiency parameter Y. For this purpose, we split all range of changing of the efficiency parameter into intervals (grades) $r = 1, 2, \ldots, N_y$. We carry out summation of the probabilities P_{y_i} of values of the parameter on the chosen intervals and also construct the integrated discrete distribution for the efficiency parameter. Now we can calculate the admitted value Y_{ad} for the given risk $Risk$, by using the formula for linear interpolation.

2. The method of sorting. The simple and accurate algorithm of calculation of the admitted value of the efficiency parameter Y_{ad} is the method of sorting. Really, arrays of values of the parameter Y_i and their probabilities P_{y_i}, $i = 1, 2, \ldots, N$, can be sorted by the value of the efficiency parameter Y_i in the ascending order. Then, for the sorted arrays we should sum up the probabilities P_{y_i} of values of the parameter Y_i until we get the given $Risk$. The last term in the sum from the array of probabilities will correspond with the value of the efficiency parameter, which should be taken as the admitted value Y_{ad}. Complexity of the process of sorting depends on the number N of conditions of the efficiency parameter, and in practice the time of repeated fulfilment of sorting in the process of optimization is quite acceptable.

3. The method of half division. The method of half division consists in the fact that the interval $[Y_{min}, Y_{max}]$ is consecutive divided into two equal parts $[Y_{min}, Y_{1/2}]$ and $[Y_{1/2}, Y_{max}]$. For each parts are counted by summation of probabilities $P_1 = P\{Y < Y_{1/2}\}$, $P_2 = P\{Y > Y_{1/2}\}$ and the number of objects or conditions in parts. The part in which there is the risk $Risk$, is divided again half-and-half. This procedure proceeds so long as the number of conditions in the part will equal to 1 (one object). At $N = 1000$ objects, the search Y_{ad} by the method of half division occupies in three times less time for calculation than by using the method of sorting.

The parameter N_{ad}. The parameter N_{ad} is the number of conditions of the efficiency parameter Y_{ad} in "the tail" of the contribution, that is at condition $Y_i < Y_{ad}$. This is a very important characteristic of the risk, because it is the integer number and can be calculated with accuracy until 1. As we shall see later, it is possible to solve the optimization problem, for example, of the security portfolio risk, by using not the criterion function Y_{ad} but the equivalent function N_{ad}.

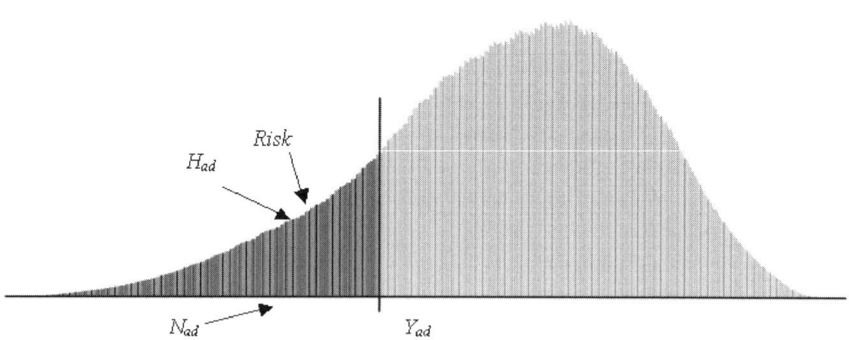

Fig. 11.4. Risk parameters in "the tail" of the distribution

The entropy H_{ad} is yet the one important characteristic for "the tail" of distribution (Shannon's entropy). The level of the heterogeneity or the probabilities variety of the set of state-events in the "tail" depends on the number of states and their probabilities.

For measuring variety of probabilities in the set, we will use the entropy, calculating from expression

$$H_{ad} = -\sum_{i=1}^{N_{ad}} P_i \cdot ln\ P_i, \tag{11.20}$$

where H_{ad} is the entropy, P_i is the probability of the object or the object state in "the tail" of the distribution, N_{ad} is the number of objects or states of the object in "the tail." Summing is making for all the objects of "the tail."

The entropy is absolutely warranted itself as the diversity measure in common case, because it has the following properties:

1. It is equal to zero, when one event appearing from the set is a certain event and other events appearing is impossible;
2. It has maximum, when this events appearing is equiprobable.
3. It increases at increasing the number of events in the set.
4. It has the additivity property.

It is to note here that the risk LP-theory can be stated as the theory of integer numbers with arithmetical operations of addition and division of integer numbers. The expressions for calculation of probabilities (11.2) of grade-events and calculation of the object risk (11.11), which are given above, confirm this idea. Hereinafter, we shall show that the risk LP-theory with GIE also uses the results of the Weil's theorem on division of integer numbers and Bayes's formula on conditional probabilities.

Below, at the description in details of risk LP-models with GIE in various problems and object fields, we shall use more than once the arithmetical operations with the integer numbers. And it will be right back to basics, logic and arithmetic, to solve complex problems. This situation is not surprising, because arithmetics has already taken the worthy place in digital communication and digital photo-equipment. Computers also are based on use of logic and arithmetics (binary).

Measure of risk and cost of risk. Let us introduce an admitted risk P_{ad} separating the objects into good and bad: if $P_i > P_{ad}$, then the object is bad; if $P_i < P_{ad}$, the object is good (Fig. 11.5 a). If the objects are classified into a greater number of classes, then a corresponding number of admitted risks: $P_{ad1}, P_{ad2}, \ldots$, is introduced (Fig. 11.5 b).

Let us assume that the probabilities of grade-events $P_{jr}, j = 1, 2, \ldots, n$; $r = 1, 2, \ldots, N_j$ are known. Then, from the risk P-model we calculate risks of all N objects of Table 11.1. We plot these risks on the risk axis. If the resulting event Y has two grades (Fig. 11.5 a), we choose the admitted risk

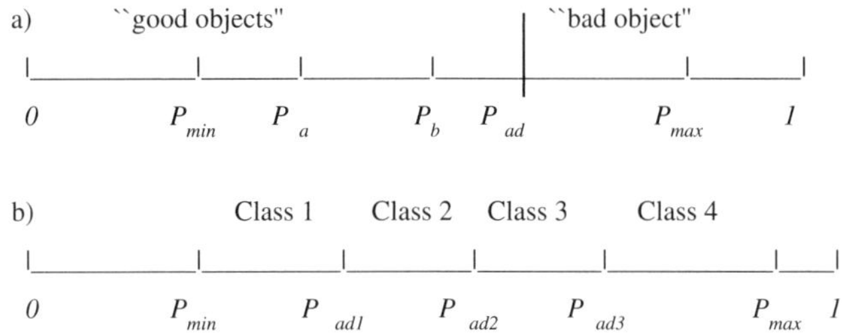

Fig. 11.5. The scheme of classification: (a) into two and (b) into several classes

P_{ad} so that N_b from N objects are bad and N_g are good. For the object i, the distance between the risk P_i and the admitted risk P_{ad} is a natural measure of its being good or bad:

$$d_i = |P_i - P_{ad}|. \tag{11.21}$$

The object risks can be represented in a different manner. We calculate the numbers of objects N_{ad} and N_i having risks, respectively, smaller than the admitted risk P_{ad} and smaller than the risk P_i of the object i and establish the following risk measures:

(1) The relative number of objects having risks, respectively, smaller than a_i and greater than the risk b_i of the object i under consideration:

$$a_i = N_i/N; \quad b_i = 1 - a_i; \tag{11.22}$$

(2) The relative numbers of good, f_i, and bad, e_i, objects having risks greater than the considered object i among the good and bad objects:

$$f_i = (N_{ad} - N_i)/N_{ad}; \quad e_i = (N_i - N_{ad})/(N - N_{ad}). \tag{11.23}$$

The above measures are used to calculate the cost of risk, for example, rate on credit. The simplest formula of the risk cost is as follows:

$$Ci = C_{ad} + C \cdot (P_i - P_{ad}), \tag{11.24}$$

where the cost of the admitted risk C_{ad} and the coefficient C are chosen by the bank on the basis of the market conditions.

11.6 Applications of the risk LP-theory with GIE

The LP-calculus is applied to estimate the risk in complex technical systems [1, 2]. It is based on the logic representation of development of dangerous conditions and mathematical methods of calculation of the truth functions of logic algebra.

The risk structural model represents the graph; its nodes are connected by connection of types *AND*, *OR*, *NOT*. Each graph node can accept value 1 or 0. Some nodes of the graph are random events with known probabilities (initiating events), other nodes are derivative events. The probabilities of initiating events are known. The probabilities of derivative events are to be calculated.

The risk logic function (L-function) is composed according to the graph, by finding risk shortest ways, or with the help of the minimal cross-sections of risk prevention. We obtain the risk P-function after orthogonalization of the risk L-function. The risk computation is made on the risk P-polynomial by substituting suitable probabilities of the initiating events.

The LP-calculus allows us to estimate numerically the object risk and to analyze contributions of initiating events to the object risk [1, 2]. However, direct application of the LP-calculus is impossible for estimation of the non-success risk in banks, business, and quality of production. In development of the LP-calculus of reliability and safety by I. Ryabinin [1, 2], we introduce new concepts and risk problems [3, 4, 30]:

1. One considers a set of homogeneous risk objects or conditions of the object in different moments of time.
2. Initiating and final events are considered on many levels.
3. One considers the incompatible events for grades of signs (as in the Bayes' formula), and not just for the different signs (as in A. S. Mozhaev).
4. One considers also the associative risk LP-models constructed for all events or for the limited set of events;
5. Problems of parametrical and structural identification of risk LP-models from the statistical data are solved;
6. New problems of the risk analysis on the basis of calculation of contributions of initiating events into the mean risk of an object set and into the accuracy of the risk LP-model are solved.

Let us discuss in more detail the groups of incompatible events (GIE). A. S. Mozhaev suggests to link the groups of incompatible events with rows of Table 11.1 on horizontal (see also Fig. 11.6). For this purpose, the algebra of GIE is used, which was specially developed by him and which is represented by expressions (11.3) and (11.4). The GIE's by A. S. Mozhaev appear in risk problems to denote special dependence between parameter-events.

We additionally introduce another type of GIE, namely, GIE for grade-events for each parameter-event (Fig. 11.6). The introduction of such GIE's, for which expressions (11.3) and (11.4) are also valid allow us to formulate problems of LP-modeling and LP-analysis of risk in the fields of classification, investment, and efficiency. Earlier in the fields, only the normal laws for distributions of random variables were considered, and now we have the opportunity to use discrete distributions constructed for arbitrary law of distribution, or, more precisely, on statistical data.

Below the risk LP-modeling and LP-analysis in systems with GIE is considered in detail for the following problems with different statements, methods of calculation of the admitted risk P_{ad}, and optimization criteria:

1. Risk in problems of classification (credit risks, ratings of banks, etc.);
2. Risk in problems of investment (security portfolio);
3. Risk in problems of efficiency (quality, accuracy);
4. Risk in problems of management (non-success risk models of companies);
5. Risk in problems of bribes and corruptions.

The non-success risk is a generalized term that we will use for all these problems and tasks in order to avoid such various conceptions as risk of accidents and crashes, of securities portfolio, of frauds and bribes, etc.

Statistical information in these problems are equally well given, namely, by Table 11.1 except for the last column of the efficiency parameter Y.

In the risk classification problem, the efficiency parameter Y in the last column of Table 11.1 is given and has two or more grades. It is necessary to construct the risk LP-model, i.e., to determine probabilities of grade-events, that would classify new objects with the least error. The admitted risk P_{ad} or several admitted risks $P_{ad1}, P_{ad2}, \ldots, P_{adn}$ are also to be determined; the risks are the thresholds for division of objects into classes by the risk value.

In the risk investment problem, the efficiency parameter Y is the yield of portfolio in the last column of Table 11.1. The optimum relative shares of securities in the portfolio are calculated from the condition that the maximum of the minimal admitted yield of portfolio Y_{ad} at the given *Risk* has the yield smaller than the admitted yield Y_{ad}.

In the risk efficiency problem, the parameter Y or its grades in the last column of Table 11.1 are given. It is necessary to determine weights of the influence of parameter-events, placed in other columns of the table, to the risk distribution of the efficiency parameter, having the admitted value Y_{ad}.

The special questions of the risk LP-modeling and analysis in systems with GIE, including the statements and solutions of optimization problems, are considered in detail for the above mentioned problems of classification, investment, efficiency, management, and bribes in Chapters 12–21, respectively.

11.7 Procedures of construction and use of the risk LP-model with the GIE

The technology of construction and use of the risk LP-models with GIE in the problems of classification, investment, efficiency, and bribes includes the following procedures:

1. Tabular representation of the statistical data.
2. Transition from DB to KB.
3. Construction of scenario and structure risk models.

4. Definition of parameter-events and grade-events.
5. Definition of groups of incompatible events.
6. Quantization of distributions of random grade-events.
7. Generation of arbitrary discrete distributions.
8. Construction of the logic risk model.
9. Orthogonalization of the logic risk model.
10. Construction of the probabilistic risk model.
11. Normalization of probabilities of events.
12. Identification of the LP-risk model on the statistic date with taking into account the GIE and Bayes' formula.
13. Computing the admitted risk.
14. Computing the risk attributes Y_{ad}, $Risk$, N_{ad}, H_{ad}.
15. Risk analysis by contributions of grade-events into Y_{ad}, $Risk$, N_{ad}, H_{ad}.
16. Choosing the recognition non-symmetry factor of good and bad objects.
17. Estimation and analysis of the object risk.
18. Management of the risk.

Scenario of success or non-success risk. We can use the risk scenario for both the success and the non-success. The probabilities of the success and the non-success are connected by the simple formula: one completes another up to 1. Constructively it is better to chouse the non-success as the basic concept and design and use the scenario and LP-model for the non-success risk. In the last case, we consider the risk. The scenario may have the physical base (for example, the electric circuit) or be associative and define the complete or limited number of danger states of the system or object. The scenario may be always represented in the form of the graph.

Logic model of non-success risk is written in the form of the disjunctive (DNF) or conjunctive (CNF) normal form, that is in the form of the logical expression with operations OR, AND, NOT, cycles, and GIE, but without parentheses. The risk logic model can also be written in the form of the orthogonal disjunctive normal form ($ODNF$) or the perfect disjunctive normal form ($PDNF$). The risk logic model may determine the complete or limited number of danger states of the system or object.

Probabilistic model of non-success risk is designed after the orthogonalization of the risk logic model in the form of DNF or at once for $ODNF$ and $PDNF$. The risk probabilistic model may determine the complete or limited number of danger states of the system or object. This model allows quantitatively to estimate and analyze the risk.

Probabilities of sign-events and grade-events is given on the statistical data by the frequency of occurring grade-events in the states, or is determined in result of solving task of the identification by statistical data.

Quantization of the random variable may be natural or artificial. For example, the random variable Z_j of the credit application sign has the natural dividing into grades meaning the credit application for buying house Z_{j1}, buying car Z_{j2}, on traveling Z_{j3}, etc., but the random variable of the security yield sign Z_j, dividing into intervals $Z_{j1}, Z_{j2}, \ldots, Z_{jNj}$, has the artificial quantization. In a number of tasks, for example the credit risk estimation, both natural and artificial quantization (the credit sum is divided into several intervals) is used. In all cases, the grade-events for one sign-events or the security yield form the group of incompatible events, in which the sum of probabilities of grade-events is equal to 1.

Normalization of probabilities of events (states) initiate from that their sum must be equal to 1 by the sense. It is carried out in the following cases:

- At the identification (optimization) of the risk P-model by the statistical data for grade-events in the GIE.
- In the presence of the limited statistics for the system states or the objects from the complete set of the possible states or objects.
- At using Monte Carlo method for the choice of the limited set of states (objects) from the complete set of possible states.

Normalization is carried out through the division of each event probability by the sum of probabilities of the considered set of events (states).

Orthogonalization of the non-success risk L-function is carried out by methods stated in Chapters 8, 9, and 10. The orthogonalization allows one to pass from the risk L-function to the risk P-function or, in other words, from the logic expression to the arithmetic expression. This operation allows one quantitatively to estimate and analyze the risk.

Identification. For the construction of the risk P-model in systems with GIE, we solve the optimization task on the set of objects or states of the object in Table 11.1. This task is formulated in different ways for classification, investment, efficiency, and bribes and has the different criterion functions. The identification in the classification task consists in the construction of the risk P-model, more correct, in the determining admitted risk and probabilities of grade-events by the statistical data. The optimization in the investment task consists in the determining the capital optimal shares, being invested in portfolio assets. The optimization in the efficiency task consists in the determining weights of processes influencing the final process.

Risk analysis. The analysis is carried out for "the tail" of the efficiency parameter distribution (the object risks, the portfolio yields, etc.). It consists in estimating contributions of parameter-events and grade-events in the risk of the object and in the object set, in the accuracy of the risk LP-model, in the admitted risk, in the admitted value of the efficiency parameter, in the entropy of probabilities and the number of danger states.

Generation of arbitrary distributions. For improvement of techniques of the risk LP-estimation and LP-analysis and for purposes of the examination of the risk LP-theory, it is necessary to generate the arbitrary discrete distribution event-grades of signs. For example, the probability distributions of event-grades of credit signs, of share yields, values of the influencing parameters and the efficiency parameter.

We shall obtain the arbitrary discrete distribution by addition of a few elementary distributions, generated by different laws. As elementary distribution laws we use, for example, the normal law, the uniform law, the trapeze law, the law of slope line, the Weibull's law, etc.

The technology of obtaining the arbitrary discrete distribution is as follows:

(1) To choose the distribution elementary law of parameters Z and to generate at random N values of the parameter in the interval $\{Z_{min}, Z_{max}\}$;

(2) To divide the obtained values of the parameter into the chosen number N_j of grades (intervals);

(3) To calculate the probabilities for the grades by formula (11.2);

(4) To repeat the operations 1–3 for the generation of all chosen elementary distributions, each of which has N_j grades;

(5) To add the obtained different elementary distributions by formula:

$$P_j = x_1 \cdot P_{1j} + x_2 \cdot P_{2j} + \ldots + x_k \cdot P_{kj}, \quad j = 1, 2, \ldots, N_j, \tag{11.25}$$

where

$x_1, \ldots, x_k$ are weights of elementary events; their sum is equal to 1;
P_j is the probability of the grade j of the resulting parameter;
$P_{1j}, \ldots, P_{kj}$ are probabilities of grades j of elementary distributions $1, \ldots, k$.

11.8 The basic equations for GIE and Bayes' formulas

The basic equations for GIE. We consider values of grades as random variables or the grade-events, which distributions are given by the discrete rows. The signs are the characteristics of the object. For measurement of signs, scales are used: logical (true or false, 1 or 0), qualitative/enumeration, linear order ($a_1 > a_2 > \ldots > a_n$), numerical (intervals [a,b]), etc. The grade-events for each sign form the GIE (Fig. 11.6).

The grade-event Z_{jr}, according to a grade r for a sign j, with the probability P_{jr} leads to the non-success of event Y and with probability $Q_{jr} = 1 - P_{jr}$ leads to the success of event Y. The sign-event Z_j leads to the non-success of event Y with the same probability P_{jr} and leads to the success of event Y with probability $Q_{jr} = 1 - P_{jr}$. The vector $Z(i)(Z_1, \ldots, Z_j, \ldots, Z_n)$ describes the object i from Table 11.1. In assigning the object i instead of the logic variables $Z_1, \ldots, Z_j, \ldots, Z_n$, it is necessary to substitute the logic variable Z_{jr} for grades of sign-events of the object i. Let's write down the general form of L-function for non-success for any object

$$Y = Y(Z_1, \ldots, Z_j, \ldots, Z_n). \tag{11.26}$$

and the non-success risk P-function of any object, given by the vector $Z(i)$,

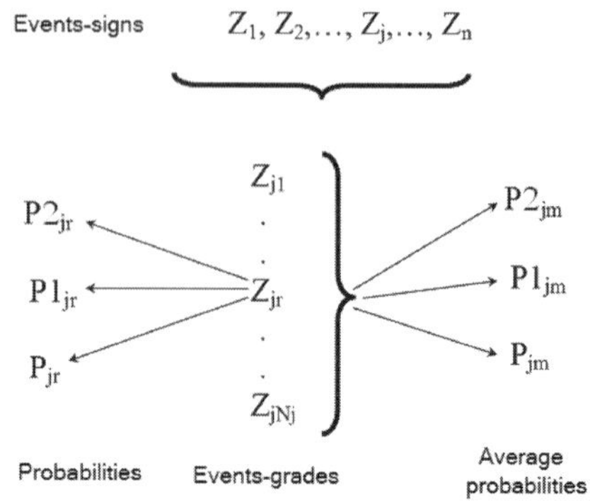

Fig. 11.6. Probabilities in GIE

$$P_i\{Y = 1|_{Z(i)} = \Psi(P_1, \dots, P_j, \dots, P_n), \quad i = 1, 2, \dots, N. \tag{11.27}$$

For each grade-event in GIE, we consider three probabilities: $P2_{jr}$ is the relative frequency of the grade in objects of Table 11.1; $P1_{jr}$ is the probability of the grade-event in GIE; P_{jr} is the probability of the grade-event to be substituted into (11.27) instead of the probability P_j. We define these probabilities for the j-th GIE as follows:

$$P2_{jr} = P\{Z_{jr}\}; \ \sum_{r=1}^{N_j} P2_{jr} = 1; \ r = 1, 2, \dots, N_j; \tag{11.28}$$

$$P_{jr} = P\{Z_{jr}|Y = 0\}; \ r = 1, 2, \dots, N_j; \tag{11.29}$$

$$P1_{jr} = P_{jr}/\left(\sum_{r=1}^{N_j} P_{jr}\right); \ \sum_{r=1}^{N_j} P1_{jr} = 1; \ r = 1, 2, \dots, N_j. \tag{11.30}$$

Here "|" is reading "at the condition."

The mean probabilities $P2_{jr}$, $P1_{jr}$ and P_{jr} in GIE are equal to:

$$P2_{jm} = 1/N_j; \quad P_{jm} = \sum_{r=1}^{N_j} P_{jr} \cdot P2_{jr}; \quad P1_{jm} = \sum_{r=1}^{Nj} P1_{jr} \cdot P2_{jr}. \tag{11.31}$$

The object risk P_i is calculated using (11.27) by replacing probabilities P_{jr} by P_j. We shall estimate probabilities P_{jr} during the process of algorithmic iterative training (identification) of the risk P-model by using the data from Table 11.1. In the beginning it is necessary to determine the probabilities $P1_{jr}$, satisfying (11.29), and further to pass from the probabilities $P1_{jr}$ to the probabilities P_{jr}. The number of the estimated independent probabilities P_{jr} is equal to:

$$N_{ind} = \sum_{j=1}^{n} N_j - n. \tag{11.32}$$

Table 11.3. Designation in the Bayes' formula and GIE

Event $A \equiv$ Sign-Event Z_j	Probability $P(H_k/A) \equiv P1_{jr}$
Hypothesis $H_k \equiv$ Grade-Event Z_{jr}	Probability $P(A/H_k) \equiv P_{jr}$
Probability $P(H_k) \equiv P2_{jr}$	Probability $P(A) \equiv P_{jm}$

The connection of the probabilities P_{jr} and $P1_{jr}$ for the grades is expressed through the mean values of the probabilities P_{jm} and $P1_{jm}$:

$$P_{jr} = P1_{jr} \cdot \frac{P_{jm}}{P1_{jm}} = K_j \cdot P1_{jr}. \tag{11.33}$$

GIE and the Bayes' formula. Connection between probabilities of grades P_{jr} and $P1_{jr}$ in GIE is expressed in terms of the mean values of probabilities P_{jm} and $P1_{jm}$ (11.31). We shall prove that this fact follows from the Bayes' formula. The condition probability $P(H_k/A)$ that a hypothesis H_k is true after the event A happens is given by the following formula:

$$P(H_k/A) = P(H_k) \cdot P(A/H_k)/P(A), \quad where \tag{11.34}$$

$$P(A) = \sum_{i=1}^{m} P(H_i) \cdot P(A/H_i),$$

and hypothesis $H_i, i = 1, \ldots, k, \ldots, m$, form a complete GIE.

There are usually many GIE in risk problems. For Z_j, each group forms a complete GIE of Z_{jr}, $r = 1, \ldots, N_j$. For simplicity, the following notation is introduced for the j-GIE (Table 11.3), equivalent to (11.28–11.31).

We are going to use Bayes' formula only for training the risk LP-model on the statistical data by solving the corresponding problem of optimization. Therefore, there is no sense in discussing here "a priori" and "a posteriori" probabilities in the real sense. The Bayes' formula can be written down formally in terms of $P1_{jr}$ instead of P_{jr} or, on the contrary, in terms of P_{jr} instead of $P1_{jr}$. For the procedure of identification of the risk LP-model, the Bayes formula (11.34) is written down in terms of probabilities P_{jr}:

$$P_{jr} = P1_{jr} \cdot \frac{P_{jm}}{P2_{jr}}. \tag{11.35}$$

It allows us to decrease by one the number of independent probabilities $P1_{jr}$ in the GIE in comparison with generation of the probabilities P_{jr}. Estimation of accuracy of the probabilities $P1_{jr}$ also becomes simpler — indeed, the sum of probabilities $P1_{jr}$ in GIE is equal to 1.

However, one meets difficulty in using (11.35) because for a limited number of statistical data, the denominator can turn to zero. Therefore, it is suggested to make use of (11.33) to relate the probabilities P_{jr} and $P1_{jr}$ in the GIE.

Table 11.4. Initial probabilities and characteristics of grade-events

Signs	Grades	P_{jr}	$P2_{jr}$	P_{jm}	$P1_{jr}$	$P1_{jm}$	K_j
Z_1	1	0.05	0.5	0.075	1/3	0.5	0.15
	2	0.10	0.5		2/3		
Z_2	1	0.15	0.5	0.175	3/7	0.5	0.35
	2	0.20	0.5		4/3		
Z_3	1	0.25	0.5	0.275	5/11	0.5	0.35
	2	0.30	0.5		6/11		

Example 1. Consider a complete set of risk objects of the "node" type with three sign-events Z_1, Z_2, Z_3 [3, 39]. Each of the sign-events has two grades, 1 and 2. The number of distinct objects in the complete set is $N = 2^3 = 8$. The risk LP-functions for each object are as follows:

$$Y = Z_1 \vee Z_2 \vee Z_3; \tag{11.36}$$
$$P\{Y = 1\} = P_1 + P_2 \cdot (1 - P_1) + P_3 \cdot (1 - P_1) \cdot (1 - P_2).$$

We fix arbitrary values of the probabilities of grade-events P_{jr} (Table 11.4), calculate first $P2_{jr}$, P_{jm}, $P1_{jm}$, K_j for all three GIE's and then probabilities $P1_{jr}$ using the Bayes' formula (11.34) (which is equivalent to (11.35) for the complete set of different objects).

Calculate the risks P_i of all objects in the complete set of different objects (Table 11.5) from the accepted values of the probabilities of the grade-events P_{jr}. As an example we choose in Table 11.4 three objects, maximum values of risk. With this aim in view, we define the admitted risk $P_{ad} = 0.462$, so that only for three objects their risk is greater than the admitted risk. These three objects are declared as "bad," that is, $Y = 0$.

The problem of identification of the risk LP-model by statistical data makes use of the signs Y of object classification and descriptions of objects by their grades. The obtained results explain the following important properties of GIE consideration [3, 30]:

1. For training the risk LP-model not taking into account GIE, one has to determine six probabilities $P_{jr}, j = 1, 2, 3;\ r = N_j = 2$. For training the risk LP-model taking into account GIE, one determines three probabilities $P1_{jr}$ (one in each GIE).

2. For uniting the grades of each sign into one grade, instead of eight different objects we get only one "averaged" object with the probabilities of events Z_1, Z_2, Z_3 being equal, we get to $P_1 = P_{1m} = 0.075$; $P_2 = P_{2m} = 0.175$; $P_3 = P_{3m} = 0.275$ respectively (Table 11.3). For this object, the risk is $P_m = 0.4522$. The introduction of grades for the signs and GIE allows us to distinguish objects in risk within the range $P = 0.394 \div 0.496$ (Table 11.4) instead of assigning them the same value of the mean risk $P_m = 0.4522$.

3. For the identification of the risk LP-model, one needs to use the modified Bayes' formula (11.33) to relate the probabilities in the GIE at the limited amount of statistical information. This makes training the LP-model possible.

Table 11.5. Non-success risks of objects

Obj-ects	Z_1 Grades	Grade-events	Z_2 Grades	Grade-events	Z_3 Grades	Grade-events	Estimation of risk $P_i = P(Y)$	The sign of class Y_i
1	1	Z_{11}	1	Z_{21}	1	Z_{31}	0.394375	1
2	1	Z_{11}	1	Z_{21}	2	Z_{32}	0.43475	1
3	1	Z_{11}	2	Z_{22}	1	Z_{31}	0.43	1
4	1	Z_{11}	2	Z_{22}	2	Z_{32}	0.468	0
5	2	Z_{12}	1	Z_{21}	1	Z_{31}	0.42625	1
6	2	Z_{12}	1	Z_{21}	2	Z_{32}	0.4645	0
7	2	Z_{12}	2	Z_{22}	1	Z_{31}	0.460	1
8	2	Z_{12}	2	Z_{22}	2	Z_{32}	0.496	0

11.9 Risk LP-models for the limited number of events

The non-success risk LP-models above describe all the possible states and are the most complete and accurate ones. In a number of cases, however, it is unnecessary to take into account all the possible states of the system. For example, it is known from the statistic data that there was non-success when one and not more than two sign-events from $Z_1, Z_2, \ldots, Z_n$ occurred. Then, using results of E. A. Losev [111], we can simplify the model. We should use the risk model for the limited number of system states.

Let we have the logical risk model of four elements

$$Y = Z_1 \vee Z_2 \vee Z_3 \vee Z_4. \tag{11.37}$$

After orthogonolization we have:

$$Y = Z_1 \vee Z_2\overline{Z_1} \vee Z_3\overline{Z_1}\ \overline{Z_2} \vee Z_4\overline{Z_1}\ \overline{Z_2}\ \overline{Z_3}.$$

Then the risk P-model is

$$P\{Y\} = p_1 + p_2q_1 + p_3q_1q_2 + p_4q_1q_2q_3,$$

where: $q_1 = 1 - p_1; \ldots; q_4 = 1 - p_4$.

For the limited number of states, when one or either one of two events occur, the non-success risk L-model is

$$\begin{aligned} Y = {} & Z_1\overline{Z_2Z_3Z_4} \vee Z_2\overline{Z_1}\ \overline{Z_3}\ \overline{Z_4} \vee Z_3\overline{Z_1}\ \overline{Z_2}\ \overline{Z_4} \vee Z_4\overline{Z_1}\ \overline{Z_2}\ \overline{Z_3} \vee Z_1Z_2\overline{Z_3}\ \overline{Z_4} \\ & \vee Z_1Z_3\overline{Z_2Z_4} \vee Z_1Z_4\overline{Z_2Z_3} \vee Z_2Z_3\overline{Z_1Z_4} \vee Z_2Z_4\overline{Z_1Z_3} \vee Z_3Z_4\overline{Z_1Z_2}. \end{aligned} \tag{11.38}$$

In the risk L-model, all the logical items are orthogonal in pairs, which allows the non-success risk B-model (B-polynomial) to be written directly:

$$\begin{aligned} P\{Y\} = {} & p_1q_2q_3q_4 + p_2q_1q_3q_4 + p_3q_1q_2q_4 + p_4q_1q_2q_3 + p_1p_2q_3q_4 \\ & + p_1p_3q_2q_4 + p_1p_4q_2q_3 + p_2p_3q_1q_4 + p_2p_4q_1q_3 + p_3p_4q_1q_2. \end{aligned} \tag{11.39}$$

Naturally, the events $Z_1, Z_2, \ldots, Z_n$ can include GIE. The non-success risk L-model (11.38) is described for sign-events. At computation of the probability of the non-success for the final event, we should put in expression (11.39) the probabilities of grade-events from GIE for corresponding sign-events.

11.10 Dynamic risk LP-models

Probabilities of initiating events in the risk LP-model change in time. That is, it is possible to set them as functions of continuous or discrete time: $P_{jr} = P_{jr}(t),\ j = 1, 2, \ldots, n;\ r = 1, 2, \ldots, N_j$. It is typical for complex technical systems — probabilities of element failures change because of deterioration, corrosion, aging, repair, replacement of components, training personnel, etc.

In many branches of engineering (nuclear, rocket, etc.), the functions of change of probabilities $P_{jr} = P_{jr}(t)$ are constructed practically for all initiating elements as functions of the time and the determining size or characteristic varied in time. For example, in the starting rocket complexes, the thickness of pipes, which are used for transmission of the acid or the fuel, decreases as a result of corrosion. The pipe thickness is measured at monitoring, and the probability of the pipe destruction is calculated by known equations for the given loading.

Actually as a result of monitoring, the technical condition of the operating the SCS, we get the information on its actual condition, estimate the risk of its operation and accept the appropriate decisions. Especially it is important for the numerical estimation of the non-success risk of the SCS with the prolonged service life.

In the course of time, the probabilities of elementary events, initiating failures or accidents, change in all other complex systems: economical, medical, social, etc. Therefore monitoring in these systems should be regular periodic and take decisions of possibility of the safe exploitation and capital investment in safety.

At the first sight, the risk LP-models are static models, because they do not include time explicitly. However, this opinion is erroneous, as actually risk LP-models are always used as dynamic models with substitution of probability values of initiating events at the real time. And the risk LP-models can be built from the initial stage as dynamic ones. Let us consider some ways of construction of dynamic risk LP-models.

As an example, we construct the risk LP-model for the classical figures of the Technical Analysis (TA) for the financial and commodity markets, represented by linear diagrams and Japanese candles (Fig. 11.7) [112].

For the classical figure "Head and Shoulders," we shall introduce the sign-events (axis X) and grade-events (axis Y), which are marked by asterisks (*) and have numbers. That is, signs and grades are used instead of values of rates (of currencies or goods) and discrete time. The total number of sign-events is equal to 7, and the total number of grade-events is equal to 26. Now it is

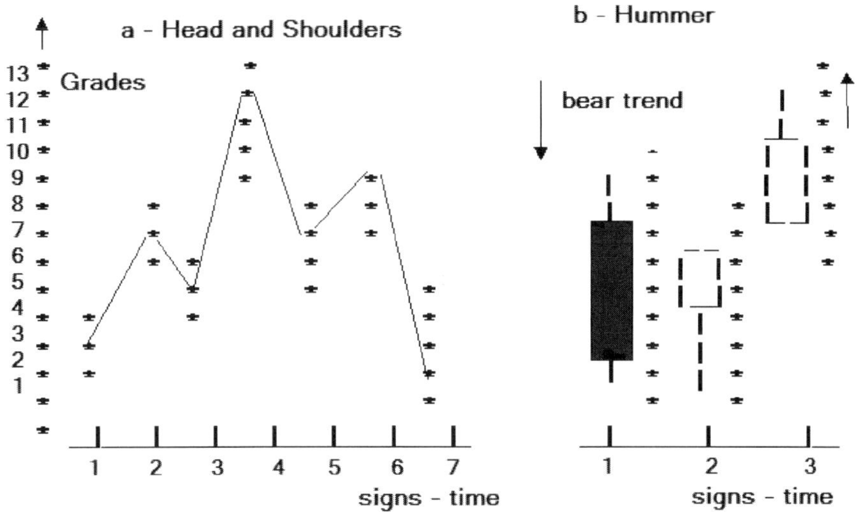

Fig. 11.7. Construction of the dynamic risk LP-model

possible to distinguish a lot of figures "Head and Shoulders," which differ by their grades, and to calculate their risks.

For training the risk P-model, it is necessary to collect statistics. For this purpose, we look through the linear diagrams of the currency rate, choose classical figures of one type, fix signs, grades, and the result of the success of the event of buying / selling ($Y = 1/0$). After the sufficient information is gathered, the risk model is trained and is used for forecasting risk of buying / selling for each new classical figure of this type.

Decisions in the risk P-model are made in the following sequence: (1) recognition of the classical figure in the dynamic mode from the diagram; (2) the decision-making for buying / selling in dependence on the risk. It is also easy to take into account the factors of "Volume of sales" and "Open interest." For this purpose, we shall introduce two signs 8 and 9 (Fig. 11.7 a). Let each of these signs have three grades. For example, sign 8 has grades: 1 means that the sales volume grows, 2 means that the sales volume does not change, 3 means that the sales volume falls. Thus, we construct the risk L-function by nine signs. Such model is hybrid, as it contains both sign-events (time) and usual sign-events.

In TA, the risk LP-model can be constructed for the Japanese candles (Fig. 11.7 b), too. On the axis X, we introduce sign-events Z_1, Z_2, Z_3 for each of three candles in the figure. Besides, we introduce element-events Z_{11}, Z_{12}, Z_{13} for description of the form of the first candle and other candles (top shadow, body, and bottom shadow), respectively. For each element-event, we shall define some event-grades (for example, $Z_{121}, Z_{122}, Z_{123}, Z_{124}$) for the body of the first candle, which are designated by asterisks * and have appropriate numbers on the axis Y. Now, after training the risk LP-model, it is possible to distinguish many figures of different types and to calculate the risk for them.

11.11 Combined risk LP-models

If we have two different scenarios and models of the risk with outputs Y_1 and Y_2, so logic criterions of the combined risk models can be written down as follows:

1. $(Y_1 \vee Y_2)$ is a realization of the criterion Y_1 or the criterion Y_2;
2. $Y_1 \wedge Y_2$ is a realization of the criterion Y_1 and the criterion Y_2;
3. $Y_1 \wedge \overline{Y_2}$ is a realization of Y_1 and not realization of Y_2;
4. $\overline{Y_1} \wedge Y_2$ is not a realization of Y_1 and realization of Y_2;
5. $\overline{Y_1} \wedge \overline{Y_2}$ is not a realization of the criterion Y_1 and the criterion Y_2.

Thus, we can combine some risk scenarios and models as logical variables, for example, all risk models of banks: a non-success risk model of credit risk for natural persons, a non-success risk model of credit risk for juridical persons, a non-success risk model for investigation, etc.

12

Identification of Risk LP-Models with Groups of Incompatible Events

The true logic of our world is calculation of probabilities.

D. K. Maxwell

We solve the task of the identification of risk LP-models on statistical data. This is the return nonlinear task of optimization. It is solved by algorithmic iterative methods by random search and gradients [3, 27–30, 114, 115].

The algorithmic iterative methods guarantee the possibility of the solving of risk problems regardless of:

- the number of objects N (or states of the object) in statistical data,
- the number of parameters n describing an object (a state),
- the number of grades N_j in every parameter,
- the complexity of logical functions of risks $Y = f(Z)$.

The following scheme of the problem solution is proposed. Suppose that first approximation for probabilities of grades P_{jr}, $r = 1, 2, \ldots, N_j$; $j = 1, 2, \ldots, n$ are known and the risks P_i, $i = 1, 2, \ldots, N$ of objects in Table 11.1 are computed.

We shall determine the admitted risk P_{ad} (Fig. 11.5a) so as to have the given number of good objects N_{gc} with the risk less than the admitted one. On the step of optimization, it is necessary to change probabilities P_{jr}, $r = 1, 2, \ldots, N_j$; $j = 1, 2, \ldots, n$, in such a way that the number of correctly recognized objects F increases.

It is notable that variables N_{gc} and P_{ad} are one-to-one related. In the problem solution algorithm, it is more convenient to set N_{gc} and to determine the admitted risk P_{ad}, because the latter would have to be set with the precision 6–7 digits after the decimal point.

E.D. Solojentsev, *Scenario Logic and Probabilistic Management of Risk in Business and Engineering*, DOI 10.1007/978-0-387-77946-1_12,

12.1 Statement of identification problem

We have the statistical data from N objects, in which N_g is the good objects and N_b is the bad objects. We have the knowledgebase (KB) in the form of the system of the logical equations (L-models for N objects of statistical data) from (11.25) and the corresponding system of the algebraical equations (P-models for N objects of statistical data) from (11.26). We substitute in (11.25) and (11.26) the logical variables and probabilities for grades-events.

The condition $P_i > P_{ad}$ let us distinguish the following types of objects:

- N_{gg} are good by the technique and statistics;
- N_{bb} are bad by the technique and statistics;
- N_{gb} are good by the technique and bad by statistics;
- N_{bg} are bad by the technique and good by statistics.

The object risks of $N_{gg}, N_{bb}, N_{gb}, N_{bg}$ move about P_{ad} when P_{jr} change. If some objects pass to the right from P_{ad} on the value of risk (Fig. 11.5a), then the same number of objects pass to the left. Optimal change of P_{jr} will be that which moves objects of N_{gb} and N_{bg} through P_{ad} toward each other — then, the criteria F is increased by two units.

Thus, the identification of the risk P-model consists in the determination of optimal probabilities P_{jr}, $r = 1, 2, \ldots, N_j$; $j = 1, 2, \ldots, n$ and the admitted risk P_{ad}, using the statistical data on credits.

We have to minimize the risk in recognition of good and bad objects:

$$P_m = [N - (N_{bb} + N_{gg})]/N \Rightarrow \min_{P_{jr}}$$

or

$$N_{bb} + N_{gg} \Rightarrow \max_{P_{jr}}.$$

Statement of the optimization problem. The problem of identification (training) of the risk P-model is formulated as follows.

Specified data: Table 11.1 with N_g good and N_b bad objects and the risk P-model in the form of (11.17), (11.19), (11.22) are given;

Expected results: Probabilities P_{jr}, $r = 1, 2, \ldots, N_j$; $j = 1, 2, \ldots, n$ for grade-events and the admitted risk P_{ad}, separating the objects into good and bad ones based on their risk should be determined.

We need: To maximize the criterion function (CF), which is the number of correctly classified objects:

$$F = N_{bb} + N_{gg} \Rightarrow \max_{P_{jr}}, \tag{12.1}$$

where N_{gg}, N_{bb} are the numbers of objects classified as good and bad using both the statistics and the risk P-model (both estimates should coincide).

From (12.1) it follows that errors or accuracy indicators of the risk P-model in the classification of good E_g, and bad E_b objects and in classification of the whole set E_m are equal:

$$E_g = N_{gb}/N_g; \; E_b = N_{bg}/N_b; \; E_m = (N - F)/N. \tag{12.2}$$

Imposed restrictions:

(1) The probabilities P_{jr} and $P1_{jr}$ have to satisfy the condition:

$$0 < P_{jr} < 1, \quad j = 1, 2, \ldots, n; \quad r = 1, 2, \ldots, N_j; \tag{12.3}$$

(2) The average risks of objects P_m based on the risk P-model and on the table P_{av} must be equal for the preservation of the probabilistic and common sense of the task; while training the risk P-model, we should correct the probabilities P_{jr} on every step of iterative training:

$$P_{jr} = P_{jr} \cdot (P_{av}/P_m); \quad r = 1, 2, \ldots, N_j; \quad j = 1, 2, \ldots, n; \tag{12.4}$$

(3) The admitted risk P_{ad} should be determined at the given ratio of incorrectly classified good and bad objects, in view of non-equivalence of losses by their wrong classification (asymmetry of recognition):

$$E_{gb} = N_{gb}/N_{bg}. \tag{12.5}$$

Features of identification. The formulated problem of identification of the risk P-model has the following features and complications (Fig. 12.1):

- The criterion function depends from a large number of real positive arguments P_{jr} (for example, this number is equal to 94 in the credit risk problem of physical persons);
- The criterion function — the number of correctly recognized good and bad objects — accepts integer values and it is stepped;
- The criterion function, in view of the risk P-model structure, has local extreme;
- In the search for the optimal F_{max}, it is impossible to increment parameters P_{jr} by arbitrary positive or negative values because it would change the average risk;
- The derivatives of the criterion function F with respect to parameters P_{jr} cannot be computed by analytic methods.

12.2 Basic statements of identification algorithm

In view of the identification problem complexity, we shall solve it by the algorithmic iterative method. The identification algorithm of the risk P-model is proposed; it iteratively generates $P1_{jr}$, P_{jr}, so as to maximize the value of the criterion function F.

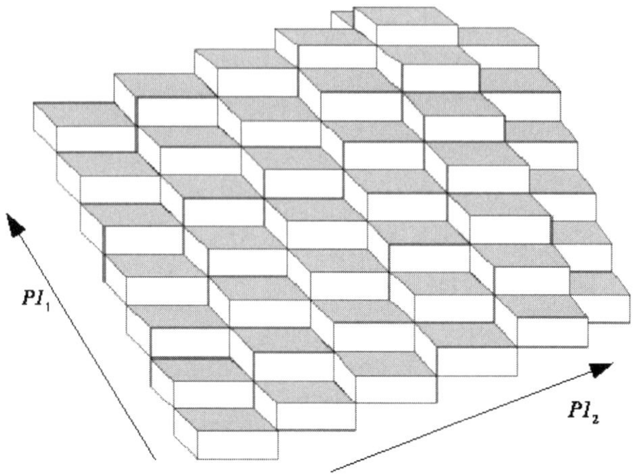

Fig. 12.1. Stepped changing the criterion function

For optimization process we give:

N_{gc} is the calculated number of good objects (the calculated number of bad objects $N_{bc} = N - N_{gc}$). This number approximately equals the number of good objects in statistics; later we consider the choice of optimal N_{gc} by condition (12.5) in details.

N_{opt} is the number of the optimization steps, on which the value of the criterion function increases (does not decreases). This number approximately equals half of the number of good objects in statistics.

We denote steps of optimization $0, 1, \ldots, v, \ldots, N_{opt}$. The step of the currently optimization is N^v.

Initial values. It is better to give initial values of probabilities P_{jr} and $P1_{jr}$, $j = 1, 2, \cdots, n$; $r = 1, 2, \cdots, N_j$ using the results of the former session of optimization. For the first session of optimization, we determined their credit risk using the next way.

(1) To determine the initial values of probabilities $P1_{jr}$ from Table 11.1

$$P1_{jr} = \frac{N_{b\ jr}}{N_b}, \tag{12.6}$$

where $N_{b\ jr}$ is the number of bad credits for the grade-event; N_b is the general number of bad credits in statistics.

(2) In response with (11.17), the initial values of probabilities P_{jr} are equal

$$P_{jr} = 1 - (1 - P_{av})^{1/n} \tag{12.7}$$

or to determine by formula

$$P_{jr} = P1_{jr} \cdot \frac{P_{jm}}{P1_{jm}} = \frac{N_{b\ jr} \cdot N_j}{N \cdot n}, \tag{12.8}$$

where $P1_{jm} = 1 \;/\; N_j;\;\; P_{jm} = N_b \;/\; (N \cdot n)$ (at arithmetic adding events); P_{av} is the average risk by statistics; n is the number of sign-events, N_j is the number of grade-events for each sign-event; N is the number of objects.

(3) To calculate P^0_{jm}, $P1^0_{jm}$ and $K_j = P^0_{jm} \;/\; P1^0_{jm}$ from (11.30).

(4) To assign a small value to the function F_{max}, say $F = P_{av} \cdot N$.

Optimization. We will optimize the criterion function iteratively for steps $1, 2, \ldots, v, \ldots, N_{opt}$ (N_{opt} cannot be more then $N/2$) until the criterion function increases.

The number of the step changes when the following condition is satisfied

$$F^v \;>\; F_{max} \;, \tag{12.9}$$

where F^v and F_{max} are the current and earlier found maximum values of the criterion function.

(5) To generate small increments $\triangle P1^v_{jr}$.

(6) To compute new values $P1^v_{jr}$ from the formula

$$P1^v_{jr} = (P1^v_{jr} + \triangle P^v_{jr}); \tag{12.10}$$

(7) To normalize the new values of probabilities $P1^v_{jr}$

$$P1^v_{jr} = P1^v_{jr} \;/\; \sum_{r=1}^{Nj} P1^v_{jr}. \tag{12.11}$$

(8) To compute P^v_{jr} from the expression

$$P^v_{jr} = K_j \cdot P1^v_{jr}; \tag{12.12}$$

(9) To compute risks of all objects $P^v_i = P_r\{Y = 1|Z(i)\}$ using P^v_{jr};

(10) To compute the mean risk for objects P^v_m using P^v_i, $i = 1, \ldots, N$ and the mean risks $P1_{jm}$ and P_{jm}, using $P2_{jr}, P_{jr}, P1_{jr}$;

(11) To determine P_{ad} from P^v_i so as to obtain N_{gc} and N_{bc};

(12) To compute F^v with P_{ad} and P^v_i, $i = 1, 2, \ldots, N$;

(13) If $F^v > F_{max}$, then to set

$$F_{max} = F^v;\; P1_{jr} = P1^v_{jr};\; P_{jr} = P^v_{jr};$$
$$K_j = P^v_{jm}/P1^v_{jm};\; K^{v+1}_j = K^v_j \cdot (P^v_m/P_{av}).$$

12.3 Identification by methods of random search

In the random search method, on each optimization step we do up to N_{mc} attempts to increase the criterion function value.

In the random search for finding $\triangle P1_{jr}$, we use the following formula

$$\triangle P1_{jr} = K_1 \cdot \frac{N_{opt} - N^v}{N_{opt}} \cdot K_3 \cdot P1_{jr}, \quad (12.13)$$

where K_1 is the coefficient ($\approx$0.05); N_{opt}, N^v are the maximal allowed and the current number of optimization steps for the criterion function; K_3 is a random number from the interval $[-1, +1]$.

The obtained values $P1_{jr}$ are normalized by (12.11). New values $P1_{jr}$ obtained with $F^v > F_{max}$, are regarded as optimal and saved.

If some values $P1_{jr}$ become negative or exceed 1, then we set them to 0 or 1 respectively. The convergence of the optimization method is guaranteed by the second factor in (12.13), which works for 0 as N^v grows.

Way output of impasses during optimization. The criteria function F_{max} has stairs, equaled to 2, and platforms. At approaching to global extreme, the sizes of platforms are decreased.

If the criterion function does not strictly increase after a chosen number attempts N_{mc} attempts (dead-lock), then F_{max} is reduced

$$F_{max} = F_{max} - \triangle F, \quad (12.14)$$

where $\triangle F$ is the value of the deviation $\triangle F = 2 \div 4$.

The meaning of this operation is the following. The discrete criterion function cannot increase to the higher level, as we can change $P1_{jr}$ only by the value determined by (12.13). Therefore we reduce the obtained value of F_{max}. When the optimization continues, the criterion function rushes to reach the former value F_{max}. However, the new calculation will give other values of probabilities $P1_{jr}$ and P_{jr} of grade-events. Therefore the optimization trajectory will be crocked and, as a result, the criterion function can increase. The deviation from the purpose is typically for optimization steps beginning with the middle of the optimization process.

Choice of coefficient K_1. Using (12.13), in the optimization beginning, the maximum amplitude of the probability increment equals

$$\triangle P1_{jr} = K_1 \cdot P1_{jr}^{beg}, \quad (12.15)$$

In the optimization finish, the amplitude of the probability increment equals zero. Let us mark the current amplitude of the increment $\triangle P1$. There is the optimal area A for $\triangle P1$. We do not know the spot and size of this area (Fig. 12.2).

For large $\triangle P1$, the probability of increasing F_{max} is small, but for small $\triangle P1$ there is a high probability that the local extremum is at the attained F_{max} (see Fig. 12.11). Optimization process of the risk LP-models must be confined within the optimal domain A for a sufficiently long time. The duration of optimal domain $\triangle N_{opt}$ is the bigger for greater number of optimizations N_{opt} and smaller K_1.

In results of computer investigations, we suggest the effective technology for determining the global extremum of the criterion function. It allows to

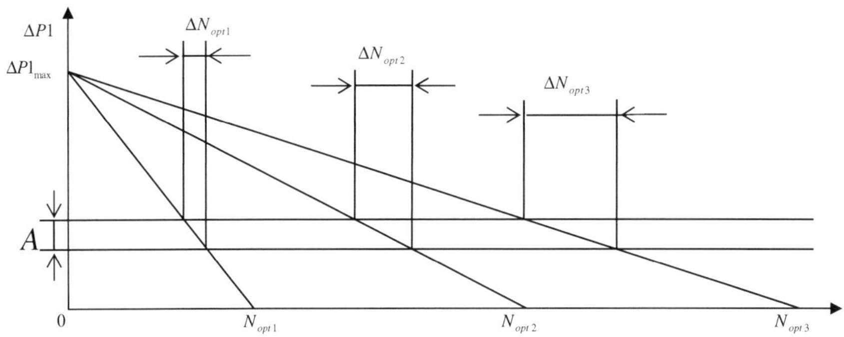

Fig. 12.2. Dependence of the optimization number $\triangle N_{opt}$ from $\triangle P1$ and N_{opt}

solve the problem of the multidimensional multicriterion optimization with the integer criterion function within reasonable computation time (on order less than it was before).

Illustrations for the identification of the risk LP-model. Let us show the graphic illustrations of the identification process of the risk P-model by the random search method. The calculations were carried out by PC for the credit risk LP-model of the physical persons. The risk LP-model has 20 sign-events (correspondingly GIE) and 94 grade-events. The credit risk L-function in the disjunctive normal form [3, 30] is as follows:

$$Y = Z_1 \vee Z_2 \vee \ldots \vee Z_{20}. \quad (12.16)$$

By words, it can be formulated as follows: non-success occurs, if any one, or any two, ... or all initiating events happen. The credit risk L-function in the orthogonal form can be written in the following form (after (12.16)):

$$Y = Z_1 \vee Z_2\overline{Z}_1 \vee Z_3\overline{Z}_2\overline{Z}_1 \vee \ldots$$

The credit risk P-model is given by the formula:

$$P = p_1 + p_2q_1 + p_3q_1q_2 + \ldots . \quad (12.17)$$

The investigations were carried out in a set of $N = 1000$ credits of the standard package, $N_g = 700$ of which were good and $N_b = 300$ were bad [97]. The credit is described by $n = 20$ signs, which in the sum have 94 grades.

Distribution of objects by risk. Looking at the histogram (Fig. 12.3) of the risk distribution for all objects and objects for which the classification by the risk LP-model and the statistics do not coincide, it is possible to make the following conclusions:

1. The distribution of the object risk is not submitted to the normal or any known law of distribution;

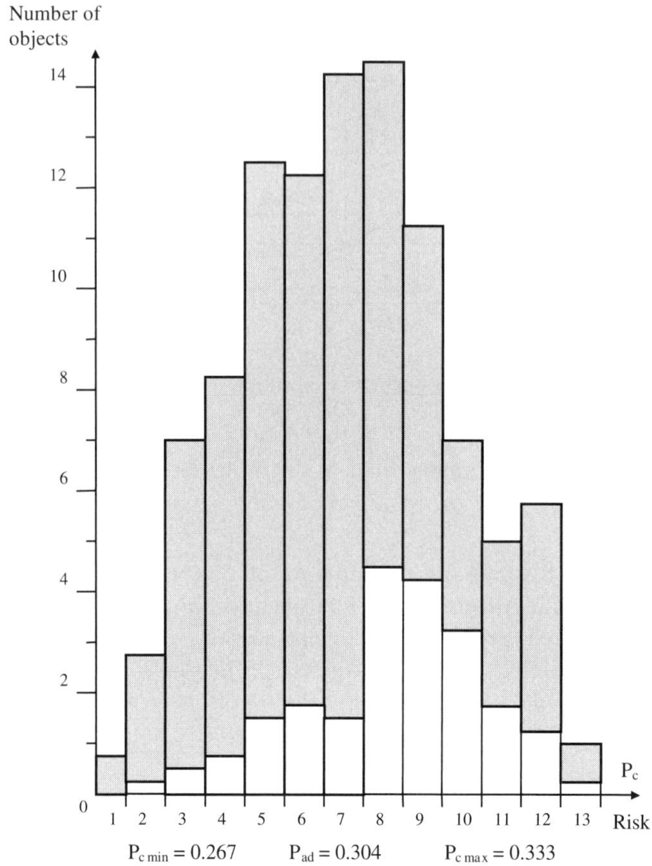

Fig. 12.3. Histograms of distributions of objects on risk

2. The ratio of the numbers of the objects N_{gb} and N_{bg} depends on the values N_g and N_{gc}; at $N_g = N_{gc}$ the N_{gb} and N_{bg} are equal.

The process of training is not monotonically growing. This is confirmed by the change diagrams of estimations of probabilities P_{jr} (Fig. 12.4), the average risk P_m, and the criterion function F_{max} (Fig. 12.5) in function of the number of optimization N_{opt}. At the final stages of the optimization process, when the increments of estimations $\triangle P1_{jr}$ tend to zero, the optimization process asymptotically approaches the extremum (Fig. 12.6). The stepped criterion function during the optimization is increased on an optimization step by an integer number (more often it is two). The section of the criterion function, constructed after end of the optimization process by changing only one parameter P_{jr} with a small step $\triangle P_{jr}$ (Fig. 12.7), shows that the height of steps is also equal to 2. The width of steps increases when P_{jr} go to their optimum values.

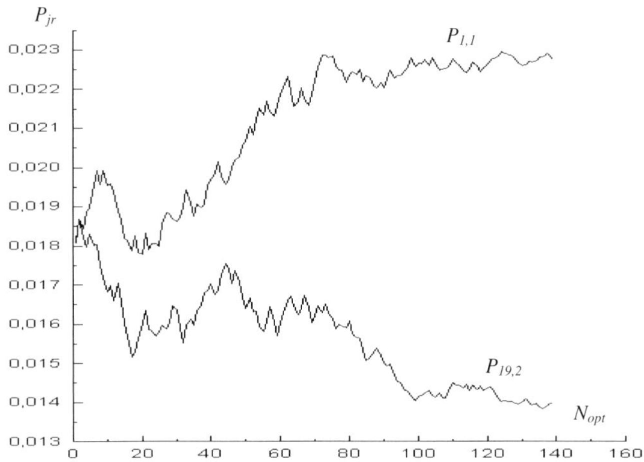

Fig. 12.4. Diagrams of change of estimations of probabilities $P_{1,1}$ and $P_{19,2}$ for grade-events in function of optimization number N_{opt}

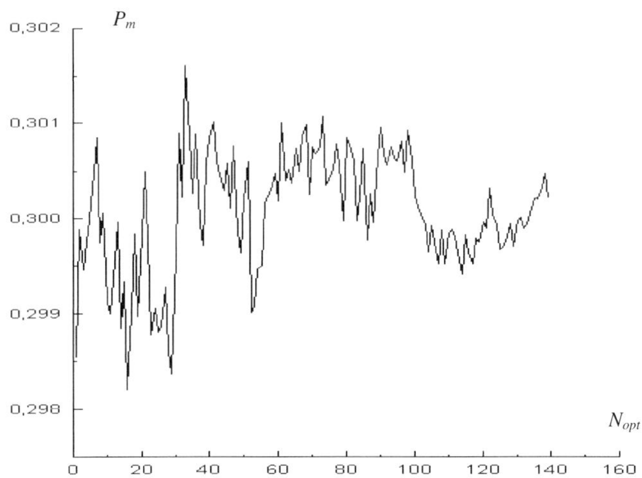

Fig. 12.5. Changing the average risk P_m against the optimization number N_{opt}

Determination of the calculated number of good objects N_{gc}. The calculated number of good objects N_{gc} and the admitted risk P_{ad} are determined from the given ratio E_{gb} of incorrectly classified good and bad objects due to non-equivalence of losses by their wrong classifications (12.5). The ratio of these mistakes is prescribed, in the case of the credit risks its value E_{gb} is equal to $2 \div 4$. The needed value N_{gc} is determined after several calculations for different N_{gc}.

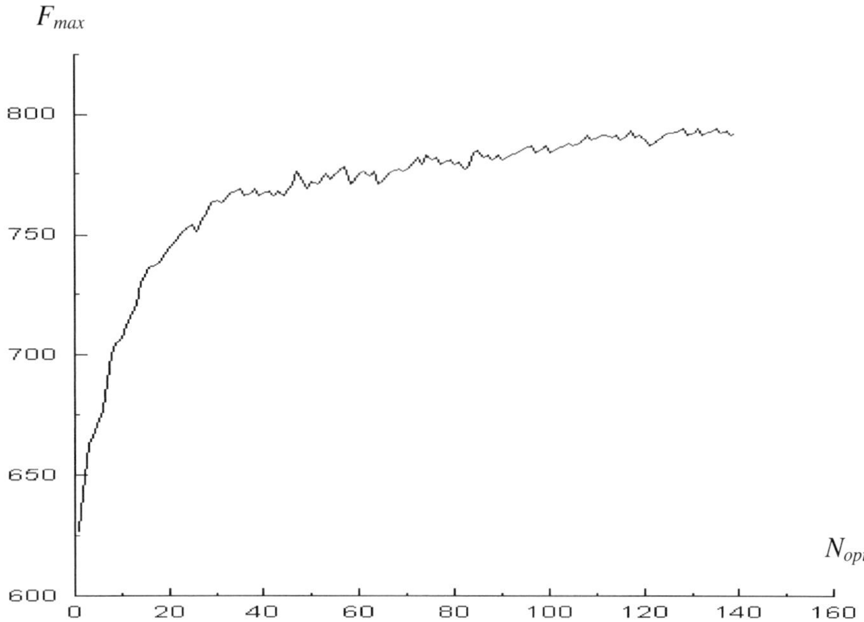

Fig. 12.6. Changing criterion function F_{max} against the optimization number N_{opt}

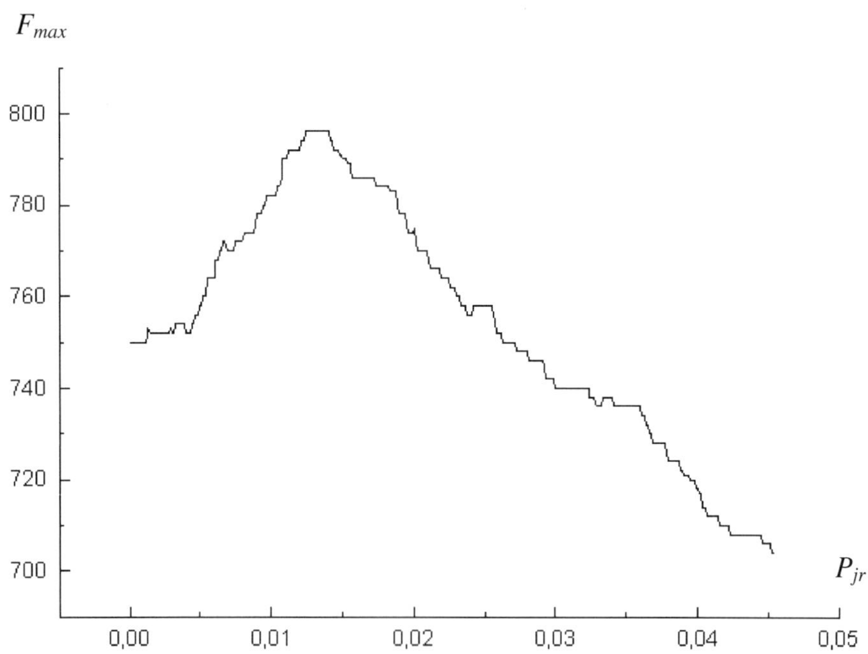

Fig. 12.7. The section of criterion function F_{max} by changing one probability P_{jr}

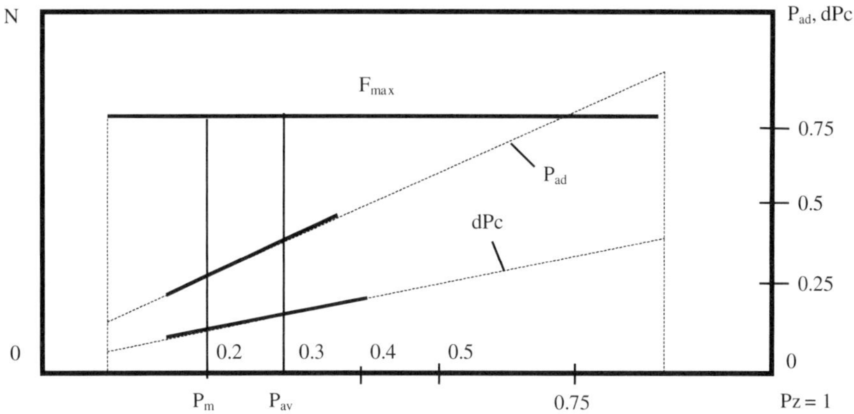

Fig. 12.8. Updating F_{max} in function of the average risk

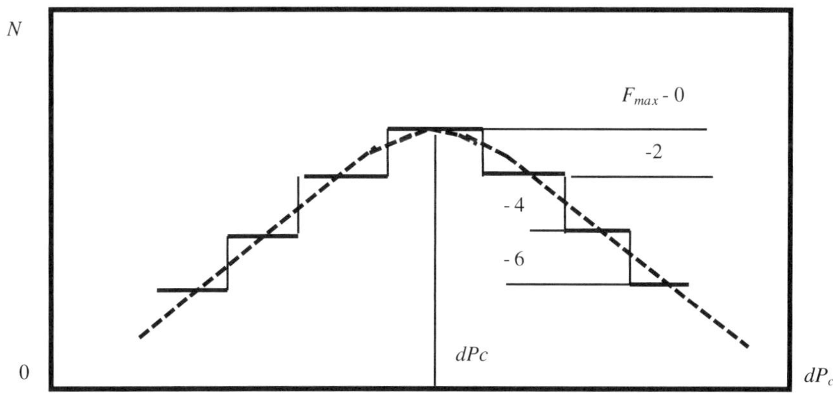

Fig. 12.9. Determining the global extremum in the optimization task

In detail, results of investigations on the choice of the calculation number of good objects N_{gc} for the credit risk LP-model of natural persons in Chapter 15 is stated.

Choice of the average risk P_z. The risk average statistical value P_{av} is known from Table 11.1. During training the risk P-model, we obtain a computed average value of the average risk P_m. Questions arise about which average value of the risk P_z should be chosen to train the risk P-model and how the value P_z influences the criterion function F_{max}. The identification results of the risk P-model for different values P_z, shown in Fig. 12.8, allow us to make the following conclusions:

- The maximal value of the criterion function F_{max} formally does not depend on the value P_z;

- For different P_z, naturally, we obtain different values of probabilities P_{jr}, the admitted risk P_{ad}, and the object risk distribution characteristic $\triangle P_c = P_{i\,max} - P_{i\,min}$;
- The risk P-model obtains the probabilistic sense for $P_z = P_m \div P_{av}$ only, because in this case all results agree with the average risk by the real data.

Determination of global extremum of the criterion function F_{max}. Depending on the parameters K_1 for chosen N_{gc} and P_z, we obtain different values of the criterion function F_{max} (different local extremum). The solution in each of the local extremes we shall characterize by F_{max} and by the difference of the maximal and minimal credit risks of objects $\triangle P_c = P_{i\,max} - P_{i\,min}$. The stepped dependence of F_{max} on the parameter $\triangle P_c$ is shown in Fig. 12.9. It has an extremum for some value $\triangle P_c$. The solutions $P1_{jr}$ and P_{jr} in the extremum of the criterion function are accepted as optimal.

Accuracy of calculation of probabilities of grade-events. Using the example of the risk calculation for $N = 1000$ credits, we shall estimate the necessary accuracy of calculations of probabilities $P1_{jr}$, P_{jr}, $r = 1, 2, \ldots, N_j$; $j = 1, 2, \ldots, n$ and the admitted risk P_{ad} [113].

In the example, the credits have different sets of grades and the classes of good and bad credits do not intersect. Using results of investigations, we set the parameter $\triangle P_c$ to be approximately equal to:

$$\triangle P_c = P_{i\ max} - P_{i\ min} = 0.1.$$

Then adjacent (in risk) credits differ on average by the value:

$$\triangle P_{im} = \triangle P_c / N = 0.1/1000 = 0.0001.$$

In the interval $[P_m, P_{ad}]$, the density of risk distribution is approximately 100 times higher than on average (see Fig. 12.3). Credits, which are neighboring in risk, should differ in the interval $[P_m, P_{ad}]$ by the value:

$$\triangle P_i = \triangle P_{im}/100 = 0.0001/100 = 0.000001.$$

In the algorithm of risk P-model training, the number of good credits N_{gc} is given and the admitted risk P_{ad} is selected by an iterative process. The first approximation of the admitted risk is set to be equal to the average risk $P_{ad} = P_m$. Then we compute the calculated number of good credits N_g. If N_{gc} is not equal to N_{gc}, we increase the admitted risk $P_{ad} = P_{ad} + \triangle P_{ad}$. After that, we compute N_g again, etc., until condition $N_{gc} = N_g$ is satisfied. It is clear that the value of P_{ad} should not contain risks for more than one credit. The accuracy of calculation of the admitted risk is equal to:

$$\triangle P_{ad} = \triangle P_i = 0.000001.$$

Now we can evaluate the calculation accuracy of the probabilities P_{jr}, $r = 1, 2, \ldots, N_j$; $j = 1, 2, \ldots, n$. The accuracy of calculation of P_{jr} is equal to:

$$\triangle P_{jr} = \triangle P_{ad}/n = \triangle P_{ad}/20 = 0.00000005.$$

Accuracy assessment of the trained risk LP-model. Traditionally in the recognition theory for training the model the training set is used, but the trained model accuracy checking is made by test set [116, 117]. It should be done for the neuron nets too, in case when due to a large number of estimated coefficients, we can train the classification model with zero values of recognition errors, but on the test set we will have the big errors of recognition. The risk LP-models accuracy checking using training and testing sets shows the high accuracy of these models. Divergence in accuracy assessment for training and testing sets was not more than 1% for different methods of forming sets [3, 27].

However for identification problems, one uses other methods of checking accuracy and adequacy of model. They use calculation of dispersion of assessments of parameters (probabilities). We think that for the risk LP-model, using the test set is unnecessary. In the identification task, all statistic data should be used for identification. The more data are used for training, whether they are training or testing, the more exact are assessments of parameters $P1_{jr}$ and $P_{jr}, r = 1, 2, \ldots, N_j; j = 1, 2, \ldots, n$ for risk LP-model. Assessments $P1_{jr}$, P_{jr} asymptotically strive to true values with increasing a number of objects N.

In technology of searching global extreme we use the following determined regularities:

- The criterion function asymptotically increases with increasing number optimizations N_{opt};
- The minimal admissible amplitude $\triangle P1_{min}$ of increments of probabilities $P1_{jr}$ is determined by two or three computations: for the values of $\triangle P1_{min}$ less than $0.0025 \cdot P1_{jr}$ (0.25 %), optimization does not go.
- The initial criterion function F^{beg} should not be set too small because this invariably yields small final values for F_{max} due to the unsatisfactory optimization trajectory;
- The maximal amplitude of increments of the probabilities $\triangle P1_{max}$ should not be taken greater than $0.05 \cdot P1_{jr}$ (5 %), as training dynamics worsens and the criterion function F decreases.

12.4 Identification by the gradient method

The identification method of the risk P-model on the basis of the random search of optimal parameters $P1_{jr}$, $r = 1, 2, \ldots, N_j$; $j = 1, 2, \ldots, n$ is very labor-intensive, and calculations by the method are non-repeatable. Therefore we develop some more the determinate gradient method of identification with calculation of the increment $\triangle P1_{jr}$, using the sign and the value of the increment of the criterion function $\triangle F_{jr}$ [30].

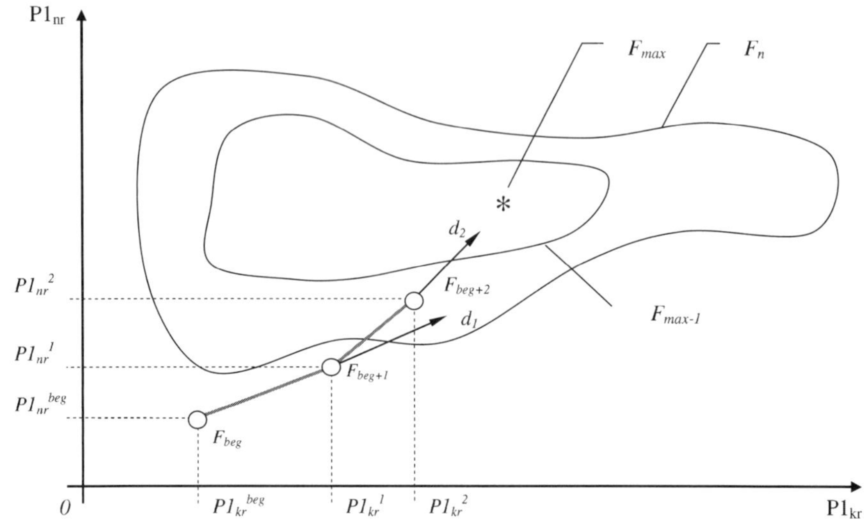

Fig. 12.10. Scheme of optimization in the gradient method

Scheme of optimization by gradient method. At the identification of the credit risk LP-model, it is necessary to estimate the probabilities of grade-events using the statistic data. Following classical scheme of non-linear optimizations, the task is decided by the algorithmic method for several steps. At each step (Fig. 12.10) it is necessary:

- To determine the increment $\triangle P1_{jr}^{(1)}$ of eachprobability $P1_{jr}$ for computing the gradient of the criterion function;
- To determine the increment $\triangle P1_{jr}^{(2)}$ of each probability $P1_{jr}$ in the gradient direction $\triangle F_{jr}$ and compute the new value of function F.

Computing difference in GIE. Probabilities P_{jr} and $P1_{jr}$ in GIE are connected by the following formulas:

$$P_{jr} = P1_{jr} \cdot \frac{P_{jm}}{P1_{jm}}; \; j = 1, 2, \ldots, n; \; r = 1, 2, \ldots, N_j. \tag{12.18}$$

Here probabilities $P_{jr}, P1_{jr}, P_{jm}, P1_{jm}$ are determined from (11.28–11.30).

For optimization of the risk LP-model, it is necessary to compute the differences in GIE by the algorithmic method. We present (12.18) as the function

$$Y = f(X), \tag{12.19}$$

which difference equals to

$$\triangle Y = f(X + \triangle X) - f(X). \tag{12.20}$$

For computing the difference, we transform the expression (12.18)

$$\frac{P_{jr}}{\sum_{r=1}^{Nj} P_{jr} P2_{jr}} = \frac{P1_{jr}}{P1_{jm}} = A. \tag{12.21}$$

Let us write this expression in detail

$$P_{jr}/(P_{j1}P2_{j1} + \cdots + P_{jr}P2_{jr} + \cdots + P_{jN_j}P2_{jN_j}) = A. \tag{12.22}$$

Let us do the following transformations

$$P_{jr} - AP_{jr}P2_{jr} = A(P_{j1}P2_{j1} + \cdots + P_{jr-1}P2_{jr-1} + P_{jr+1}P2_{jr+1} + \cdots).$$

From received expression, we determine

$$P_{jr} = A \cdot \frac{P_{j1}P2_{j1} + \cdots + P_{jr-1}P2_{jr-1} + P_{jr+1}P2_{jr+1} + \cdots + P_{jN_j}P2_{jN_j}}{1 - A \cdot P2_{jr}}.$$

At changing $P1_{jr}$ on $\triangle P1_{jr}$, probability P_{jr} is computed by (12.18), with taking into account changing both numerator $P1_{jr}$ and denominator $P1_{jm}$.

From above given expression we also see, as the number of grades for signs is different and changing in a wide range from 2 to 11, so non-correctly computing difference (differential) can cause the big error, particularly for signs with the little number of grades.

Computing gradient $\triangle F_{jr}$. In the beginning of the step of optimization (identification), the parameters P_{jr}^{beg}, $P1_{jr}^{beg}$, P_{jm}^{beg}, $P1_{jm}^{beg}$, F^{beg} are known. For each optimization step, we compute gradients of the criterion function $\triangle F_{jr}$ for each probability $P1_{jr}$ in the following sequence:

(1) To compute $\triangle P_{jr}^{1}$ by formula

$$\triangle P1_{jr}^{(1)} = K_1 \cdot P1_{jr}^{beg} \cdot \frac{N_{opt} - N^v}{N_{opt}}, \tag{12.23}$$

where K_1 is the coefficient of changing probabilities.

(2) To compute the new value of probabilities

$$P1_{jr}^{(1)} = P1_{jr}^{beg} + \triangle P1_{jr}^{(1)}. \tag{12.24}$$

(3) To normalize probabilities in GIE

$$P1_{jr}^{(2)} = \frac{P1_{jr}^{(1)}}{\sum_{r=1}^{N_j} P1_{jr}}; \; r = 1, 2, \cdots, N_j. \tag{12.25}$$

(4) To compute the new value of the mean probability $P1_{jm}^{(2)}$ in GIE

$$P1_{jm}^{(2)} = \sum_{r=1}^{N_j} P2_{jr} \cdot P1_{jr}^{(2)}. \tag{12.26}$$

(5) To compute the new value of probabilities P_{jr} in GIE

$$P_{jr}^{(2)} = P1_{jr}^{(2)} \cdot \frac{P_{jm}^{beg}}{P1_{jm}^{(2)}}. \tag{12.27}$$

(6) To compute the F_{jr} at changing one probability in the GIE only.
(7) To compute the gradient of the criterion function for one probability

$$\triangle F_{jr} = F^{beg} - F_{jr}; \; r = 1, 2, \cdots, N_j, \; j = 1, 2, \cdots, n. \tag{12.28}$$

Computing differences of probabilities in GIE. The new values of probabilities in GIE are computed by the gradients of the criterion function:
(8) To compute differences for all probabilities using formula

$$\triangle P1_{jr}^{(2)} = K_2 \cdot \triangle P1_{jr}^{(1)} \cdot \triangle F_{jr}, \tag{12.29}$$

where K_2 is the coefficient of changing the step.
(9) To compute all probabilities in GIE using formula

$$P1_{jr}^{(3)} = P1_{jr}^{beg} + \triangle P1_{jr}^{(2)}; \; r = 1, 2, \cdots, N_j; \; j = 1, 2, \cdots, n. \tag{12.30}$$

(10) To normalize the received probabilities in GIE

$$P1_{jr}^{end} = \frac{P1_{jr}^{(3)}}{\sum_{r=1}^{N_j} P1_{jr}^{(3)}}. \tag{12.31}$$

(11) To compute the new values of the mean probabilities in GIE

$$P1_{jm}^{end} = \sum_{r=1}^{N_j} P2_{jr} \cdot P1_{jr}^{end}. \tag{12.32}$$

(12) To compute the values of probabilities P_{jr} in GIE in the step end

$$P_{jr}^{end} = P1_{jr}^{end} \cdot \frac{P_{jm}^{end}}{P1_{jm}^{end}}; \; r = 1, 2, \cdots, N_j; \; j = 1, 2, \cdots, n. \tag{12.33}$$

(13) To compute the new values of mean probabilities in GIE

$$P_{jm}^{end} = \sum_{r=1}^{N_j} P2_{jr} \cdot P_{jr}^{end}; \; j = 1, 2, \cdots, n. \tag{12.34}$$

(14) To compute the values of the criterion function F^{end} in the step end.

Choice of coefficients by the method of planning experiments. The increments of probabilities $\triangle P1_{jr}^{(1)}$ for computing gradients of the criterion function and the step of changing probabilities $\triangle P1_{jr}^{(2)}$ in the direction

Table 12.1. Planning experiments

No	K_1	K_2	N_{opt}	F	N^{end}
1	+	+	+	834	860
2	–	+	+	830	1100
3	+	–	+	834	1100
4	–	–	+	830	718
5	+	+	–	798	752
6	–	+	–	830	900
7	+	–	–	842	900
8	–	–	–	830	900
9	0.15	0.4	1000	834	1000

of the gradient are decreasing with increasing the optimization number N_{opt}. Coefficients K_1 and K_2 for computing $\triangle P1_{jr}^{(1)}$ and $\triangle P1_{jr}^{(2)}$ are chosen by method of optimal planning experiments. Coefficients K_1, K_2 and the number of optimization N_{opt} change in following intervals: $K_1 = 0.13 \div 0.17$; $K_2 = 0.3 \div 0.5$; $N_{opt} = 900 \div 1100$.

Results of computed experiments are presented in Table 12.1. The last column of the table is the optimization number N^{end}, at which the optimization process is ended, as all gradients $\triangle F_{jr}$ are equal to zero. After processing the matrix of planning experiments, we get the following expression:

$$F = a_0 + a_1 \cdot \frac{K_1 - 0,15}{0,02} + a_2 \cdot \frac{K_2 - 0.4}{0.1} + a_3 \cdot \frac{N_{opt} - 1000}{100}, \tag{12.35}$$

where:

$$a_0 = \frac{\sum_{i=1}^{8} F^{(i)}}{8} = 828.5; \; a_1 = \frac{\sum_{i=1}^{8} K_1^{(i)} \cdot F^{(i)}}{8} = -1.5;$$

$$a_2 = \frac{\sum_{i=1}^{8} K_2^{(i)} \cdot F^{(i)}}{8} = -5.5; \; a_3 = \frac{\sum_{i=1}^{8} N_{opt}^{(i)} \cdot F^{(i)}}{8} = 3.5.$$

With increasing the optimization number, increments of probabilities $\triangle P1_{jr}^{(1)}$ for computing the criterion function gradient are decreased. Increments of probabilities $\triangle P1_{jr}^{(2)}$ in the gradient direction are decreased, too. We begin to calculate gradients of the criterion function at changing probabilities $P1_{jr}$ to $13 \div 17\%$, whereas in determined direction we do 1/3 or 1/2 step from $13 \div 17\%$.

Proceed from (12.35) and values of coefficients a_0, a_1, a_2, a_3, the maximum value of the criterion function cannot be more than $F_{max} = 842$ in the area of possible values of parameters.

Investments on the standard statistic data. Let us bring the illustration of the identification process of the risk LP-model by the gradient method. Computing made for the credit risk LP-model for natural persons.

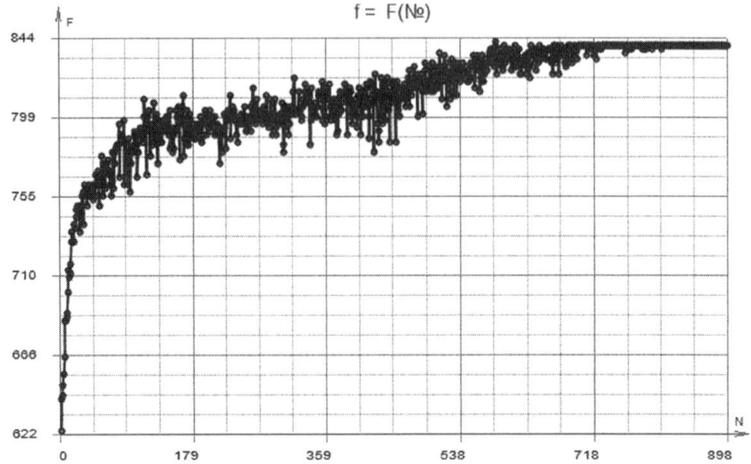

Fig. 12.11. Changing criterion function at optimization

As the statistic data we use standard Western packet of credits. The packet had $N = 1000$ credits, $N_g = 700$ of which were good and $N_b = 300$ were bad. The credit description has $n = 20$ signs, which have in the sum 94 grades.

We processed the statistic data and determine the initiating approximate probabilities P_{jr}, $P1_{jr}$ and the relative frequencies of grades $P2_{jr}$.

The optimization process of the credit risk LP-model is illustrated in Fig. 12.11 and Fig. 12.12. We got $F_{max} = 842$, $N_{opt} = 898$. The accuracy of classification (the relation of the number of correctly recognized objects to the global number of objects) is equal to 84.2%. That is the good result. The credit risk LP-model has substantively less errors in classification of credits $E_m = 0.158$; $E_g = 0.125$; $E_b = 0.262$, than known methods, which have $F_{max} = 750 \div 720$; $E_m = 0.25 \div 0.28$.

On basis of stating we made the following conclusions:

1. The identification method of the credit risk LP-model with GIE by the gradient method shows the value of the criterion function, coincided with Monte Carlo method. However, the computing time is substantively less and results are stable and repetitious.

2. In identification by the gradient method, the differences $\triangle P1_{jr}^{(1)}$ and $\triangle P1_{jr}^{(2)}$ for probabilities $P1_{jr}$ should be determined by (12.23) and (12.29).

3. The initiating values of probabilities P_{jr}, $P1_{jr}$, $j = 1, 2, \cdots, n$; $r = 1, 2, \cdots, N_j$ should give by formulas (12.6–12.8).

4. It is necessary to normalize $P1_{jr}$ in GIE at computing gradients of the criterion function by formulas (12.25, 12.31).

5. It is necessary to recompute the mean probabilities $P_{jm}, P1_{jm}$ in GIE by (12.26, 12.34) at computing gradients of the criterion function.

6. Coefficient $K_1 = 0.13 \div 0.17$ means the relative difference of probabilities $P1_{jr}$ in choosing direction of the gradient; coefficient $K_2 = 0.3 \div 0.5$ decreases this difference for computing the gradient.

7. The optimal values of coefficients K_1 and K_2 and the number of optimization N_{opt} should estimate by the experiment planning method.

8. Convergency of the identification provides the cofactor $(N_{opt} - N^v) / N_{opt}$ in the formulas for computing difference $P1_{jr}, j = 1, 2, \cdots, n; r = 1, 2, \cdots, N_j$.

At last, we state the sequence of changing formulas for the identification of the risk LP-model during investments (Table 12.2). We used the formula construction idea for identification of the risk LP-model by the statistic data (12.13) and (12.29) from publishing by training the neuron nets [114]. In beginning for the errors estimation of coefficients-wights at the training the neuron nets by Monte Carlo the normal distribution lows are used. However it caused large computing time (days and weeks) because the interpolation of the tabular values of the normal distribution function. Therefore the normal distribution low is replaced by Cauchy' distribution, and the function $tg(K_3)$ appeared in the training formula. Here K_3 is the random number from interval $[-\pi/2, +\pi/2]$. The random function $tg(K_3)$ changed in the interval $[-\infty, +\infty]$. The training time of the neuron net, or the number of the optimization steps, in advance does not set.

The first formula, which is used for training the risk LP-model, is in Table 12.2 in line 1. From the essence of the criterion function F_{max}, one also inputs the given number of the optimization steps N_{opt}. It is obvious that in any case, probabilities $P1_{jr}$ must belong to the interval $[0, 1]$.

Further it became clear that for decreasing computing time the function tg should not be used, and for the convenience estimation of the probability

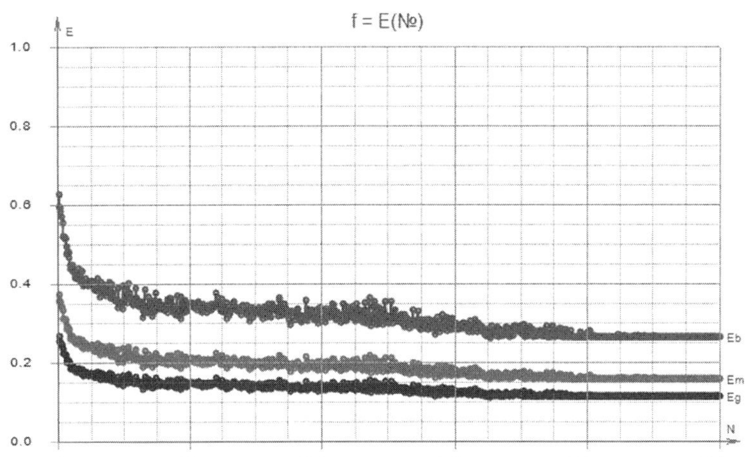

Fig. 12.12. Changing classification errors at optimization

Table 12.2. Developing formulas for identification

No	Formula for training	Values of coefficients	Method
1	$\Delta P1_{jr} = K_1\ (N_{opt} - N^v)\ tg\ K_3$	$K_3 = [-\pi/2, \pi/2]$ $K_1 = f(N_{opt})$	Monte Carlo
2	$\Delta P1_{jr} = K_1\ (N_{opt} - N^v)\ K_3$	$K_3 = [-1, +1]$ $K_1 = f(N_{opt})$	Monte Carlo
3	$\Delta P1_{jr} = K_1\ \frac{(N_{opt} - N^v)}{N_{opt}}\ K_3$	$K_3 = [-1, +1]$ $K_1 = f(N_{opt})$	Monte Carlo
4	$\Delta P1_{jr} = K_1\ P1_{jr}\ \frac{(N_{opt} - N^v)}{N_{opt}}\ K_3$	$K_3 = [-1, +1]$ $K_1 = [0.025; 0.05]$	Monte Carlo. Take into account $P1_{jr}$
5	$\Delta P1_{jr}^{(1)} = K_1\ P1_{jr}\ \frac{(N_{opt} - N^v)}{N_{opt}}$ $\Delta P1_{jr}^{(2)} = K_2\ \Delta P1_{jr}\ \Delta F_{jr}$	$K_1 = [0.13; 0.17]$ $K_2 = [0.3; 0.5]$	Gradient method. Take into account $P1_{jr}$

errors of $P1_{jr}$ the random number K_3 it should be generated from the interval $[+1, -1]$ (line 2).

There are difficulties in the choice of the training coefficient K_1, so its value depends on the optimization number N_{opt}, and probabilities $P1_{jr}$ must belong to the interval $[0, 1]$. Because of this we input in the training formula the relation $(N_{opt} - N^v)/N_{opt}$ (line 3).

The important guess: the input in the training formula of the probability $P1_{jr}$. Now, at last, the coefficient K_1 gets the transparency sense — it is maximum relative amplitude of changing probabilities $P1_{jr}$ (line 4). It simplified also the estimation task of the errors of probabilities $P1_{jr}$, as the number of the late optimization N_v is known.

The training formula of the risk LP-model by the gradient method (line 5) is constructed by analogy of Monte Carlo method (line 4). The optimal values of coefficients K_1 and K_2 easily choose, as their sense is transparency.

12.5 Identification criteria of the credit risk LP-models

There is the problem of the criteria chosen for identification of the credit risk LP-model by the statistic data of the bank. It is actual, as the identification

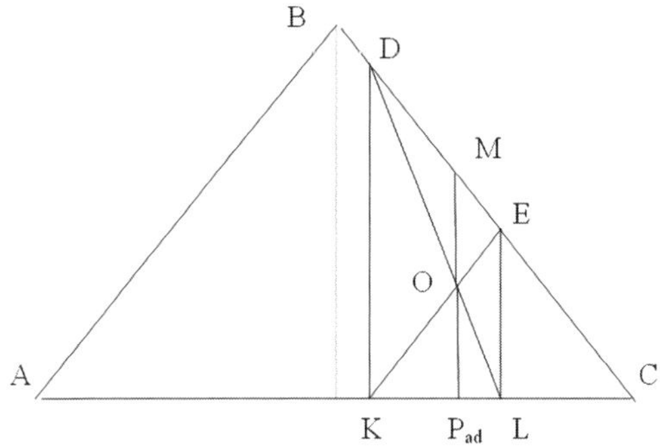

Fig. 12.13. Schema of distribution, good and bad credits by risk

task is complex. One must solve the multiparametric (about 100 parameters) optimization task with the integer criterion function and with a great set of the local criterions. The task can be solved in several hours on modern computers.

Before we used only one identification criterion: the sum of the correctly recognized bad and good credits must be maximum [3, 30]. However the discrete integer criterion function does not allow us to apply the known non-linear methods of optimization.

Let us consider the other criterion functions, which can be used for the identification of the risk LP-model.

1. The number of the correctly classified credits

$$F = N_{bb} + N_{gg} \Rightarrow \max_{P_{jr}}, \tag{12.36}$$

where N_{gg}, N_{bb} are, respectively, the numbers of credits, classified as good and bad by statistics and P-model (coincident estimations).

2. The entropy of probabilities of the correctly classified credits

$$H = -\sum_{k=1}^{N_k} P_k \cdot Ln\ P_k \Rightarrow \max_{P_{jr}}, \tag{12.37}$$

where P_k are probabilities of the correctly classified credits.

3. The sum of probabilities of the correctly classified credits

$$S = \sum_{k=1}^{N_k} P_k \Rightarrow \max_{P_{jr}}, \tag{12.38}$$

Let us use the conventional schematic representation of credit risk distribution (Fig. 12.13). Here the areas of the triangle are the numbers of corresponding credits. Then, for example, the criterion function for the entropy of credit risks can be written down for one group of criterions (H-criteria):

Table 12.3. Criteria of identification (optimization)

Criteria by order	Probabilities entropy of credits (H-criteria)	Number of credits (F-criteria)	Sum of probabilities probabilities of credits (S-criteria)	Notes
1	H_{gg}	F_{gg}	S_{gg}	Good and correctly identified
2	H_{bb}	F_{bb}	S_{bb}	Bad and correctly identified
3	H	F	S	Good and Bad correctly identified
4	H_{gb}	F_{gb}	S_{gb}	Good and incorrectly identified
5	H_{bg}	F_{bg}	S_{bg}	Bad and incorrectly identified

- H_{gg} is the number of correctly identified good credits (ABMPad–KOPad),
- H_{bb} is the number of correctly identified bad credits (CMPad–LOPad),
- H is the number of correctly identified good and bad credits)(ABC–KOL),
- H_{gb} is the number of incorrectly identified good credits (KOPad),
- H_{bg} is the number of incorrectly identified bad credits (LOPad).

Different possible criterions of optimization are tabulated (Table 12.3). Their sense is clear from the above-mentioned description; it is also explained in the column notes in the table. One should note that the criteria by the incorrectly recognized credits complement the criteria by the correctly recognized credits. It is also notable, that from the point of view of the algorithmic calculations, they may have the different characteristics.

The optimization dynamic for the criterion function F and the criterion function H is presented in Fig. 12.14 and Fig. 12.15. Investments are conducted by the random search method with use of the statistical data.

Investments for different criteria of optimization

Characteristics of criteria. The criterion F is integer, it changes discretely and equals the number of the correctly recognized good and bad credits (12.36). The criteria H is continue-discrete and equals the entropy of probabilities for the correctly recognized good and bad credits (12.37). The criterion S is continue-discrete and equals the sum of probabilities for the correctly recognized good and bad credits (12.38). The criteria H and S are continue-discrete, so their values depend on the number of the correctly recognized credits and on their probabilities (from the non-success risk).

Conditions of optimization. We use the initiate values of probabilities $P1_{jr}$ and P_{jr}, computed at the optimization by the gradient method for the criterion $F = 820$ on in-between step. Then we made the optimization for all three criterion F, H, S by the Monte Carlo method. At the optimization by one criterion, we fix values of other criterion, too.

Features of optimization. For way output from the local extremes, we use the procedure "deviation" (12.14), if for the given number of attempts $N_{mc} \approxeq 150$ on the optimization step the criterion function does not increase. The optimal deviation values for optimization criterions were equal: $\triangle F = 4$; $\triangle H = \triangle S \approxeq 0.1125$.

Results of investments of the above named optimization criterion, resulting in Table 12.4, allow us to do the following conclusions:

1. Optimization by criteria H and S is convenient, so we can use the whole arsenal of non-linear optimization methods.

2. The biggest value of the number of the correctly recognized credits (good and bad) or, the same, the biggest accuracy of the risk LP-model, is achieved at the optimization with use of the direct integer criteria F. The indirect continue-discrete criteria H and S correctly recognize the less number of credits. The criteria H and S give the non-optimal and biased estimates for a number of correctly recognized credits F.

3. The difference of optimization processes with different criteria can be estimated by deviation values in case of unsuccessful optimizations attempts in the step. The optimal deviation for the criterion F equals $\triangle F = 4$ and equals approximately $4/800 = 1/200$ from the optimal value of the criterion function. The optimal deviation for criteria H and S equals $\triangle H = \triangle S \approxeq 0.1125$ and

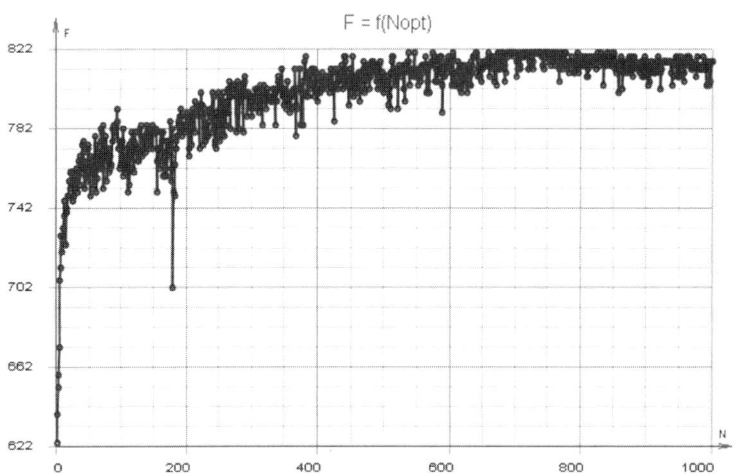

Fig. 12.14. Changing the criterion function F in optimization

Table 12.4. Optimization results by different criteria

Criterions of optimization	F	H	S
$F \rightarrow max$	842	223.35	182.84
$H \rightarrow max$	844	225.21	190.34
$S \rightarrow max$	830	223.98	197.13

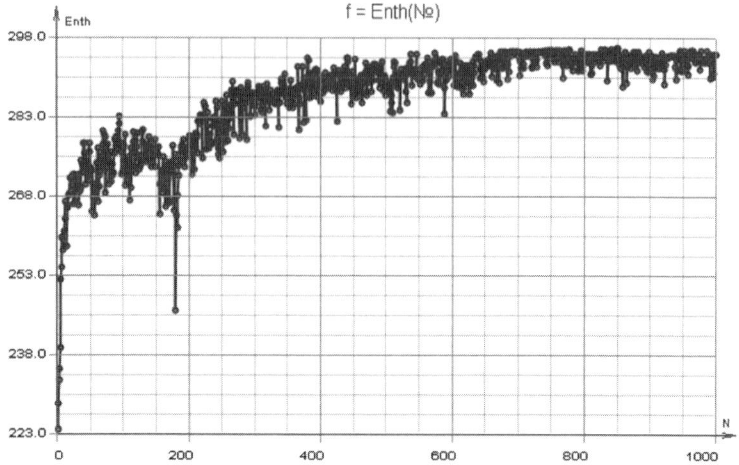

Fig. 12.15. Changing the entropy H in optimization by F

equals approximately $0.1125/225 = 1/2000$ from the optimal value of criteria function (Fig. 12.2), i.e. for the continuous-discrete criteria the height of stairs are decreased ten times.

4. The indirect continue-discrete optimization criteria H and S should be used to determine the initial values of probabilities $P1_{jr}$ and P_{jr} for the optimization by the criterion F and for checking solutions for this criterion.

5. The indirect continue-discrete optimization criterion H is better than the criterion S, so at this criterion we have the higher value of the criterion F and the less translocate it from the optimal value.

12.6 Investigations on identification of risk LP-models

Signs of the credit are the following: Y is the sign of credit success (2 grades), Z_1 is Balance of Account in Bank, Z_2 is Duration of Loan, Z_3 is Credit History, Z_4 is Intended Use of Loan, Z_5 is Amount of Loan, Z_6 is Securities Account, Z_7 is Duration of Employment, Z_8 is Payment to Partial Discharge (in % of available income), Z_9 is Marital Status and Gender, Z_{10} is Common Liability or Guarantee, Z_{11} is Time Spent in Current Residence, Z_{12} is a Type of Guaranty, Z_{13} is Age, Z_{14} is Other Loans, Z_{15} is Accommodation, Z_{16} is Number of Loans in Bank Including new Loan, Z_{17} is Profession, Z_{18} is Number of Relatives Dependent for Support, Z_{19} is Phone, Z_{20} is Citizenship.

The fragment of the statistical data on credits, given by their grades, result in Table 12.5.

It is known that multiparametric multicriterion optimization is an extremely difficult problem [114–117] and has some features and complications [3,28,29]: the criterion function F is the number of correctly recognized

Table 12.5. Fragment of statistical data on credits

Y	Z_1	Z_2	Z_3	Z_4	Z_5	Z_6	Z_7	Z_8	Z_9	Z_{10}	Z_{11}	Z_{12}	Z_{13}	Z_{14}	Z_{15}	Z_{16}	Z_{17}	Z_{18}	Z_{19}	Z_{20}
1	1	3	5	3	3	1	2	4	2	1	4	2	1	3	1	1	3	1	1	1
1	1	2	5	1	5	1	3	2	3	1	2	1	2	3	1	2	3	2	1	1
1	2	3	3	10	2	2	4	2	2	1	4	1	1	3	1	1	2	1	1	1
1	1	1	1	5	4	4	1	4	1	2	1	4	3	2	3	2	1	3	1	1
1	2	8	4	11	7	2	1	2	3	1	4	4	1	3	3	1	4	1	2	1
1	1	4	3	4	4	5	4	2	4	1	4	3	1	3	1	2	2	1	1	1
0	2	7	3	6	4	1	2	4	3	1	1	4	2	3	1	1	2	1	1	1
1	2	3	5	5	3	1	4	4	3	1	3	2	2	3	2	1	3	1	1	1
1	1	1	5	1	5	5	2	1	3	1	3	1	3	3	2	2	2	2	1	1
1	2	2	4	4	5	1	4	2	3	1	4	2	3	3	2	1	3	1	1	1
1	1	3	3	3	2	1	5	4	2	1	4	2	1	3	1	1	3	1	1	1
1	2	7	3	2	6	1	3	2	3	1	2	3	2	3	1	1	4	1	2	1
1	2	8	1	10	8	5	3	2	3	1	2	3	3	1	2	1	4	1	2	1
0	1	6	3	3	5	1	5	4	3	1	2	3	3	3	2	1	3	2	1	1
0	1	2	3	7	2	1	3	4	3	1	4	3	3	3	1	1	2	2	1	1
1	1	4	3	2	4	1	4	2	3	1	1	4	3	3	3	1	4	1	2	1
1	1	3	5	1	6	1	5	2	3	1	4	4	3	3	3	3	4	1	2	1
1	1	6	3	3	5	1	3	1	3	3	2	2	2	3	2	1	3	2	1	1
0	1	4	3	5	2	2	1	4	2	1	3	2	1	3	2	1	3	1	1	1
0	1	2	3	3	7	1	5	4	3	1	4	4	3	3	3	1	4	1	2	1

good and bad objects, i.e., it accepts the integer values and it is a stepped function; the criterion function has local extreme and it depends on a large number of real-valued positive parameters; derivatives of the criterion function with respect to the probabilities $P1_{jr}$ cannot be computed.

The criterion function F shown in Fig. 12.1 is a function of two parameters only. It is a stepped function. The steps have height equal to 2 and different width. The parameters $P1_1$ and $P1_2$ belong to the interval [0,1], but may differ in magnitude by one order. The weight of steps decreases when approaching to the extremum.

Optimization may be delayed on any step without attaining the maximum or crossing it. The variation of the criterion function in a multidimensional space remains the same. We recall that the dimension of the space of optimization parameters for the logical-probabilistic risk model of the credit risk is equal to 94.

Writing object and the risk model. We conduct the model investment on PC for the credit risk LP-model. We used the structural risk model of the type “knot” (Fig. 11.3a). The credit risk LP-model has 20 sign-events accordingly GIE) and 94 grade-events. The logical and probabilistic risk function are given by expressions (12.16), (12.17).

We use the standard Western packet of the statistical data on credits as in former examples [97]. The packet has the data on $N = 1000$ credits of natural persons, of which $N_g = 700$ was good and $N_b = 300$ was bad. Every credit is described by $n = 20$ signs, which have 94 grades.

Identification of the risk LP-models, as above noted, consist of determination of optimal probabilities P_{jr}, $r = 1, 2, \ldots, N_j$; $j = 1, 2, \ldots, n$ of grade-events. The risk of each credit is computed on each step of optimization. This risk is compared with the admitted risk P_{ad}. The object refers to bad or good one. Let us recall that the object function of the optimization task is formulated so: the number of the correctly classified credits must be maximum (12.1).

Below we bring the investigation results on identification, getting by the random search method.

Choice of parameters N_{opt}, K_1, F^{beg}. Probabilities $P1_{jr}$ for the initial variant were taken without the last four digits in comparison with those for the optimal variant $F = 824$. Therefore, the optimization began with $F^{beg} =$ 690–760. That reduced the computation time. The computations were made for two maximal increment amplitudes: $K_1 = 0.05$ (5%) and $K_1 = 0.1$ (10%) and number of optimizations N_{opt} equal to 150, 300, 500, 750, 1000, 2000, 3000, 4000, 5000, 6000, 7000, 8000.

Using results in Table 12.6 (Var. 1–21), we make the conclusions:

Table 12.6. Results of the choice of optimization parameters

No	N_{opt}	K_1	F^{beg}	F	$\triangle P_c$	$\triangle P1_{min}$	N^{end}
1	2	3	4	5	6	7	8
1	300	0.05	756	794	0.197	0.002	289
2	300	0.1	712	790	0.221	0.0043	288
3	750	0.05	756	802	0.164	0.0054	669
4	750	0.1	692	790	0.205	0.0131	652
5	1000	0.05	750	802	0.187	0.0035	931
6	1000	0.1	708	792	0.217	0.0158	843
7	2000	0.05	776	808	0.159	0.0075	1702
8	2000	0.1	724	798	0.180	0.0140	1720
9	3000	0.05	748	806	0.186	0.007	2581
10	3000	0.1	708	806	0.186	0.0050	2849
11	4000	0.05	744	812	0.194	0.0079	3368
12	4000	0.1	740	802	0.212	0.0086	3656
13	5000	0.05	754	806	0.166	0.0055	4445
14	5000	0.1	738	803	0.158	0.0040	4801
15	6000	0.1	710	810	0.16	0.006	5610
16	7000	0.05	764	810	0.209	0.0041	6430
17	7000	0.1	734	810	0.169	0.0074	6479
18	8000	0.05	764	810	0.175	0.0098	6425
19	8000	0.1	718	814	0.180	0.0028	7772
20	8000	0.03	780	820	0.153	0.0025	7662
21	8000	0.07	744	814	0.173	0.0025	7801

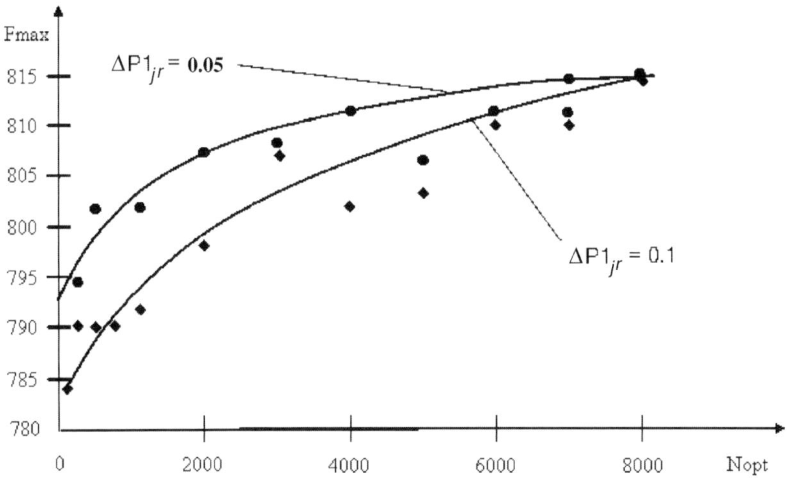

Fig. 12.16. Changing criterion function F from the optimization number N_{opt}

(1) The criterion function F (column 6 in Table 12.6 and Fig. 12.14) asymptotically increases with the number of optimization N_{opt};

(2) The minimum amplitude $\triangle P1_{min}$ (column 9) is approximately equal to 0.0025 (0.25%). Optimization does not take place for less values of $\triangle P1_{min}$ and the number of the last optimization N_{end} (column 10) is less than the given number of optimizations N_{opt}.

(3) The initiating value of criterion function F^{beg} (column 4) we must understate, as the low values bring to the low finished values F in considered case $F^{beg} = 750 \div 760$.

Discussion of results. In results of investment on identification of the risk LP-model, the effective technology of searching global extreme is supposed. It allows to decide the task of parametric multicriteria optimization with integer CF for acceptable computation time.

In the training formula, it should generate the random number K_3 in the interval $[-1, +1]$; in this case, the absolute values of the increment of probabilities $\triangle P1_{jr}$, multiplied to 100, pass to per cent (%) and give the assessment of accuracy of probabilities $P1_{jr}$ by value $\triangle P1_{min}$.

In the technology of search of the global extreme it uses the regularity:

- Criterion function asymptotically increases with the number optimizations N_{opt} (Fig. 12.16);
- Minimal amplitude $\triangle P1_{min}$ of increments of probabilities $P1_{jr}$ is the assessment of accuracy of determination of probabilities $P1_{jr}$; for the less values $\triangle P1_{min}$, optimization does not go (less than 0.25%);
- The initial criterion function F^{beg} should not be set too small because the low values lead to small final values for F_{max} due to the unsatisfactory optimization trajectory;

- Maximal amplitude of increments of the probabilities K_1 should not be taken greater than $0.02 \div 0.05$ ($2 \div 5\%$), as training dynamics worsens and the criterion function F decreases;
- Determination of the global extremum of the criterion function should be verified with the help of the graph of variation of F in the function of the difference $\triangle Pc$ between the maximal and minimal risks of objects in statistics. The function F has an extremum at a certain value of $\triangle Pc$.

12.7 Accuracy and robustness of risk LP-models

We carried out the estimation of the accuracy and robustness of the LP-model for the credit risk of natural personals. The structural risk LP-model of the type "unit" (Fig. 11.3a) is used. The credit risk L-model in the disjunctive normal form is presented by the formula (12.16), and the P-model is presented by the formula (12.17). The scenario is that non-success occurs when any one, two, ..., or all initiating events take place.

For comparison of the different risk evaluation methods, we use the standard statistical packet of credits [97]. The packet has $N = 1000$ credits of natural persons, of which $N_g = 700$ are good and $N_b = 700$ are bad. The credit has 20 sign-events (corresponding with GIE) and 96 grade-events. The essence of signs has been described above.

Accuracy is the basic attribute of the risk estimate method. The accuracy is estimated by the relative errors in recognition of bad and good credits and in the average (12.2). Usually we want to recognize the bad credits better than good credits. The relation of the non-corrected recognized good and bad credits is chosen to equal 2–10.

The estimates of the accuracy of the risk LP-models have the optimal properties: efficiency, consistency, and unbiasedness. This follows from the direct criterion function (the maximum number of the correctly recognized credits) and from its accurate determination as the integer number. Because of this, we think that the statistical data should not be divided between training and testing random samples.

It is also known the estimates of the accuracy of probabilities $P1_{jr}; r = 1, 2, \ldots, N_j; j = 1, 2, \ldots, n$ of grade-events. They have the optimal estimates too. This follows from the correct determination of the maximum error $\triangle P1_{jr}$ and its possible values on the interval $[\triangle P1_{min}, 0]$.

The results of accuracy comparison of different methods on the same statistical data (Table 12.7) show that the risk LP-model is almost 2 times more accurate than other classification methods based on the linear (LDA) and quadratic (QDA) discriminant analysis, on the cluster analysis (CARD), and neuron networks (NN) [97].

Robustness (stability) is also an important attribute of risk estimate method. The different methods of risk estimation (or one method at different algorithm of identification on the statistical data) classify differently the

Table 12.7. Parameters of classification accuracy of credits by different methods

Used method	Error of classification of bad objects, E_b	Error of classification of good objects, E_g	The average error, E_m
LDA Resubtitution	0.26	0.279	0.273
LDA Leaving-one-out	0.287	0.291	0.29
QDA Resubtitution (QDA)	0.183	0.283	0.253
QDA Leaving-one-out	0.283	0.34	0.323
CART	0.273	0.289	0.285
Neuron networks 1 (NN)	0.38	0.24	0.282
Neuron networks 2 (NN)	0.24	0.312	0.29
LP-model without GIE (Var. 1)	0.167	0.201	0.191
LP-model with GIE (Var. 2)	0.1433	0.190	0.176
LP-model with GIE and after structure identification (Var. 3)	0.126	0.174	0.155

credits into good and bad ones. One of two models may classify a credit as good, whereas another may classify it as bad. The non-stability (non-robust) estimation may be 20% from the global number of credits.

The estimate of the robustness (stability) of the risk LP-model was carried out by the pairwise comparison method of different variants of the solution in the credit classification. In the variants, difference of the criterion functions F reach 10 units. The comparison was based on the inconsistencies number of the estimates for the good n_g, bad n_b credits.

Our investments show, that different methods for classification (Table 12.7) differ on the robustness seven times. The obtained result can be generalized to the instability of the risk models based on the neuron networks, where a large number of the net weights links is introduced without any restrictions. The non-robustness of the methods of the risk estimation, based on neuron networks, has been noted in numerous papers [97].

In detail, we state the investment results on accuracy and robustness of the credit risk LP-model in Chapter 15.

13

LP-Analysis of Risk in Systems with Groups of Incompatible Events

Without the quantitative analysis of risk it is impossible to operate risk.
A. S. Mozhaev

The technique of risk analysis should be transparent and provide calculation of the contributions of sign-events and grade-events in risk of each object, in the mean risk of object set, and in accuracy of the risk model. It is important for management of risk. For various purposes of modeling and estimation of risk, it is necessary to develop different methods of the risk analysis, namely statistical, combinatorial, and logic-probabilistic methods of the risk analysis. We consider these methods of analysis denoting indexes in variables so it is made in the computing program.

13.1 Statistical risk analysis

The statistical analysis of risk (S-analysis) is necessary for calculation of the grade-events probabilities at the first, but rather good approximation. It is necessary for the beginning training process of the risk LP-model.

By using Table 11.1, it is possible to determine the number NS_{jr} of objects with the given grade, the number $NS0_{jr}$ of bad objects with the grade; $NS1_{jr}$ is the number of good objects with the grade. Then the probabilities of event-grades $P1_{jr}$ in GIE are equal to

$$PS1_{jr} = NS0_{jr}/NS_{jr}. \tag{13.1}$$

It is to note that the obtained probabilities of grade-events, when being substituted into the risk LP-model without any optimization, give higher value of the criterion function $F_{max} = 776$, than other known risk models after optimization ($F_{max} = 725 \div 750$).

E.D. Solojentsev, *Scenario Logic and Probabilistic Management of Risk in Business and Engineering*, DOI 10.1007/978-0-387-77946-1_13,

13.2 Combinatorial risk analysis

The risk combinatorial analysis (C-analysis) explains the high accuracy of the risk LP-models. In works [3, 30, 118], it is shown that the risk LP-models trained by methods of parametrical identification on the statistical data have almost two times higher accuracy and seven times larger robustness in classification of objects than do known methods of classification on the basis of the discriminant analysis and neuron networks.

However, explanation of high accuracy of the risk LP-model risk by its logic structure does not satisfy all experts. Really, the common sense of the risk model, which states that failure occurs if there is any one, any two, or all sign-events, requires other confirmations.

The researches of the risk LP-model with one more sign-event having only two grades, the same as those of the final event, have shown 100% recognition of objects [3]. Each grade of such sign is projected to grades “1” or “0” of the final event (success or failure). If the sign has a few grades, theoretically a part of grades in GIE should project to “1” and rest of grades to “0.” These ideas form the basis of the risk combinatorial analysis [118].

We describe formulation of the combinatorial analysis problem by an example. Let the grades $r = 0, 1, 2, 3, 4$ of the sign $j = 12$ be investigated. The calculation scheme for grade $j = 0$ with relative frequency $P2_{12\ 0}$, is as follows:

- we regard all objects with the grade 0 as the bad, that is this grade is projected to value $Y = 0$ of the failure sign. We get estimations $NC00$ (the correct bad objects) and $NC01$ (the incorrect bad objects);
- we regard all objects with grade 1 as the good, that is this grade is projected to $Y = 1$. We get estimations $NC11$ (the correct good objects) and $NC10$ (the incorrect good objects).

In the combinatorial analysis for each grade, only two types of such objects $NC00, NC01$ or $NC10, NC11$ are calculated, that is, the ideal participation of the grade in classification is determined in view of its frequency $P2_{jr}$.

Let us write out possible combinations for the first branch from 15 variants, in which numbers before the hyphen denote grades for the good objects, and numbers after the hyphen specify grades for the bad objects:

0–1234; 01–234; 12–034; 23–014; 34–012;
1–0134; 02–134; 13–024; 24–013;
2–0124; 03–124; 14–023;
3–0123; 04–123;
4–0123

Let us consider the combination: 04–123; N=1000; N_b=300; N_g=700.

For the first branch we have:

$NC^b_{04} = 250; NC00 = 150; NC01 = 100;$
$NC^g_{123} = 750; NC11 = 550; NC10 = 200;$
$ECm = (N - (NC00 + NC11))/N = (1000 - (150 + 550))/1000 = 0.3;$

For the second branch we have:

$NC_{04}^{g} = 250; NC11 = 100; NC10 = 150;$
$NC_{123}^{b} = 750; NC00 = 200; NC01 = 550;$
$ECm = (N - (NC00 + NC11))/N = (1000 - (100 + 200)/1000 = 0.7.$

Thus, in the problem of combinatorial analysis the following tasks can be determined:

- to calculate numbers $NC00$, $NC01$, $NC10$, $NC11$ for each grade in GIE;
- to calculate numbers $NC00, NC01, NC10, NC11$ for all possible combinations of grade in GIE on the basis of the obtained results;
- to determine the variant with optimum combination of grades and to fix its parameters of classification accuracy.

We carried out numerical investigation in the field of the combinatoric analysis. There were used "standard" statistical data for 1000 credits, from which 700 were good and 300 were bad. The structural LP-model of credit risk for physical persons is given in Fig. 11.3a. It has 20 sign-events (accordingly, GIE) and 94 grade-events. We used the risk L-function (11.15) and the risk P-function (11.17).

The following variants of C-model were built: with the number of good objects $N_{gc} = 610$, with $N_{gc} = 700$, and with the maximal value of the criterion function F_{max}. The results of calculations are given in Table 13.1. For each, from 20 signs for optimal division of their grades into good and bad, we calculated the following extra parameters:

$EC0$ is an error in classification of bad objects;
$EC1$ is an error in classification of good objects;
ECm is an error in classification of objects.

For the sign 0, effective combinations belong to the range $ECm = 0.304 \div 0.696$. Two most effective combinations have $ECm = 0.304$ and 0.335, that is, differ not much. For the sign 8, efficiency of combinations are in the range $ECm = 0.310 \div 0.690$. Two most effective combinations have $ECm = 0.310$ and 0.342 that differ not much, too.

Let us construct one more table based on the research results (Table 13.2), where for each sign we show: the number of grades N_j, value of the average error of classification ECm, the ratio EC_{max}/EC_{min} for different accepted number of good objects N_{gc}, and the decrement of the criterion function $\triangle F_j$, when this sign is eliminated.

We consider in more detail the results of calculations for the signs 0 and 8, having four grades. We write in Table 13.3 the values of efficiency of each grade of the sign 0 at its projection to 0 or 1 of the final event.

We present the results of Table 13.3 for the optimal combination with $ECm = 0.304$ (1, 2, 3 are good grades; 0 is bad grade) in the graph form (Fig. 13.1). The normalized probabilities of grade-events in GIE is as follows:

$$P1_{0r} = EC_r \Big/ \sum_{r=0}^{3} EC_r \qquad (13.2)$$

Table 13.1. Results of the combinatorial analysis

Signs	$N_{gc} = 610$				$N_{gc} = 700$				N_{gc} for F_{max}			
	Combi- nations	E_m	E_1	E_0	Combi- nations	E_m	E_1	E_0	Combi- nations	E_m	E1	E0
0	+0,1,2 -3	.60	.5	.84	+1,2,3 -0	.30	.2	.55	+1,2,3 -0	.30	.20	.55
1	+1,2,3,4 +6,7,9 -0,5,8	.49	.42	.67	+0,2,3,4 5,6,8,9 -1,7	.42	.30	.69	+0,1,2,3, +4,5,6,8 -7,9	.29	.04	.87
2	+2,3 -0,1,4	.47	.4	.65	+0,1,2,3 -4	.49	.35	.83	+1,2,3 -0,5	.28	.05	.82
3	+1,2,3,4 +5,7,10 -0,6,8,9	.4	.35	.51	+0,1,3,4 +6,8,10 -2,5,7,9	.40	.28	.66	+0,1,2,3, +4,5,6,8,9 -10	.30	.0	1.0
4	+0,1,2,4 +5,8,9 -3,6,7	.45	.38	.60	+1,3,4,6 +7,8 -0,2,5,9	.43	.31	.72	+0,1,2,3, +4,5 -6,7,8,9	.29	.01	.93
5	+0 -1,2,3,4	.53	.45	.72	+0,1 -2,3,4	.5	.4	.83	+0,2,3,4 -1	.33	.1	.88
6	+1,2,3 -0,2	.45	.39	.57	+1,2,3 -0,4	.44	.32	.71	+1,2,3,4 -0	.32	.06	.92
7	+0,3 -1,2	.47	.40	.64	+1,3 -0,2	.43	.31	.74	+0,1,3 -2	.37	.16	.85
8	+0,2 -1,3	.43	.38	.55	+0,2,3 -1	.39	.28	.64	+1,2,3 -0	.31	.04	.93
9	+0 -1,2	.34	.09	.91	+0 -1,2	.34	.09	.91	+0,2 -1	.30	.030	.95
10	+0,1,2 -3	.46	.41	.58	+0,2,3 -1	.41	.30	.67	+1,2,3 -0	.36	.13	.9
11	+0,2 -1,3	.41	.35	.54	+1,2,3 -0	.46	.32	.80	+0,1,2 -4	.3	.12	.77
12	+1,3,4 -0,2	.44	.39	.58	+0,1 -2,3,4	.44	.31	.74	+0,1,2,3 -4	.32	.03	.98
13	+2 -0,1	.33	.16	.74	+2 -0,1	.33	.16	.75	+0,3 -1	.31	.04	.936
14	+1 -0,2	.36	.25	.62	+1 -0,2	.36	.25	.62	+0,1 -2	.32	.09	.85
15	+0 -1,2,3	.47	.38	.66	+0,2,3 -1	.45	.34	.69	+0,1,2 -3	.30	.01	.99
16	+2 -0,1,3	.44	.36	.62	+0,2 -1,3	.43	.34	.64	+1,2,3 -0	.31	.02	.97
17	+0 -1	.36	.15	.85	+0 -1	.36	.15	.84	+0 -1	.36	.15	.85
18	+0 -1	.47	.41	.62	+0 -1	.47	.41	.62	+0 -1	.48	.41	.62
19	+0 -1	.33	.05	.98	+0 -1	.33	.05	.98	+0 -1	.33	.05	.98

Table 13.2. Influence of the number of grades to the recognition error

Signs, j	N_{gc}=610 ECm	N_{gc}=700 ECm	F^{abs}_{max} ECm	Number of grades, N_j	EC_{max} EC_{min}	$\triangle F_j$
0	0.602	0.304	0.304	4	1.95	- 64
1	0.496	0.419	0.290	10	1.71	- 27
2	0.476	0.493	0.283	5	1.74	- 18
3	0.398	0.400	0.301	11	1.33	- 26
4	0.452	0.430	0.288	10	1.57	- 20
5	0.531	0.496	0.335	5	1.59	- 20
6	0.447	0.441	0.316	5	1.41	-6
7	0.474	0.435	0.367	4	1.29	-5
8	0.434	0.392	0.310	4	1.4	-11
9	0.337	0.337	0.305	3	1.11	-10
10	0.465	0.414	0.358	4	1.3	0
11	0.41	0.462	0.32	4	1.44	0
12	0.445	0.443	0.316	5	1.41	-16
13	0.334	0.334	0.309	3	1.08	-2
14	0.359	0.359	0.32	3	1.12	-8
15	0.467	0.449	0.302	4	1.54	-2
16	0.442	0.434	0.308	4	1.43	0
17	0.363	0.363	0.363	2	1.0	0
18	0.478	0.478	0.478	2	1.0	0
19	0.329	0.329	0.329	2	1.0	-2

Table 13.3. Individual efficiency of grade-events

Grades	0	1	2	3
Good	0.696	0.641	0.665	0.398
Bad	0.304	0.359	0.335	0.602

After calculations we get:

$P100 = 0.151; \quad P101 = 0.319; \quad P102 = 0.331; \quad P103 = 0.198.$

These probabilities should be used as initial approximation when solving the problem of identification.

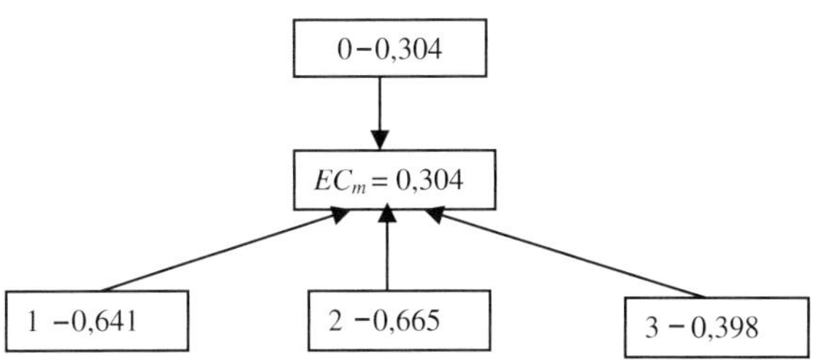

Fig. 13.1. The scheme of probability calculation of grade-events for the optimum combination of grades

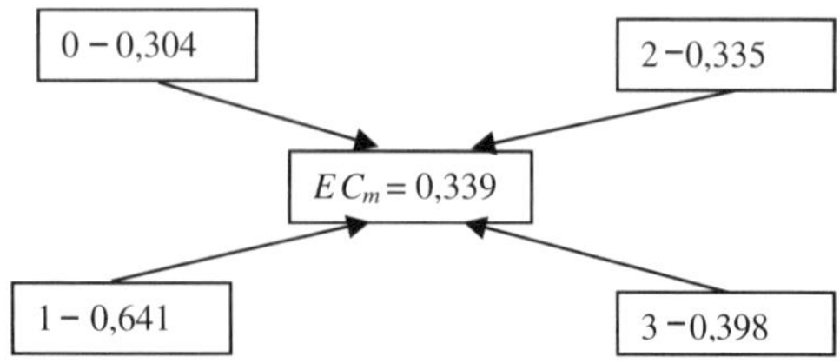

Fig. 13.2. The scheme of probability calculation of grade-events in GIE for any combination of grades

If, for example, as optimum we fix the combination with $ECm = 0.339$ (1, 3 are good grades; 0, 2 are bad grades), we get the graph shown in Fig. 13.2. The normalized probabilities of grade-events in GIE are as follows: $P100 = 0.181$; $P101 = 0.382$; $P102 = 0.199$; $P103 = 0.237$.

As a result of the model research by the combinatorial analysis (C-analysis), we obtained the following basic outcomes:

- The C-analysis allows us (unlike the S-analysis of risk) to estimate the importance of sign-events and grade-events without solution of the difficult problem of optimization.
- The C-analysis allows us to estimate the average recognition error of each sign. These average errors for different signs can differ almost in two times. Any of the signs cannot provide the average recognition error, obtained by all signs ($ECm = 0.174$ on the risk LP-model and $ECm = 0.3$ on the statistical data).
- The signs with a small number of grades in them ($N_j = 2, 3$) practically do not change the average classification error when the calculated number of good objects N_{gc} changes. The signs with the number of grades $N_j \geq 4$ substantially change the average error of object classification when N_{gc} changes. At the same number of grades in signs, their influence to the classification error depends on frequencies of grades $P2_{jr}$ or, which is the same, from lengths of intervals, on which the grades are constructed.
- Large number of grades in a sign is not only the measure of error in the object recognition. The error also depends on frequencies of grades $P2_{jr}$ of signs or on lengths of intervals, on which the grades are constructed.
- Large number of grades in a sign leads to "saturation" of recognition error, i.e., further increasing the number of grades does not decrease the error of object recognition.

The analysis of risk is carried out by algorithmic numerical computation on PC; this allows us to determine characteristics of system risk and characteristics of each element risk. The analysis of efficiency of each sign is done on the basis of the analysis of efficiency of its grades, which constitute group of incompatible events. Some signs of risk object have grades constructed on intervals of sign values. For example, in the problem of credit risk as the signs with grades we can name the sum of the credit, duration of the credit, age of the client, etc. For such signs with continuous value, it is possible to set the problem of optimal division of the sign value area into intervals of different optimal length with the number of interval corresponding with the number of sign grades. Similarly, the problem of optimal choice of grades number for a sign with discrete value by merging or splitting separate grades.

13.3 Logical-probabilistic risk analysis

Let the risk P-model be trained and the probabilities of grade-events P_{jr} be known. In order to carry out analysis, we determine contributions of sign-events and grade-events in the object risk and mean risk of a set of objects, as well as the accuracy of the risk LP-model. Using ideas of works [2, 3, 119], this task can be readily computerized by calculating the differences between the values of the above mentioned characteristics in the optimal mode and those obtained (|) for the zero probabilities of the grade-events.

The contribution of a sign (all grades of the sign) to the risk of the object i is as follows:

$$\triangle P_j = P(i) - P(i)|_{P_j=0}; \quad j = 1, 2, \ldots, n. \tag{13.3}$$

The contribution of a sign to the mean risk P_m of the set of objects is given by the formula

$$\triangle P_{jm} = P_m - P_m|_{P_j=0}, \quad j = 1, 2, \ldots, n. \tag{13.4}$$

The contributions of grades to the mean risk P_m of the set of objects is as follows:

$$\triangle P_{jrm} = P_m - P_m|_{P_{jr}=0}, \; j = 1, 2, \ldots, n; \; r = 1, 2, \ldots, N_j. \tag{13.5}$$

The contribution of a sign to the criterion function F_{max} is as follows:

$$\triangle F_j = F_{max} - F|_{P_j=0}, \quad j = 1, 2, \ldots, n. \tag{13.6}$$

For the contributions of grades to the criterion function F_{max} we have:

$$\triangle F_{jr} = F_{max} - F|_{P_{jr}=0}, \quad j = 1, 2, \ldots, n. \tag{13.7}$$

We note that the calculation of the contributions of the grade-events to the mean risk and the criterion function ($\triangle P_{jrm}$ and $\triangle F_{jr}$) is incorrect because it is not known how to correct the relative frequencies of other grades W_{jr} in GIE if one of them is set to zero. Therefore, by analogy with (12.2), instead of the contributions $\triangle F_{jr}$, one should calculate the errors of object classification for each grade-event:

$$E_{jrg} = N_{jrgb}/N_{jrg}; \quad E_{jrb} = N_{jrbg}/N_{jrb};$$
$$E_{jrm} = (N_{jrgb} + N_{jrbg})/N_{jr}, \tag{13.8}$$

where the N_{jrg}, N_{jrb}, N_{jr} are the numbers of good, bad, and all objects with grade r, respectively, and N_{jrgb} and N_{jrbg} are, respectively, the numbers of objects with the grade r and non-correct classification.

Calculation of the above contributions of the sign-events and grade-events underlies structural identification of the risk LP-model. The latter implies varying the L-function, and variation of the number of the signs and grades in the signs, which enables one to tackle the important applications of risk control. The performance of the LP-analysis of the risk will be discussed in detail in Chapter 15 by an example of the bank credit activity analysis.

13.4 Transparency of risk LP-models

Transparency is the important characteristic of methods of estimation of credit risks. Transparency of the method, as we will understand is accuracy of used mathematical methods, the absence of subjective expert estimations, the obviousness of results of estimation and analysis of the risk for bank clerks and for controlling agencies.

Transparency of the credit risk is determined by knowledge of quantitative attributes of the risk for every grade, every sign, every credit, and the set of all credit of the bank.

Quantitative attributes of the grade-event risk of the sign are the non-success probability of the credit, the relative non-success probability among grades of the sign, the probability-frequency in the set of credits, the contribution to the accuracy exactness of the model.

Quantitative attributes of the sign-event risk are the non-success average probability, the structural weight and significance in the risk model risk, the contribution to the credit risk, the contribution to the average risk of the credit set.

Quantitative attributes of the credit risk are the non-success risk, possible losses, the price for the risk, the contribution to the risk of the credit set.

Quantitative attributes of the risk of the credit set are the admitted risk, the average risk, the average losses, the admitted losses, the number of credits, the number of hazardous credits, the entropy of hazardous credits risk.

For analysis, forecasting, and management of the credit risk, every bank must be able to determine the named attributes, to analyze the risk, and to do the constant monitoring attributes of the credit risk. Using results of analysis of risk attributes of grades, signs, credits, and the set of credits, we can optimize credit risk for increasing its accuracy and robustness.

In common case, using Fig. 11.4, the contributions of grade-events into the admitted value of the output parameter Y_{ad} are equal to

$$D_{jr} = N_{jr}/N_{ad}, \quad j = 1, 2, \ldots, n; \quad r = 1, 2, \ldots, N_j, \tag{13.9}$$

where N_{ad} and N_{jr} are the numbers of all conditions of the output parameter and conditions of the output parameter with the grade r of the parameter j, such that the conditions satisfy the inequality

$$Y_i < Y_{ad}. \tag{13.10}$$

Contributions of grade-events to $Risk$ are equal to:

$$C_{jr} = P_{jr}/Risk, \quad j = 1, 2, \ldots, n; \quad r = 1, 2, \ldots, N_j, \tag{13.11}$$

where P_{jr} is the total probability of conditions of the output parameter with the grade-event r of the parameter j, satisfying (13.10).

On the basis of the given above expressions, contributions of the group of grades-events for one or a few different securities [2, 3] can be calculated. Grades or their groups, having the maximum contributions, are the best indicators, showing the opportunity of non-success for the output parameter.

Some signs of the risk object have grades, constructed on intervals of sign values. For example, in the task of the credit risk these are the signs of sum and term of credit, the client's age, etc.

For such signs with continuous values it is possible to use the task of optimal subdivision of sign values area not uniformly as usual but to intervals of different optimal length, the quantity of which corresponds to the number of sign grades. Similarly we may put the task of choice of optimal number of grades for a sign with discrete values by means of combination/subdivision of separate grades.

13.5 Management of risk

Let's consider management of risk by the example of the credit risk. The purpose of management of the credit risk is a decrease in financial losses of the bank and an increase of an accuracy of recognition of the bad and good credits. Parameters of risk management of the credit and the credit activity of the bank are

- The risk of the credit. We compare the risk of the credit with the admitted risk and take decision on delivery of the credit (Fig. 11.5a).
- The sum and term of the credit are established by the bank depending on the risk of the credit. A number of the countries give out the credit practically to everything, but its sum, term, and price depends on values of the parameters describing the credit and client. The maximal sum and term are limited to norms of the regulating body (the central bank).
- The price (percent) for the credit depends on the risk value of the credit according to the formula (11.24). This dependence can be more complex, and not just linear.
- The number of the attributes describing the credit and the client.

- The number of grades for each attribute.
- The recognition asymmetry factor of good and bad credits. Losses of bank are nonequivalent at erroneous recognition of the good and bad credit.
- The width of intervals at allocation of grades for such attributes as the sum and term of the credit, the age of the client, etc.
- The use of the non-success risk model of the credit with full (11.16) or the limited (11.38) set of events.

14

Software for Assessment, Analysis, and Management of Risk

The computer allows one to solve all those problems which up to the invention of a computer did not exist.
Computer News

In this chapter we shall consider intellectual work station and software for management of the safety of SCS. It was indicated already that the problems of safety and risk are characterized by extreme computing complexity. Thus, to speak about their solution is only possible in presence of program means. These means are not standard and cannot be joined in packages, such as $MATLAB$, etc.

The risk LP-theory with GIE and corresponding special logic software are tools of the new generation for modeling, analysis, and management of risks in business, economics, and engineering.

14.1 Intellectual work station for safety management

The necessity of a new approach to automation of design, tests, and maintenance of safe operation of SCS is caused by the fact that computer-aided design (CAD) systems are ineffective [3, 7]. They solve utilitarian problems: calculations are carried out, design drawings and technological processes are prepared. However, there are no means for the senior staff of the design office to be directed to safety maintenance of the created SCS.

Possible ways and methods of the practical solution of these problems should be considered in a context of complex automation, for example, of machine-building enterprise, including (Fig. 14.1) the automated design office or the system CAD (computer-aided design), automated manufacture or the system CAM (computer-aided manufacturing), and automated testing benches or system CAT (computer-aided testing).

E.D. Solojentsev, *Scenario Logic and Probabilistic Management of Risk in Business and Engineering*, DOI 10.1007/978-0-387-77946-1_14,

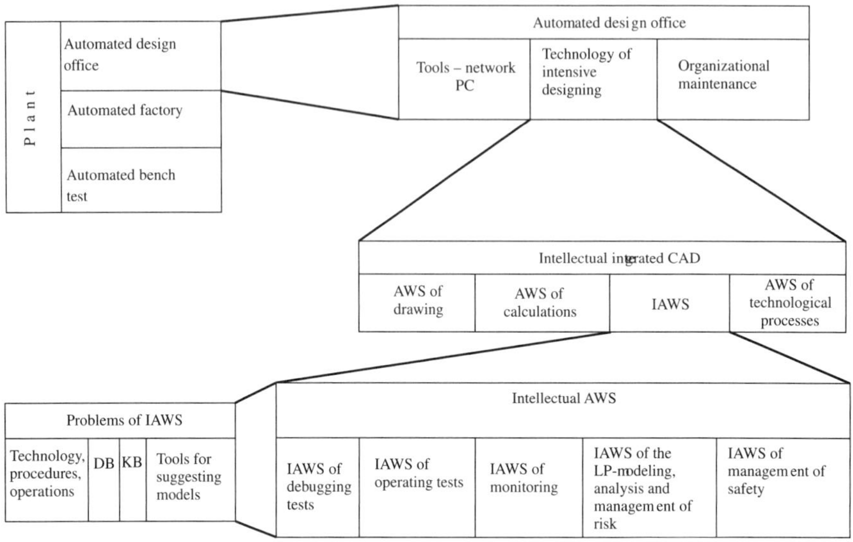

Fig. 14.1. CADS with Intellectual Automated Work Station (IAWS)

The automated design office includes software tools (local network of computers), technology of intensive development of objects and organizational maintenance. In its turn, the technology of intensive development of objects is presented as an intellectual integrated CAD system. This means that in the *CAD* system at the design stage, the problems of debugging and operational tests, monitoring and diagnosing, modeling and analysis of risk, management of safe operation are solved, too [7, 67]. It is suggested to equip the integrated CAD system by the certain set of problem-oriented automated work stations on the base of computers with intellectual software.

Such integrated *CAD* system, along with the work station for calculation, drawings, and making technological processes, includes also the following intellectual work stations for:

- debugging tests;
- operational tests and estimation;
- monitoring;
- logic and probabilistic modeling and analysis of risk;
- management of safe functioning.

The problems of development of these intellectual work stations include:

- Development of technology of the work station as a set of procedures and operations;
- Development of a database (*DB*);
- Development of a knowledgebase (*KB*);
- Development of means for support of models (graphical, logic ones, etc.).

If we do translucence by "floodlight" the automated manufacture and automated testing benches, then we see the same intellectual work stations and problems.

Let us give engineering interpretation of knowledge of the intellectual work station. The concept of knowledge of the intellectual work station is usually related to the documents, models, and technology [3, 67]. For example, in development of the intellectual work station for debugging tests, the engineering interpretation of knowledge is used (shown in Fig. 14.2). Among the mentioned components of the intellectual work station, the leading position belongs to the technology, which is built on the basis of experience of experts and achievements in information technologies. The technology defines the requirements to toolkit and work organization. It is supported by appropriate program-methodical means and provides the disciplining and organizing foundation. It is also one of the most valuable component knowledge, as it is invariant to different objects.

To the basis of technology of the intellectual work station, invariant for different objects, it is necessary to put any comprehensive ideas, which give a direction and common philosophical approach. For this, as it was marked earlier in Chapter 3, we can use the following principles of:

- management of complex object,
- knowledge management,
- recognition images,
- training,
- structural design.

Consider development of technology of intellectual work station, for example for planning and the realization of debugging tests of machines and man–machine–environment systems. All the above mentioned principles are used in the development, but the leading role belongs to the scheme of management of complex object with movement along a chosen trajectory and

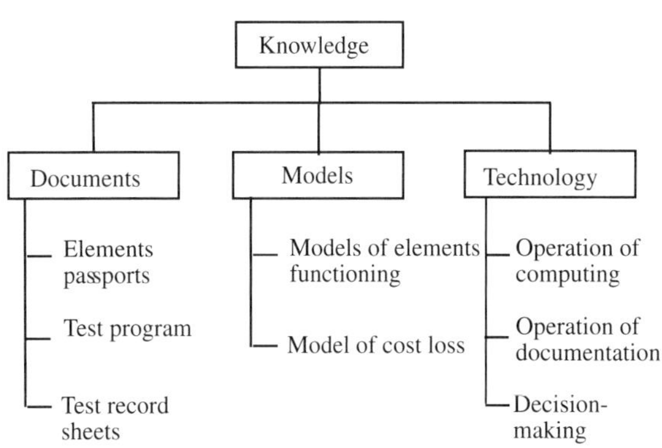

Fig. 14.2. Engineering interpretation of knowledge of IAWS

correction of deviation from the way and the scheme of knowledge as the logically closed sequence of procedures. In technology of this intellectual work station, the following procedures are input:

- forecasting debugging process;
- technical and economic modeling debugging process;
- planning debugging process;
- decision making during tests;
- processing test protocols and improving models.

In their turn, the procedures of the intellectual work station consist of elementary operations. More than 100 operations can be named in total. Approximately one third of operations carry out computations on models, approximately one third of operations deal with documents; and approximately one third of operations are related to decision making in conditions of uncertainty with use of productive rules "if–then" and "on analogy," formulated on the basis of judgments of experts.

14.2 Software for risk LP-models with the GIE

The software for identification and analysis of risk LP-models (software for "LP-estimation of risk") is the so-called "know-how." It uses the mathematical risk LP-theory with groups of incompatible events. The LP-model of credit risk of physical persons has shown almost two times higher accuracy and seven times higher robustness in classification of credits to good and bad than other used methods (see Tables 12.4 and 12.5). The decrease of risk almost to half allows us to reduce losses of a bank and essentially to lower the rate for the credit, to attract more clients, and to raise competitiveness of the bank.

Functions of Software "Risk LP-estimation." The software provides the solution of the whole complex of new important problems for estimation and analysis of risk. For example for credit risk, three groups of problems are distinguished:

1. Determination of the credit risk

(1) Estimation of the credit risk;

(2) Classification of the credit (good or bad);

(3) Determination of the price for risk of the credit;

(4) Analysis of the credit risk;

2. Analysis of the credit activity of bank

(5) Determination of the contributions of signs (characteristics) and of sign grades of a credit in the average credit risk of bank;

(6) Determination of the contributions of signs (characteristics) and of signs grades of a credit in accuracy of classification of the credits;

(7) Optimization of number of signs and grades, splitting into intervals (sum and period of the credit, age of clients) for higher of accuracy of classification of the credits;

3. Identification and estimation of quality of risk LP-models

(8) Statistical analysis of the risk model;

(9) Combinatorial analysis of the risk model;

(10) Probability estimation of grade-events and the admitted risk;

(11) Estimation of accuracy of the risk LP-model;

(12) Estimation of robustness of the risk LP-model;

(13) Choice of the ratio of the incorrectly classified bad and good credits.

The users of software "LP-estimation of risk". Any bank in any country can use the given software. For a specific bank, it is only necessary to train the risk LP-model on the statistical data. In the absence of statistics (the least necessary number of the credits in statistics is equal to 400), it is necessary to use statistics on the related bank. The software "LP-estimation of risk" can be easily adapted for solving risk problems in other data domains — for example, in business for estimation and analysis of risk of:

- Credits of juridical persons;
- Ruin of banks;
- Swindles;
- Bribes;
- Insurance of life;
- Buying-selling in tasks of the technical analysis, etc.

The description of variables of classes OBJECT, SIGN, GRADE. Object-oriented software for identification and analysis of risk LP-models with GIE is written on Java and Visual C++ following the above stated technique (Chapter 11). Let us present appropriate descriptions of classes and variables. As the basic class-object we choose "sign-event." Such decision allows us to use this software for risk LP-models:

- with a different number of sign-events;
- with a different number of grade-events in each sign-event;
- for different data domains of risk (business, engineering, insurance);
- for models of risk of different logic complexity.

Let us describe variables of classes *OBJECT, SIGN, GRADE* of the object-oriented program in the order of their subordination.

Variables for the OBJECT class, common for classes SIGN and GRADE:

$S[N][n]$ are statistical data (the file, Table 11.1);
n is a number of signs;
N is a number of objects;
$NS0$ is the number of bad objects in the statistical data;

$NS1$ is the number of good objects in statistic data;
$ESm = NS0/N$ is the average error in classification of objects.
$YL[N]$ is the array of final estimations of the non-success of objects;
$NL0$ is the number of bad objects by L-model;
$NL1$ is the number of good objects by L-model;
Pad is the admitted risk;
$NL00$ is the number of correctly recognized bad objects;
$NL11$ is the number of correctly recognized good objects;
$NL01$ is the number of incorrectly recognized bad objects;
$NL10$ is the number of incorrectly recognized good objects;
$Fmax$ is the common number of correctly recognized objects;
ELm is the average error in classification of objects;
$EL0$ is the average error in classification of bad objects;
$EL1$ is the average error in classification of good objects.

Variables for the SIGN class:
Nj is a number of grades in GIE;
Kj is the factor of transition from $P1jr$ to Pjr in GIE;
$PLm, PL1m$ are the average probabilities for $P1jr$ and Pjr in GIE;

Variables for the GRADE class:
$NS[\,]$ are numbers of objects with the given grade;
$NS0[\,]$ are numbers of bad objects with the grade by statistics;
$NS1[\,]$ are numbers of good objects with the grade by statistics;
$NL0[\,]$ are numbers of bad objects with the grade by model;
$NL1[\,]$ are numbers of good objects with the grade by model;
$NL00[\,]$ are numbers of correct bad objects with the grade by model;
$NL01[\,]$ are numbers of incorrect bad objects with the grade;
$NL10[\,]$ are numbers of incorrect good objects with the grade;
$NL11[\,]$ are numbers of correct good objects with grade;
$EL0[\,]$ are the errors of classification of bad objects on grade;
$EL1[\,]$ are the errors of classification of good objects on grade;
$ELm[\,]$ are the average errors in classification on grade;
$PL1[\,]$ are estimations of probabilities of grade-events $P1jr$ in GIE;
$PL[\,]$ are estimations of probabilities of grade-events Pjr in GIE.

For use in programming, the described variables are given in a more obvious and compact form in Table 14.1. Here we give extra indication of using variables in the statistical analysis (S-analysis), in the combinatorial analysis (C-analysis), and in the logic analysis (L-analysis).

The block diagram of identification and analysis. Let us name the basic actions and modules of the object-oriented program:

1. Input of $N, n, DS[\,][\,]$;
2. Computation of $YS[\,], NS0, NS1, PSm, ESm$;
3. Input of the array of a number of grades for signs $FNJ[\,]$;

Table 14.1. Object-oriented programming for risk LP-models in Java

Models	Variables
	Variables for the class "Set of risk objects"
S-model:	$DS[N][n], N, n, PSm, YS[N], NS0, NS1, ESm$;
L-model:	$PL[\,], Pad, \triangle PL, PLm, YL[N], NL0, NL1, NL00, NL01,$
	$NL10, NL11, Fmax, EL1, EL0, ELm$;
C-model:	$EC1, EC0, ECm$
	Variables for the classes "Sign-events": $1, 2, \ldots, n$
S-model:	Nj;
L-model:	$Kj, PLm, PL1m, \triangle Fmax, \triangle EL1, \triangle EL0, \triangle ELm$;
C-model:	$NC0, NC1, NC00, NC01, NC11, NC10, EC0, EC1, ECm$
	Variables for the classes "Grade-events"
S-model:	$NS0[\,], NS1[\,], P1S[\,]$;
L-model:	$NL[\,], NL0[\,], NL1[\,], NL00[\,], NL01[\,], NL10[\,], NL11[\,],$
	$EL0[\,], EL1[\,], ELm[\,], PL1[\,], PL[\,], \triangle FL[\,]$;
C-model:	$NC[\,], NC1[\,], NC00[\,], NC01[\,], NC10[\,], NC11[\,],$
	$EC0[\,], EC1[\,], ECm[\,], PC1[\,]$

4. Computing attributes of the grade: $NS[\,]$, $NS0[\,]$, $NS1[\,]$, $PS1[\,]$, $W[\,]$;
5. Input of $NL1, NL0$;
6. Input or computation of initial values of PLm for signs;
7. Input or computation of initial values of $PL1[\,]$ for grades;
8. Computation $PL1m$, Kj for signs;
9. Computation $PL[\,]$ for grades;
10. Input of parameters for the training formula: $Nopt, K1$ and $Fbeg$;
11. Beginning the cycle of optimization while $Nopt < Ncur$;
    Else STOP;
12. Beginning the cycle of optimization by Monte Carlo method;
13. Generation of $\triangle PL1[\,]$ by Monte Carlo method;
14. Computation of $PL1r[\,]$, standardization of $PL1r[\,]$, computation of $PLr[\,]$ for the grades;
15. Computation of the risk of objects in statistical data and the average risk of objects;
16. Computation of the average risk for objects and signs;
17. Computation of Pad, $Fcur$, $NL11$, $NL00$, $NL01$, $NL10$, $YL[N]$;
18. If $Fcur < Fmax$, then algorithm goes to item 12, else it goes to 19;
19. Assignment $Fmax = Fcur$; $Ncur++$;
20. If $Fmax > Fabs$, then $Fabs = Fmax$;
21. Output to the screen of $Ncur, Fmax$
22. Writing to the file of $Noptcur$, $Fmax$;
23. Computation of the new $PLmr[\,], PLm[\,]$ for signs and their correction by the condition $PSm = PLm$;
24. Computation of $PL1[\,] = PL1r[\,]$;
25. Computation of $PL1m, Kj$ for signs;
26. Ending the cycle of Monte Carlo method;

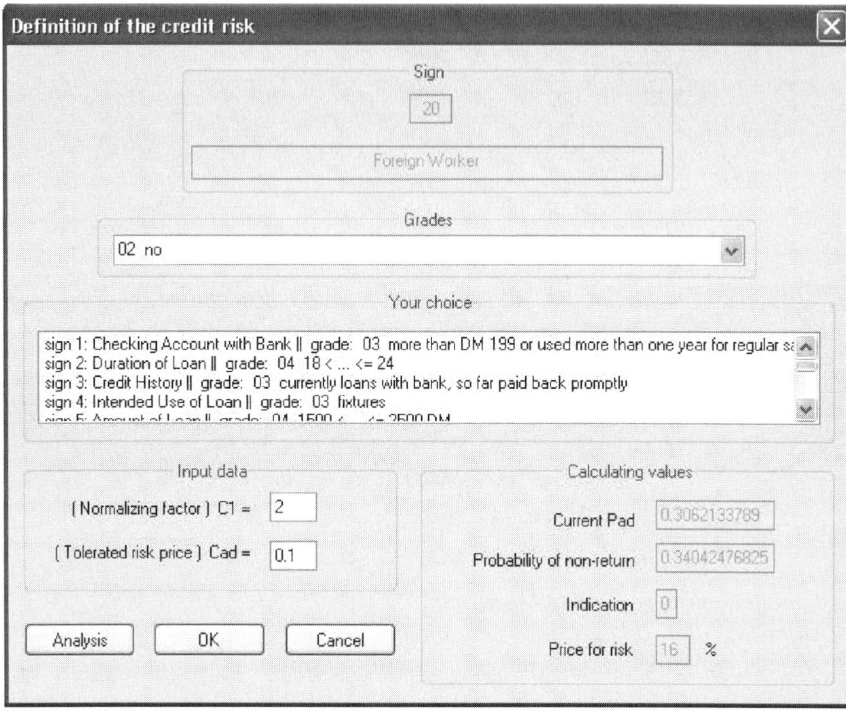

Fig. 14.3. The dialogue window for data input and estimation of the credit risk

27. If there were no optimizations in the cycle by M-C, then deviation $Fmax = Fmax - \triangle F$ is applied and the algorithm goes to 12;
28. End of the cycle of the number of optimizations.

Examples of realization. The software "LP-estimation of risk" consists of a number of modules written in C++ language. The program core of software has approximately 3000 operators. There are modules of different functional purpose written on Delfi-4, C++, and Java. The fragments of realization of the package can be seen by the demo-version on website: http://www.ipme.ru/ipme/labs/iisad/soft.htm. In Figs. 14.3–14.5, the screen shots demonstrating work of software are given.

In Fig. 14.3, the form for filling the application for credit is shown. For each of 20 signs of the credit the list of grades is given ($\vee$ is prescribed), and it is only necessary to specify values of grades for the sign. After filling the form, the risk, which is probability of non-return of the credit "Probability of non-return" and the attribute "Indication" of classification of the credit as good and bad (0 or 1), is calculated.

In Fig. 14.3, the screen form for the analysis of the credit after an estimation of its risk is shown. The probabilities of grade-events of signs result, which describe the given credit. The contributions of grades to risk of the credit are

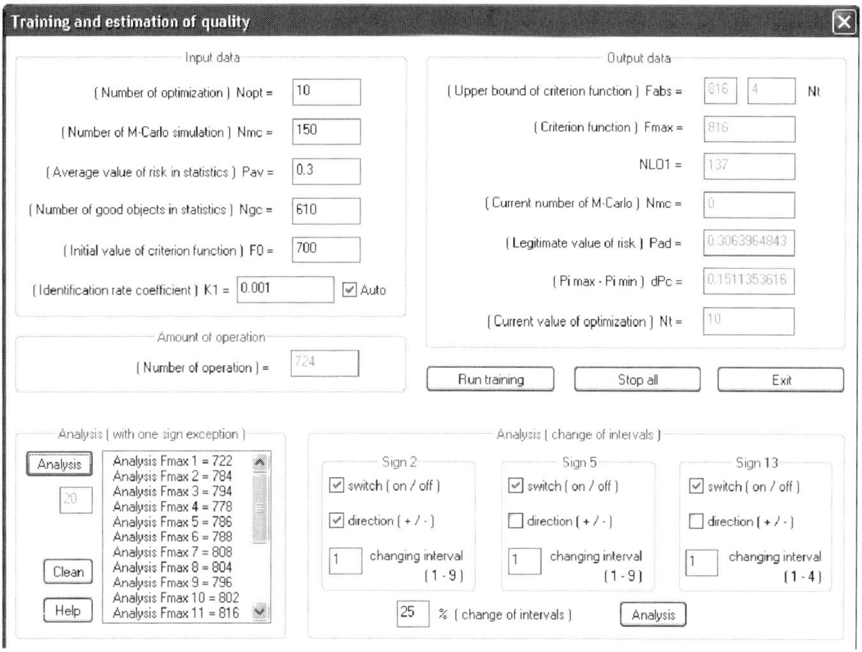

Fig. 14.4. The dialogue window "Training and estimation of accuracy"

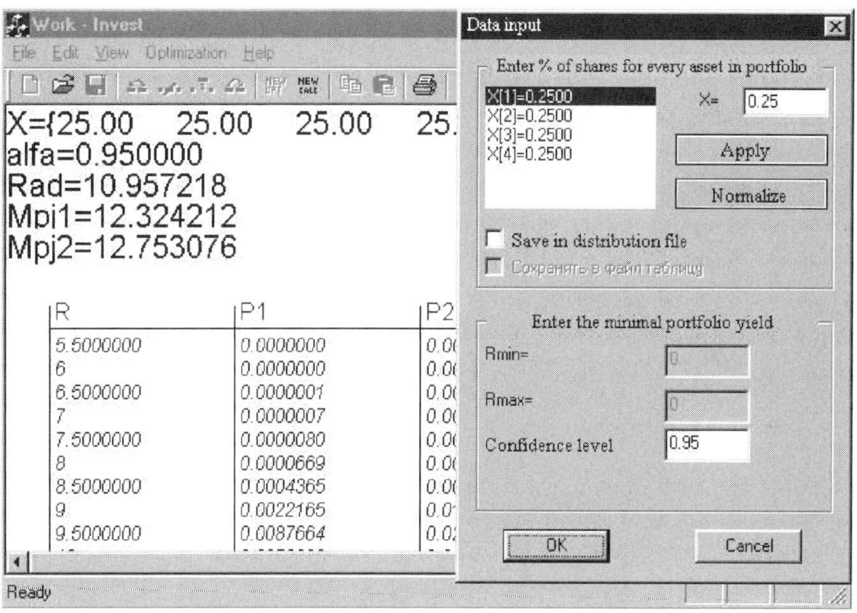

Fig. 14.5. The dialogue window "LP-selection of the optimal security portfolio"

directly proportional to these probabilities. In the same screen form, the meaning of the admitted risk "admitted risk" and the average risk of bank "average bank risk" is shown.

In Fig. 14.4, the screen form for training risk LP-model is presented. On the left side of the screen form, the parameters for the formula and process of training are given. They are set before the beginning training by the program itself, but can be changed by the user. On the right part of the screen form, the parameters of trained model in dynamics of its training are given. Values of probabilities of grade-events (their number is equal to 96) are written to a special file after each successful attempt of training. At the end of training in this file, we find optimum estimations of probabilities of grade-events in the form, which allows us to use them in the subsequent sessions of optimization.

In Fig. 14.5, the screen form for optimization of the security portfolio is presented.

14.3 Software for structural and logic modeling

In papers [106–108], the structure and characteristics of the base version of the computer program ACM-2001 is considered. The program allows one to build automatically analytical, statistical, Markovian, and network mathematical models for computation of parameters of reliability, survivability, safety, efficiency, and risk of functioning CS of the large dimension. Results of this paper by A. C. Mozhaev are stated below.

The theory of the automated structural and logic modeling. The theory is based on methods that allow one to automate processes of construction of mathematical models for calculation of stability (readiness, survivability, safety), efficiency, and risk of functioning for complex systems of any structure and organization of functioning. The actuality of this scientific direction is caused by the fact that in many applied areas, the systems risk analysis is not implemented. Construction of only the risk LP is impossible by manual methods.

Technology of the automated modeling. The technology represents such form of the practical system analysis, in which at first the block diagram of researched system is developed, and the regimes (criterions) of its work and parameters of elements are set. Then, by a computer, completely automatically, precisely and operatively the necessary mathematical models for the given regimes of operations of researched system under consideration are constructed. The system can have very large dimension and complexity. Further, on the basis of the obtained models, the machine calculations of the system characteristics are carried out and various tasks of analysis, optimization, and synthesis of system can be solved for making scientifically reasonable decisions at stages of its research, designing, operation, and management.

The greatest effect from application of technology of the automated modeling is achieved when (due to the large complexity and high dimension of system) construction of the mathematical model by old, hand-operated methods becomes impossible. Necessary conditions for realization of the given technology are creation of the appropriate theory and development of the program complex for the automation of the structural and logic modeling (SL-modeling) of complex systems of any structure.

Logical-and-probabilistic modeling. ACM-2001 is based on the LP-method of system analysis [5, 6, 106] described also in Chapter 10. It is known that in the classical LP-method, the graphic means of statement of problems (trees of events and connectivity graphs) represent only two logic operations AND, OR. This basis of operations is not functionally complete and allows one to build only a limited subclass of so-called monotonic models of systems. The means of graphic statement of problems in the program system are the special schemes of functional integrity (CFI), which by construction can represent the functionally complete set of logic operations AND, OR, and NOT. On this basis, first complete realization in the program system of all opportunities of logic algebra was achieved. Thus, with the help of ACM-2001, it is possible to build automatically both all known kinds of monotonic models of systems of any structure and essentially new class of non-monotonic models of systems, in which it is possible to take into account correctly the influence of harmful (damaging, emergency, etc.) events on processes of functioning.

The latter is especially important for construction of complex models of safety, danger, and risk of functioning various complex systems.

Now considerable scientific and practical experience of development and use of the computer programs of type ACM-2001 is accumulated. A lot of scientific research has been done, more than 40 dissertations were successfully held up, and the 100-hour educational course "Automated modeling systems" was given. The ACM-2001 was applied in various areas of system analysis, including the probabilistic analysis of safety and risk of technical, organizational, and banking systems.

The basic characteristics of *ACM-2001* are as follows:

1. *The complex is developed* by the modular principle in the program system Delphi 5.
2. *All positive properties of the previous versions are kept*, their flaws are taken into account and corrected, and new results of development of the theory and technology of the automated structural and logic modeling systems are introduced.
3. *The given version of ACM-2001 is directly intended* for maintenance of educational process in the course "Automated modeling systems" for performance of scientific work in various areas of system research.
4. *ACM-2001 is the base system* and covers four classes of tasks of the automated modeling: analytical, statistical, Markovian, and network. On the basis of this system, various specialized systems of the automated modeling (from probabilistic analysis of safety of nuclear stations to risk in business) can be developed.

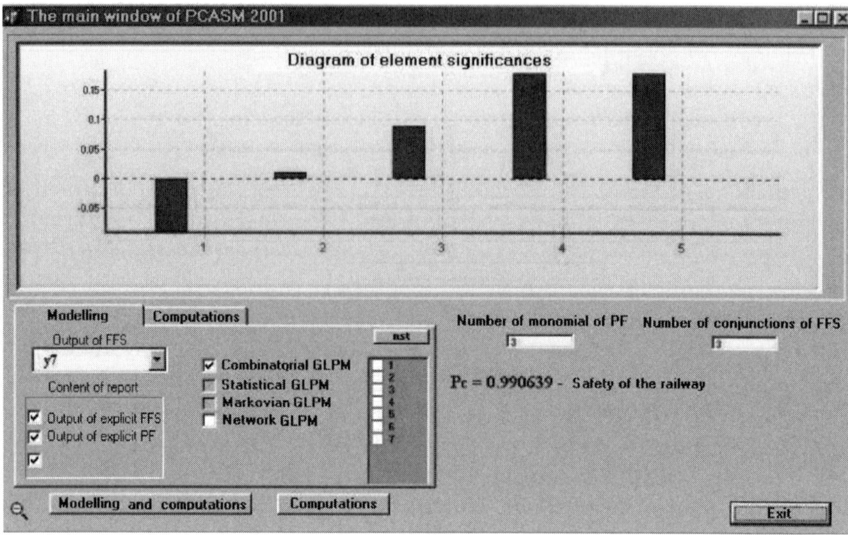

Fig. 14.6. Program modules and the structure of Software "ACM 2001"

The common structure of this complex is given in Fig. 14.6.

The graphical interface of the input-output provides graphic display of the initial data and results of automated structural and logic modeling. In Fig. 14.7, the window of preparation of the block diagram of the system under consideration is shown, and in Fig. 14.8, the window of the necessary regimes and output results of automatic modeling is given.

Modules of construction, storage, and transformations of the scheme of functional integrity (SFI) provide the evident graphical input of new and used before developed structural models of the systems. In Fig. 14.7, a variant of the window of preparation of SFI for the educational task of the safety analysis of a railway segment is given. After development of the graph, the SFI is automatically formed and stored in the computer memory as the work file of the appropriate system of L-equations (see the button Gb.dat of the tool panel in Fig. 14.7).

Modules of preparation, storage, and transformations of parameters of elements allow to introduce new and use former prepared probabilistic and other parameters of elements (see the button Harel.dat of the tool panel in Fig. 14.7). The basic parameters of simulated systems are: the static probabilities of elementary events; the intensity of elements failure, the average times of restoration of elements, the signs of two kinds of groups of incompatible events, the signs of duplication of functional nodes of SFI, the codes of four laws of distribution of non-failure operation time of elements, own operating times of elements, characteristics of efficiency or risk of various regimes of system operation.

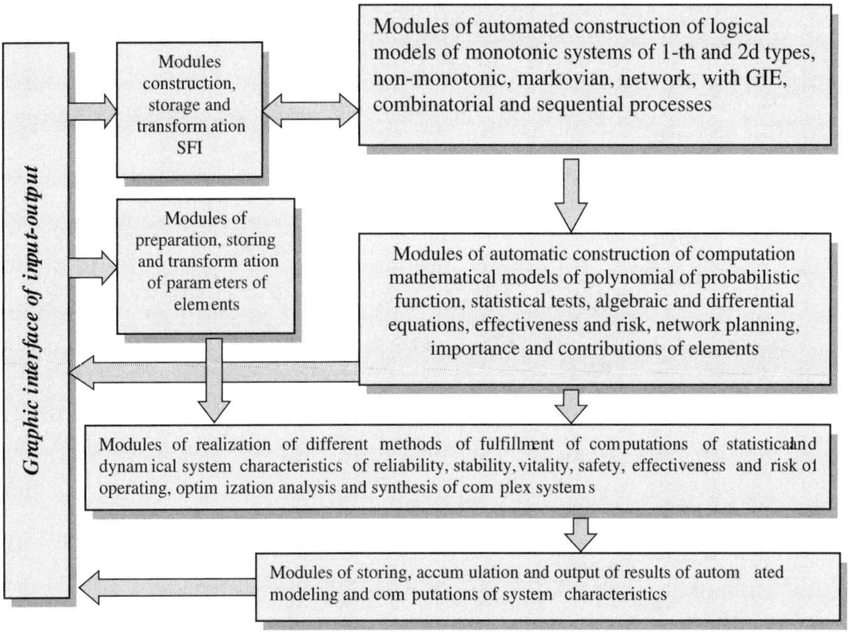

Fig. 14.7. The input window of structural model in the PC ACM 2001

The modules of automatic construction of logic models allow us to form several kinds of logic functions of serviceability of systems (FSS) and logic functions of transitions (LFT). These functions are monotonic ones (in basis of operations AND, OR) for systems of the first and second types, and any non-monotonic ones (in basis of operations AND, OR, NOT); FSS with taking into account any initial condition of a system, groups of incompatible events, multi-functional elements, elements with number of condition more than two, combinative, consecutive, and network processes.

The logic modeling in ACM-2001 realizes all combinatorial opportunities of classical logic algebra and takes into account dependence represented with the help of logic of groups of incompatible events and logic of sequences of events in complex systems and processes.

The modules of automatic construction of computed mathematical models carry out formation of the polynomials of probabilistic functions (for independent and several kinds of dependent events); imitating models for realization of statistical computations; Markovian schemes and corresponding matrixes of transitional probabilities for calculation of the conditional laws of survivability (construction of corresponding differential and algebraic equations is possible); logic sequences for calculation of various variants of the network work plans of elements of simulated systems. The dimensions of logic and calculated mathematical models in the ACM-2001 are limited only by the volume of fast memory of the computer and now achieve several tens of thousands of

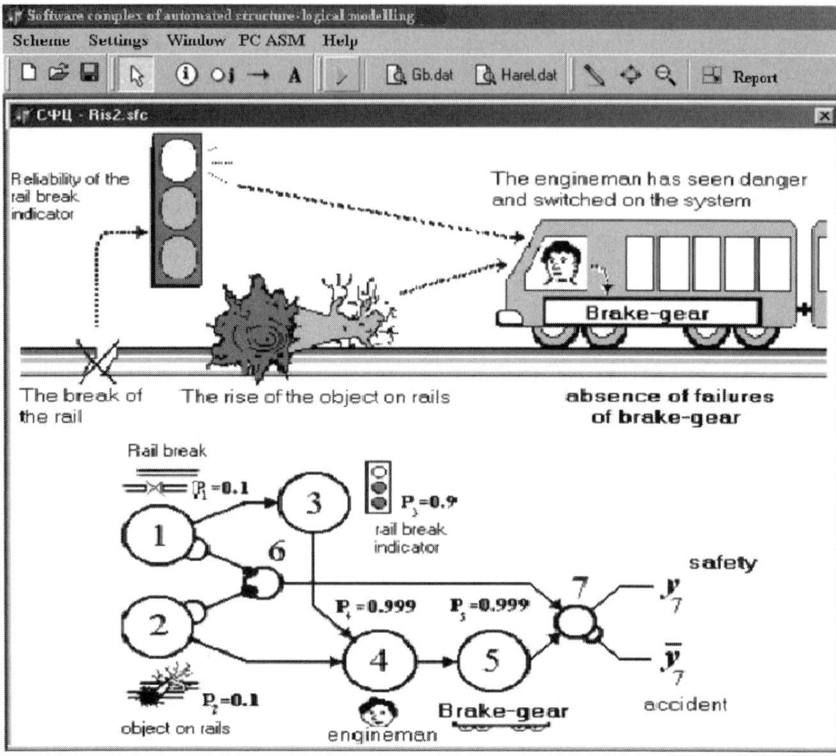

Fig. 14.8. The window of choice of simulation modes and output of results of calculations by ACM 2001

terms. By realization in ACM-2001 of methods of structural decomposition, the restrictions on dimension of formed models practically will be taken off.

The modules for realization of techniques of calculations performance are intended for quantitative estimation of various properties of systems (non-failure operation, readiness, stability, survivability, safety, efficiency, risk of functioning) on the basis of the mathematical models obtained at the previous stages. Simultaneously with calculation of the general-system parameters, the corresponding characteristics of the role of separate elements (importance and contributions) are defined. It provides an opportunity of automation of the decision processes for many special problems of optimization, distribution of resources, target planning, and management of systems. The automatic construction of models makes it possible to organize the solution of the mentioned problems in a real-time scale of the system functioning. In the ACM-2001, the inclusion of various techniques of modeling and calculation are carried out with the help of managing elements of the Basic window shown in Fig. 14.8.

The modules for preservation, accumulation, and output of results allow one to form in memory of the computer the libraries of initial structures of the systems under consideration, parameters of their elements, along with all results of automatic modeling and calculations of the system characteristics. The part of the most important information appears on-the-fly on the display panels of the Basic window (see Fig. 14.8) during automatic modeling and calculations.

With the help of this program complex, it is possible to solve practically all the problems mentioned in the published works of analytical logical and probabilistic modeling systems, as well as any problems of the given class suggested by the user.

14.4 Software for LP-models on the basis of the cortege algebra

This section follows works of B. A. Kulik [102, 103] to demonstrate the method of mathematically strict analytical logic and probabilistic modeling. The method makes it possible to obtain finite analytical expressions for the risk and to perform some additional research on the analytical risk model.

On the basis of cortege algebra, the programs for computation by a computer can be constructed. However, the orthogonalization of logic system on the basis of cortege algebra is only possible if the number of condition of elements and system is within the limits defined by resources of the computer. For example, for such problems as the credit risk and the security portfolio risk, the risk logic function and appropriate risk polynomial cannot be placed in the memory of computer.

Therefore, for solution of real problems of non-success risk, the stated earlier risk LP-theory with groups of incompatible events is used, which realization is algorithmic and is not limited by resources of the computer.

Risk analysis of systems with many conditions

Statement of the problem. Many modern methods of analysis of the risk and safety of technical systems are based on LP-methods developed by the scientific school of I. A. Ryabinin [1, 2, 25]. The application of these methods makes it possible to estimate the risk and safety of many structurally complex systems. At the same time, these methods have a number of methodological problems, which are not resolved yet. One of them is the risk modeling and analysis for systems with many conditions. Below, the approach to solution of this problem is suggested, based on the cortege algebra [102, 103].

If a system and its elements have two conditions ("work–failure," "on–off," "dangerous–safe," etc., then it is quite natural to interpret these systems in the terms of calculus of the statements (or Boolean algebra); all structures of the calculus are mapped to the set of two elements: "false," "true" or 0, 1. It is the model that is used frequently in the logic and probabilistic methods. However, as soon as we pass to logic modeling technical systems with many conditions, we need to pass to logic systems with more than two values of the logic atom. And it means the transition to multiple-valued logic, which is inevitably related to invalidating some Boolean algebra laws and, as a consequence, of classical laws of the theory of probability [128].

The way out from this "deadlock" situation nevertheless exists, and it consists in synthesis of algebra of sets and multiplace relations. This synthesis results in the mathematical model including as a special case both structures of mathematical logic, and structure of some variants of non-classical logic. Thus, in all cases the laws of algebra of sets and, correspondingly, Boolean algebra remain true.

Solution of the problem is based on the linguistic transformation — at the first stages we digress from the logic terminology and state the problem in the technical and algebraic terms. Let us begin with the first (technical) ones. Consider a system S with a set $Y = (y_1, y_2, \ldots, y_r)$ of conditions. Let the structure of conditions include some set $V = (V_1, V_2, \ldots, V_n)$ of nodes (or subsystems). Each node in its turn can also be in one of sets of conditions, i.e., each V_i has a corresponding set $X_i = (x_{i1}, x_{i2}, \ldots, x_{ik_i})$ of conditions. Here k_i is a number of possible conditions for the node V_i, and the sets X_i are arbitrary. The latter sets can also be infinite continuous; in this case, separate conditions can be given by points or intervals.

The exact model of the system S is the mapping between all possible sets of nodes conditions and conditions of the system. Mathematically, this relation can be displayed as follows:

$$S:\ D \longrightarrow Y, \tag{14.1}$$

where D is the Cartesian product $X_1 \times X_2 \times \ldots \times X_n$.

It is to note that the model (14.1) loses practical sense with increase of number of nodes, and also with increase of number of conditions of nodes and system, because even in rather small systems, the number of all elementary sets of conditions can exceed computing resources of modern computers. Besides, at the moment it is not absolutely clear how to apply to mathematical model (14.1) the powerful analytical means of the mathematical logic and the logic and probabilistic methods. The suggested approach to solution of these problems is considered in the following sections.

Logic interpretation of the model. Let us first formulate the above stated formal description of technical system in the algebraic terms of the multiplaced relations (keeping our notations). Let $D = X_1 \times X_2 \times X_n$ be the many-dimensional space of coordinates. In the space using only some of these coordinates, it is possible to choose separate subspaces or projections, and to set the multiplaced relations in the space D or any its projections. The list of the coordinates determining the given projection is referred to as the scheme of the relation. As is known, any set of relations on D, or on a fixed projection of D forms a system, isomorphic to algebra of sets, in which elements are elementary corteges, i.e., sets of n elements from various

X_i, if the complete space D is considered, or sets of m elements ($m < n$), if the system is considered in some fixed projection D. At the same time, if we consider totality of the relations, given on the projections of D with the different schemes of relations, then the representation of such system as a system, isomorphic to algebra of sets, seems to be problematic.

However, the given problem has a solution with the help of structures and methods of the cortege algebra (CA) [102, 103]. In the given system, the basic structural unit is the C-cortege, which differs from an elementary cortege — C-cortege is constituted by subsets of the sets X_i, not by their elements. The C-cortege is the set of elementary corteges, as it is interpreted as the Cartesian product of contained in it components. Association of some set of the C-corteges given in a certain projection form the C-system. Thus, it is possible to present any relation given in a certain projection as C-system.

In order to combine in a uniform algebraic system, isomorphic to algebra of sets, relations given in different projections, the fictitious coordinates are introduced in cortege algebra for formation of the fictitious coordinates and the fictitious components are used. A fictitious component "*", added to C-cortege or to C-system and taking the place i, is the set equivalent to all values area of the corresponding coordinate X_i. Thus, using fictitious components, it is possible to define any sets of the relations, given in different projections, to the dimension of the complete space or a certain generalized projection and to apply to them all operations of algebra of sets. In this form, the system of all possible relations of the complete space D, determined in different projections, becomes isomorphic to algebra of sets [128].

Essential difference of the suggested method of introducing fictitious coordinates from the known ones is that the new coordinates are introduced in the multi-placed relations expressed by C-corteges or C-systems as sets, not element-wise. The latter essentially reduces complexity of computing algorithms and volume of memory needed for storing the structures.

Along with three above mentioned structures (elementary cortege, C-cortege, and C-system in cortege algebra), two auxiliary structures are introduced: D-corteges and D-systems. More details can be found in [103].

In cortege algebra among operations of algebra of sets, the operations of formation of projections are provided, which are reduced to three elementary operations with coordinates: (1) addition of fictitious coordinate, (2) removal of any coordinate together with the components, corresponding to it, and (3) rearrangement of coordinates. The introduction of these operations allows us to add to operations of algebra of sets the operations that correspond with operations of the inversion and composition of mappings, and to operations having the same semantics as quantifiers $\forall$ and $\exists$ in the calculus of predicates. As an illustration, we shall consider an example of calculation of composition of the relations.

Let two relations be given (in brackets of names of the relations we indicate the schemes of the relations) $R_1(X_2, X_3)=\{(a, a), (a, b), (b, a), (b, c), (c, a), (c, c)\}$ and $R_2(X_1, X_2)=\{(a, b), (a, c), (b, a), (c, a)\}$. The composition of these relations (R_1R_2) is traditionally calculated by the method of comparison of pairs of elementary corteges and choice of such pairs, at which the second element of the pair from the relation R_2 coincides with the first element of the pair from the relation R_1. For each of such pairs, result of the composition is the elementary cortege containing first element from a cortege belonging to R_2, and second element from a cortege belonging to R_1. For example, for the cortege (c, a) from R_2, the suitable pair is the cortege (a, b) from R_1, and the composition of this pair is the cortege (c, b). Finally, after examination of all possible pairs of corteges from the different relations, finding suitable pairs, their compositions, and exception of repeating corteges, we obtain

$$R_1 \circ R_2 = \{(a, a), (a, c), (b, a), (b, b), (c, a), (c, b)\}.$$

Now we shall consider how this operation in cortege algebra (CA) is done. At first, the relations R_1 and R_2 are written as C-systems:

$$R_1 = \begin{bmatrix} \{a\} & \{a, b\} \\ \{b, c\} & \{a, c\} \end{bmatrix}; \quad R_2 = \begin{bmatrix} \{a\} & \{b, c\} \\ \{b, c\} & \{a\} \end{bmatrix}.$$

Further, we make calculations by the following algorithm:

(1) We shall add fictitious coordinate X_1 to R_1 and fictitious coordinate X_3 to R_2. Thus we bring these structures to the same scheme of the relation (X_1, X_2, X_3):

$$R_1 = \begin{bmatrix} * & \{a\} & \{a, b\} \\ * & \{b, c\} & \{a, c\} \end{bmatrix}; \quad R_2 = \begin{bmatrix} \{a\} & \{b, c\} & * \\ \{b, c\} & \{a\} & * \end{bmatrix};$$

(2) By using algorithms given in [89], we calculate the intersection of these C-systems:

$$R_1 \cap R_2 = \begin{bmatrix} * & \{a\} & \{a, b\} \\ * & \{b, c\} & \{a, c\} \end{bmatrix} \cap \begin{bmatrix} \{a\} & \{b, c\} & * \\ \{b, c\} & \{a\} & * \end{bmatrix} = \begin{bmatrix} \{b, c\} & \{a\} & \{a, b\} \\ \{a\} & \{b, c\} & \{a, c\} \end{bmatrix}; \quad (14.2)$$

(3) We check absence of empty C-corteges (if they are present, they should be removed). Then we delete the coordinate X_2:

$$R_1 \circ R_2 = \begin{bmatrix} \{b, c\} & \{a, b\} \\ \{a\} & \{a, c\} \end{bmatrix}.$$

It is proved that the structure of the elementary corteges, contained in the obtained C-system, up to rearrangement of elements of the obtained relation is equal to the result of calculation of the composition obtained by the traditional approach. In [102] it was established that the functionality of cortege algebra is sufficient for representation of all means of calculus of predicates.

The set of all various projections of the complete space D, including the space D itself, is defined in cortege algebra as the flexible universe. If it appears that any cortege algebra formula is equivalent to any projection (particular universe), then such formula corresponds with the general formula of many-sorted calculus of predicates. If, on the contrary, the equality of some formula of the cortege algebra to empty set is established, then the given formula corresponds with the unsatisfiable formula.

In order to specify correspondence between models of cortege algebra and various variants of logic, we shall consider a number of restrictions, imposed on the subject domain at its representation by model (14.1).

Restriction C_1: the mapping $D \longrightarrow Y$ is one-valued (or functional). It means that to any elementary cortege from D cannot correspond more than one element from Y.

Restriction C_2*:* the set Y contains exactly two elements.

Restriction C_3*:* sets $X_1, X_2, \ldots, X_n, Y$ are equivalent (on capacity).

Restriction C_4*:* restriction C_3 provided that the capacity of each of these sets is equal to 2.

It is easy to prove that

- A system, satisfying the restrictions C_1 and C_2, appears isomorphic to some model of many-sorted calculus of predicates; in this case, the relations on any projection of space D are interpreted as multi-placed predicates or logic formulae;
- A system under restrictions C_1 and C_3 corresponds with a model of multi-valued logic, in which an obligatory condition is the equivalence of value areas of the truth for all variable and for the system as a whole;
- A system under restrictions C_1 and C_4 corresponds with a model of calculus of statements. In all these interpretations, the laws of algebra of sets hold. It, seemingly, contradicts with "multi-valued" systems under restrictions C_1 and C_3, because in the multi-valued logic not all laws of Boolean algebra are true. However, this contradiction is especially terminological. In multi-valued logic, one object can have more than one negation. Certainly, it is possible to name "negations" of each other the pairs of structures with not coincident "values of the truth", if from the point of view of algebra of sets (and accordingly CA) they are not complementary. However, it is better to avoid ambiguity and to introduce for such "negations" some other term, for example "alternative." Thus, the uniqueness of negation is kept, and all laws of Boolean algebra hold.

It is possible to give one more logic interpretation for the given system by keeping only the restriction C_1 (otherwise the system appears unpredictable). It is possible to consider the united space $D \times Y$, which relations in different projections can contain (or not contain) coordinate Y. Then the set of elementary corteges of the space $D \times Y$ can be divided into two not intersecting sets: the set of admitted (true) and the set of non-admitted (false) conditions, which are determined according to constructive or technological features of simulated system. Then mapping

$$S_c : D \times Y \longrightarrow \{true, false\} \tag{14.3}$$

is isomorphic to a model of many-sorted calculus of predicates, in which there are no restrictions on number of conditions of nodes and of the system as a whole.

Immersion of a system into the probabilistic space. The success of LP-modeling and analysis of risk and safety of systems in many respects was determined by the fact that in the theory, the known algorithms were generalized and new ones were developed. These are algorithms of orthogonalization of systems allowing one to decompose the formula of calculus of the statements to a disjunction of simple conjunctions where any pair does not contain the general satisfying substitutions [102, 103]. The orthogonalization of system makes it possible to present the probabilistic measure of system as a simple analytical formula suitable for analysis and computation. However, using methods of orthogonalizations, the researchers were forced to consider one rigid restriction: each unit and system should have no more than two conditions, and the model of system should be necessarily represented as a model of calculus of the statements.

In the cortege algebra, this restriction is replaced by a weaker one: system orthogonalization is possible, if the number of condition of nodes and of the system is limited by computing resources of the computer.

Let the area of condition change for each node of systems (14.1) or (14.3) has a finite number of elements or is broken into a finite number of intervals and for each of the areas the probability distribution function is known with known parameters of the distribution. Then, if models (14.1) or (14.2) are described by a set of formulae of the cortege algebra, these formulae can be easily transformed with the help of the developed algorithms to orthogonal C-systems, which can be easily transformed to the computed formulae allowing one to calculate the probability of separate conditions of system.

The inverse problem is possible, too, when we know probabilities of the system conditions with higher accuracy. Then there is an opportunity to use the computed formulae obtained on a basis of orthogonalization to specify the estimations of the probability distribution parameters of conditions of nodes of the system.

Conclusion. On the basis of methods and means of cortege algebra, it is established that systems with many conditions can be presented not only in the terms of multi-valued logic, but also in the terms of the system, isomorphic to the algebra of sets. Then in this system, it is possible to use the laws of the classical theory of probabilities. If conditions of the system and its nodes are interpreted not as values of the truth, but as some value areas of corresponding variables, then it is possible to interpret dependence between conditions with the help of system in which the laws of Boolean algebra are true.

Description of Software. The given variant of the program has the following restrictions: the L-function should contain no more than 79 variables and consist of no more than 255 terms (disjunctions). In the program the L-model of risk is represented by matrix. It is caused by specificity of representation of the given class of problems in structures of cortege algebra and, at the same time, provides reduction of labor at input of the initial data and output of results.

In the matrix representation of the formulae, only 4 symbols are provided: "0", "1," "-" and "*." The symbols "1" and "0" in the column i replace the appropriate literals X_i and the negation X_i. Instead of the omitted literals in the conjunction we use the mark "*," in the disjunction we use the mark "-." In the mode of editing, the matrix form of the initial formula is displayed on the screen. In the program two opportunities for input of formulae are provided:

1. From a database; in this case, when choosing particular problems after pressing Enter, the user is informed how many variables and how many disjunctions the problem has;
2. The input the of formula; in this case, first we input the parameters of the problem: the number of variables and the number of disjunctions.

The program based on the original accelerated algorithm of transformation of CNF into orthogonal DNF, in which the value of conjunction of any pair of conjunction contained in it is "false." The number of conjunctions in the transformed problems is not known beforehand. Therefore, it is necessary to set this number. If it is simply required to solve the problem of feasibility, then it is enough to set the number 1. Then the program finds one conjunction, which contains some number of satisfying substitutions, or displays the message "Problem has no solution."

As a greater number of conjunctions is given, the program stops in the cases when:

1. It is the unsatisfiable formula;
2. The number of conjunctions in the completely transformed formula is less than the given number; then the number of conjunctions will be shown, equal to the given number; in this case, transformation of CNF into DNF remains uncompleted. In this case, for complete transformation of CNF into DNF in the following session with the given problem, it is necessary to set a greater number (for problems of large dimension, it is recommended to set 10,000). The examples of problems of large dimension are possible, for which in the given variant of the program the complete transformation is unattainable. It is because all obtained conjunctions are stored in fast memory (PAM) that allows to speed up performance of some operations;
3. If there is a lot of the intermediate data, the PAM can be overfilled. If it is necessary, this restriction can be taken off by allowing to keep intermediate data in disk memory, too.

In all cases, when given CNF is feasible, obtained conjunctions can be seen on the screen and written down in a file. It is also possible to write down to the file the new made problem or the problem loaded from the database if any changes were made in the mode of editing. The name of the file is chosen by the user, thus the program does not check the existence in the database of the file with the chosen name. If the name of the new problem coincides with the name of a problem already stored in the database, then the latter will be erased.

One of the essential factors of reduction of time of solution of the problem of feasibility and transformation of CNF into orthogonal DNF is the algorithm of sorting. It has polynomial complexity, but for problems of the large dimension (the number of variables is more than 30, the number of disjunctions is more than 100), for its realization some time is required (depending on the dimension of the problem and technical parameters of the computer), it takes from several seconds to several minutes. If the problem of calculation of probability is not solved, this algorithm can be switched off, but then it is possible that the process of the solution of the basic problem does not end. Besides, the use of algorithm of sorting for a problem of translation of CNF into DNF allows us to reduce general number of pairs of orthogonal conjunctions at the complete transformation of CNF into DNF.

The software can be used in any catalogue. It is desirable that this catalogue should contain a subdirectory with the name ZD, in which the texts of problems and solutions are written.

14.5 Description of software working on the ASM technology

Let us consider the history analysis, development, and the application of the theory and software, working on the technology of the automated structurally logic modeling (ASM). We use this technology for construction, estimation, and analysis of reliability of complex systems.

The ASM complexes essentially differ from other known program means of automated modeling. For the first time, we realize all opportunities of the modeling device of the logic algebra in the basis of the functionally full set of logic operations AND, OR, and NOT. It allows automatically to build all kinds of monotonous models and a new class of non-monotonic models of reliability, survivability, safety, and risk for structural-complex organizational and technical systems of the big dimension. It allows us to represent the researched system in any typical form of its structural description: the function chart, the series-parallel connection of elements, the tree of refusals, the tree events, the graph of the connectivity, etc.

At construction of the schemes of the functional integrity (SFI), the user can apply both the direct logic of reasonings and substantiations (serviceability, survivability, safety, efficiency of functioning systems) and the return logic of reasonings and substantiations (refusal, failure, risk of functioning systems).

The ASM theory was developed by A. S. Mojaev in the middle of 1980s. He offered the general logic and probabilistic method (GLPM) for the system analysis, which uses the special logically graphic device for the description of structures of properties of the reliability and the safety of researched systems. This device is the SFI.

A. S. Mojaev successfully defended the Ph.D. thesis "Development of logic and probabilistic methods for the automation of the modeling and calculation of the stability parameters of technical complexes of control systems" (1984) and the doctoral thesis "The theory of the automated structural-logic modeling of systems" (1997).

In time he accumulated the experience of the practical application of the ASM. Below we give examples of the following executed works:

1. The ASM technology was applied for the estimation of outcomes of possible emergencies during the raising of the nuclear submarine "Kursk";
2. The ASM (2001) (the prototype of ASM NWMA) is applied for:
   - the probability estimation of various consequences of nuclear and radiating failures, potentially possible at performance of the complex works on the recycling of nuclear waste products;
   - the calculation of reliability and risk at the preparation of computed-explanatory notes for the plans of the localization and liquidation of emergencies of 36 objects of the company "Kirishinefteorgsintez";
3. The first Russian ASM NWMA is applied for:
   - the parameters computing the reliability of the central storehouse of isotopes "Northern machine-building company," Severodvinsk, and also some other objects of atomic energy;
   - the definition techniques developing the residual resource of objects of the atomic energy "Northern machine-building company";
   - the estimation of reliability for six objects of "Kirishinafteorgsintez";
   - the reliability estimation of objects of the company "Mozyrsky NPZ";
   - the reliability estimation of two objects of company "Kazanorgsintez".

In 2004–2005, three companies SPb. AEP, IPU RAS, and "SPIK NWMA" executed the research work "Technology 2004," in which the opportunities of the new ASM NWMA technology have been successfully compared with two foreign program complexes:

- Risk Spectrum (Sweden), which is widely used for the probabilistic analysis of the safety of atomic power stations;
- Relex (USA), which is widely used in many countries for the estimation of the reliability.

The program complex "ARBITER" allows automatically to determine the shortest ways of successful functioning, the minimal sections of refusals and any of their non-monotonic combinations. The graphic interface realizes the principle of two-level decomposition (equivalences, aggregations). It means that the basic circuit of the SFI graph can include up to several hundreds tops, each of which, in turn, can contain up to hundred elements of circuits of subsystems of the second level. The complex automatically builds the calculated mathematical models of reliability and safety of researched systems in the form of exact polynomials of the calculated probabilistic functions.

15

LP-Model of Credit Risk for Natural Persons

The basic difficulties of introduction of the Agreement "Basel II" are connected to absence in banks of effective techniques of an estimation of credit risks of borrowers and the reservations, providing accuracy, stability and a transparency of an estimation credit risks.

Agreement "Basel II"

Credit activity is the basic form of commercial and national bank activity. Many new big, average, and small banks have appeared in Russia for crediting juridicial and natural persons. All of them are individual, because they serve various groups of society in different cities and regions of the country, and are companies of different branches and sizes with various forms of property. Competition promotes bank individuality, too.

Credit business is connected with the risk: no risk, no profits. Conditions of credit activity are changing, the level of the risk is changing also. Credit activity adapts for conditions of the developing economics of the country and for the living standards. Laws for ensuring stability and securing interests of creditors and banks are of primary importance. We would not have the successful developing economics with its circulatory system — the successful operational banks without this. Methods play the great part for estimation and analysis of the risk in conditions of the development of stability.

The base of the stable activity of the bank is the quantitative estimation and the analysis of the credit risk. Costs of the risk have to depend on the risk value of the credit. In addition to the average risk, which determines on statistic of former activity, every bank has to know the quantitative estimation and participating risk for every credit and for the average bank risk.

Every bank must have its credit risk model for the quantitative estimation of the risk of credits with the greatest accuracy. The more accuracy, the less losses of the bank, the less per cent for the credit, the more competitive ability of the bank. The whole society benefits from the increasing accuracy and transparency of the risk estimation. Creation of the effective risk model

E.D. Solojentsev, *Scenario Logic and Probabilistic Management of Risk in Business and Engineering*, DOI 10.1007/978-0-387-77946-1_15,

and optimal management of the credit risk is possible only on the basis of constant quantitative analysis of statistic data of successful credits.

These requirements to the credit risk model correspond with the risk LP-models in full measure, which have two times more accuracy, seven times more robustness, and absolute transparency in estimation and analysis of the risk in comparison with well-known models.

In the non-success risk LP-theory, in contrast with scoring methods, the risk is defined by the following quantitative attributes:

- failure probability (risk failure);
- admitted non-success probability (admitted risk);
- damage or losses (in yield or efficiency);
- admitted damage or losses (minimal admitted yield);
- global number of different states (or objects);
- number of the dangerous states (or objects).

15.1 Description of credit and data presentation

Credits of natural persons are described by features. Each has grades [3, 4, 30]. Number of features may be from 10 to 20. Number of grades in the feature may be from 2 to 11. Table 15.1 contains description of the natural personal

Table 15.1. Description of the credit for natural persons

Number of sign	Name of sign	Designation	Number of grades
0	Successful of credit	Y	2
1	Sum of account in bank	Z_1	4
2	Term of lend	Z_2	10
3	Credit history	Z_3	5
4	Purpose of lend	Z_4	11
5	Sum of lend	Z_5	10
6	Accounts on securities	Z_6	5
7	During works	Z_7	5
8	Payment in part discharge	Z_8	4
9	Family status and sex	Z_9	4
10	Joint obligations or guarantor	Z_{10}	3
11	Term of living in this region	Z_{11}	4
12	Type of guaranty	Z_{12}	4
13	Age	Z_{13}	5
14	Presence of another lend	Z_{14}	3
15	Presence of living space	Z_{15}	3
16	Number of lends with bank	Z_{16}	4
17	Occupation	Z_{17}	4
18	Number of relatives depend on	Z_{18}	2
19	Presence of telephone	Z_{19}	2
20	Foreign or native	Z_{20}	2

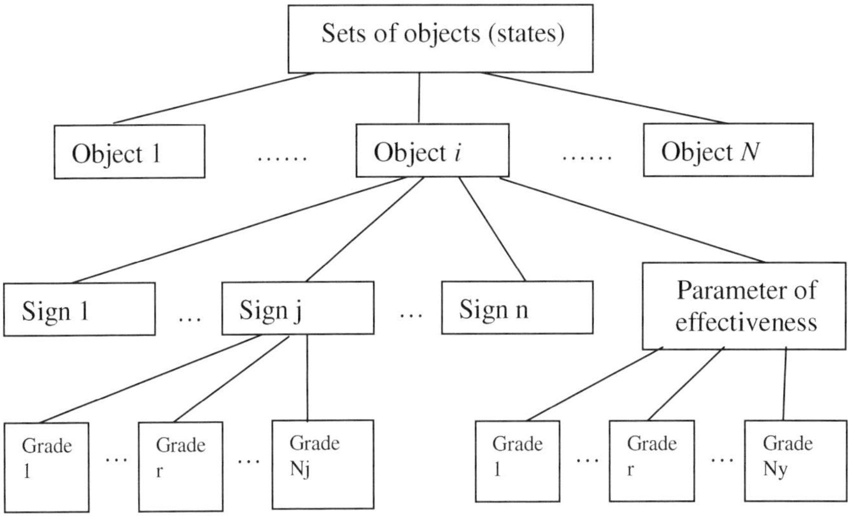

Fig. 15.1. Structure of data on credits in a bank

credit that is used in many investigations [97]. The data structure for credits is illustrated in Fig. 15.1. The basic elements of the structure are

- the set of bank credits,
- the credit,
- features for description of a credit and its efficiency,
- grades of features.

The maximum number of combinations (objects or states) is as follows:

$$N_{max} = N_1 \cdot N_2 \cdot \ldots \cdot N_j \cdot \ldots \cdot N_n \ , \tag{15.1}$$

where $N_1, N_2, \ldots, N_j, \ldots N_n$ are numbers of grades in parameters.

The statistic data for the risk analysis and evaluation are given in table form (Table 15.2) where the grades are in columns of the table.

Table 15.2. Credits and its signs

Number of credit i	Sign 1 Z_1	...	Sign j Z_j	...	Sign n Z_n	Sign efficiency of credit, Y
1						
...	...	...	...	...	...	...
i			Z_{jr}			
...	...	...	...	...	...	...
N						

15.2 Model of credit risk for natural persons and rice for risk

We adduce the regulations of the risk LP-theory with GIE, stated in Chapter 11, in conformity with the description of the credit risk model of natural persons and determining the price for risk.

The risk model. The credit risk model structure is presented in Fig. 15.2, where connections in the form of arrows are the logical one OR.

We shall consider the credit feature and its grades as random events: feature-events and grade-events. These events with certain probabilities lead to the failure of the credit. The scenario of the failure of the credit is associative and is formulated for the full set of events as follows: failure occurs if any (one or two),... or all feature-events occur. Let us mark that any scoring method cannot note this scenario with the help of mathematics.

The binary logic variable Z_j is equal to 1 with probability p_j, if the feature j leads to the failure, otherwise Z_j is equal to 0 with probability $q_j = 1 - p_j$. The binary logic variable Z_{jr}, corresponding with the grade r of the feature j is equal to 1 with probability p_{jr} , otherwise it is equal to 0 with probability $q_{jr} = 1 - p_{jr}$. Binary vector $Z(i)(Z_1, \ldots, Z_j, \ldots, Z_n)$ describes the object i from Table 15.2. In assigning the object i instead of the logic variables $Z_1, \ldots, Z_j, \ldots, Z_n$, it is necessary to substitute the logic variable Z_{jr} for grades of features for the object i.

We write down the risk failure L-function for the credit [3, 4, 27]

$$Y = Z_1 \vee Z_2 \vee \ldots \vee Z_n. \tag{15.2}$$

We write down the risk falure L-function for the credit in the equivalent orthogonalizaiton form as follows

$$Y = Z_1 \vee Z_2\ \overline{Z}_1 \vee Z_3\overline{Z}_2\ \overline{Z}_1 \vee \ldots. \tag{15.3}$$

Now we come from the logic description of the risk failure to the arithmetic one. P-model (P-polynomial) of the risk failure of credit is

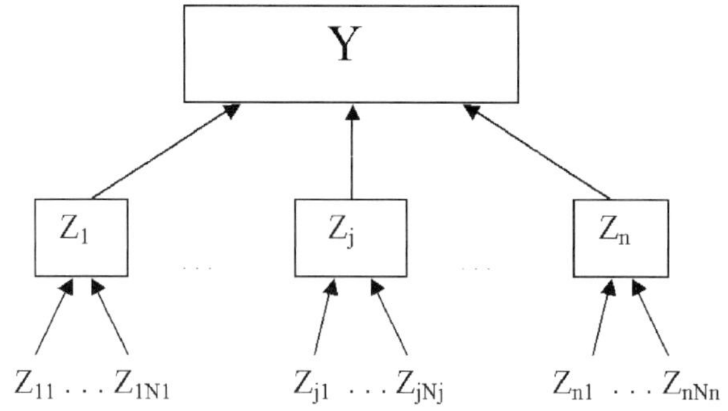

Fig. 15.2. Structural model of credit risk

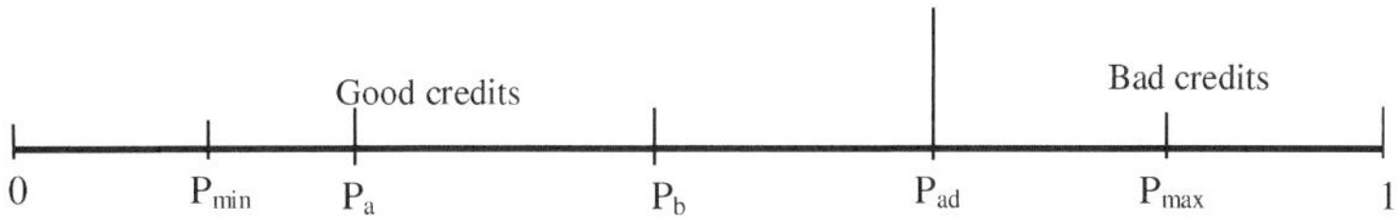

Fig. 15.3. The scheme of classification of objects by risk into two classes

$$P = p_1 + p_2 \cdot q_1 + p_3 \cdot q_1 \cdot q_2 + \ldots. \tag{15.4}$$

"Arithmetic" of risk L-model is such that for final event, the value of risk belongs to the range [0,1] at any values of probabilities of initiating events.

For all gradation-events in GIE, we shall consider three probabilities: $P2_{jr}$ is relative frequency of the grad in credits objects in Table 15.2; $P1_{jr}$ is probability of gradation-events in the GIE; P_{jr} is the probability of grad-events, substituted into the formula (15.4) instead of probability P_j. We defined these probabilities in Chapter 11.

We estimate the probabilities P_{jr} of algorithmic iterative identification of the risk P-model by statistic data in Table 15.2. In the beginning, it is necessary to determine the probabilities $P1_{jr}$ and further to pass from the probabilities $P1_{jr}$ to the probabilities P_{jr}. The number of the estimated independent probabilities P_{jr} is equal to:

$$N_{ind} = \sum_{j=1}^{n} N_j - n. \tag{15.5}$$

The connection of the probabilities P_{jr} and $P1_{jr}$ for the grades is expressed through the mean values of the probabilities P_{jm} and $P1_{jm}$ (11.32):

$$P_{jr} = P1_{jr} \cdot \frac{P_{jm}}{P1_{jm}} = K_j \cdot P1_{jr}. \tag{15.6}$$

Measure and cost of risk. Let us introduce an admitted risk P_{ad} separating the objects into good and bad: if $P_i > P_{ad}$, then the object is bad; if $P_i < P_{ad}$, the object is good (Fig. 15.3). If the objects are classified into a greater number of classes, then a corresponding number of admitted risks: $P_{ad1}, P_{ad2}, \ldots$, is introduced. Let assume that the probabilities of grade-events $P_{jr}, j = 1, 2, \ldots, n$; $r = 1, 2, \ldots, N_j$ are known. Then, from the risk P-model we calculate risks of all N objects of Table 15.2. We plot these risks on the risk axis. If the resulting event Y has two grades (Fig. 15.3), we choose the admitted risk P_{ad} so that N_b from N objects are bad and N_g are good. For the object i, the distance between the risk P_i and the admitted risk P_{ad} is a natural measure of its being good or bad:

$$d_i = |P_i - P_{ad}|. \tag{15.7}$$

The above measures are used to calculate the cost of risk, for example, rate of the credit. The simplest formula of the risk costs is

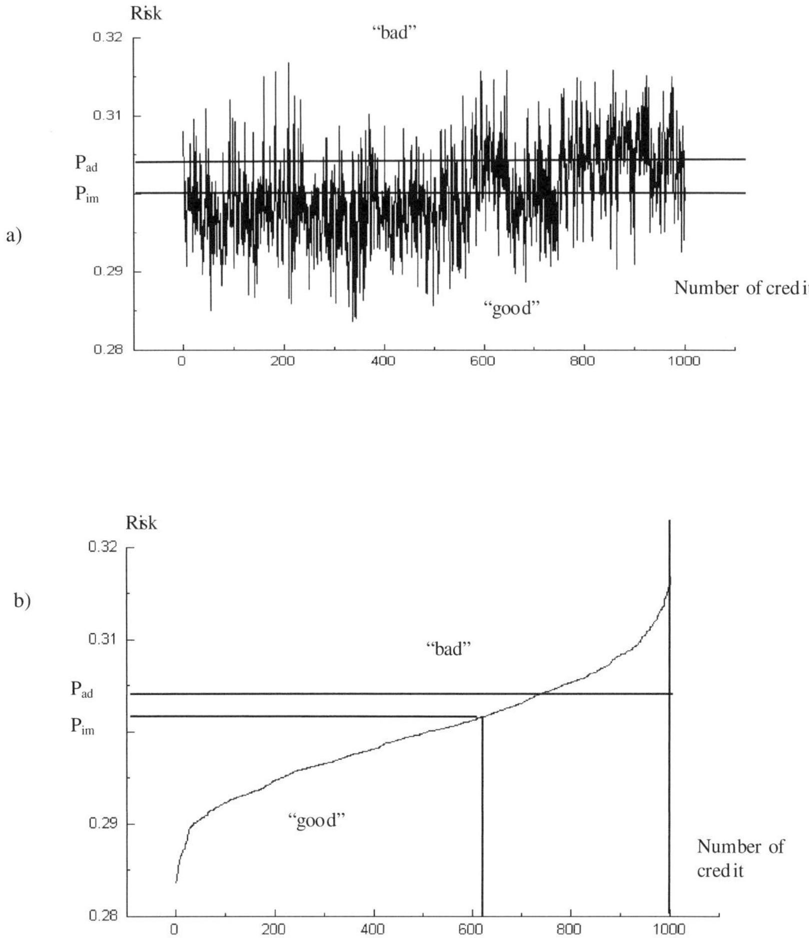

Fig. 15.4. Risks of credits, classification of credits, and price for risk

$$C_i = C_{ad} + C \cdot (P_i - P_{ad}), \tag{15.8}$$

where the costs of possible risk C_{ad} and the coefficient C are chosen by the bank on the basis of the market conditions.

The introduction of the price of credit risk can be seen if we build the graph of risks for 1000 credits before and after sorting objects by risks (Fig. 15.4). Approximately 15% of credits are very good, whereas 15% of credits are very bad, which naturally makes us believe that the price of credit should depend on its risk. Dependance must be more complex than linear one.

15.3 Identification of risk LP-model and analysis of risk

We adduce the regulations of the risk LP-theory with GIE, stated in Chapters 12 and 13, in conformation with the identification of the risk LP-model and with the analysis risk of credits. **Identification**. The following scheme of the problem solution is proposed. Suppose that the first approximation for probabilities of grades P_{jr}, $r = 1, 2, \ldots, N_j$; $j = 1, 2, \ldots, n$ are known and the risks P_i, $i = 1, 2, \ldots, N$, of objects of the Table 15.2 are calculated. We shall determine the admitted risk P_{ad} so as to have the given number of good objects N_{gc} with the risk less than the admitted one and, accordingly, the number of bad objects $N_{bc} = N_g - N_{gc}$ with risk more that the admitted one. On the step of optimization it is necessary to change probabilities P_{jr}, $r = 1, 2, \ldots, N_j$; $j = 1, 2, \ldots, n$, in such a way that the number of correctly recognized objects F increases.

It is notable that variables P_{ad} and N_{gc} are unambiguously related. In the problem solution algorithm, it is more convenient to set N_{gc} and to determine the admitted risk P_{ad}, because the latter would have to be set with the precision 6–7 digits after the decimal point.

The problem of identification of the risk P-model is formulated as follows.

Specified data: the Table 15.2 with N_g good and N_b bad credits and the risk P-model (15.4) are given;

Expected results: probabilities P_{jr}, $r = 1, 2, \ldots, N_j$; $j = 1, 2, \ldots, n$ for grade-events and the admitted risk P_{ad}, separating the objects into good and bad ones based on their risk, should be determined;

We need: to maximize the criterion function (CF), which is the number of correctly classified objects:

$$F = N_{bb} + N_{gg} \Rightarrow \max_{Pjr} , \tag{15.9}$$

where N_{gg}, N_{bb} are the numbers of objects classified as good and bad using both the statistics and the risk P-model (both estimates should coincide).

From (15.9) follows that errors or accuracy indicators of the risk P-model in the classification of good E_g and bad E_b objects and in classification of the whole set E_m are equal:

$$E_g = N_{gb}/N_g; \; E_b = (N_{bg}/N_b; \; E_m = (N - F)/N . \tag{15.10}$$

Imposed restrictions:

(1) Probabilities P_{jr} and $P1_{jr}$ have to satisfy the condition:

$$0 < P_{jr} < 1, \quad j = 1, 2, \ldots, n; \quad r = 1, 2, \ldots, N_j; \tag{15.11}$$

(2) The average risks of credits on the risk P-model and on Table 15.2 must be equal; while training the risk P-model, we should correct the probabilities P_{jr} on every step of iterative training:

$$P_{jr} = P_{jr} \cdot (P_{av} \ / \ P_m); \ r = 1, 2, \ldots, N_j; \ j = 1, 2, \ldots, n \ ; \tag{15.12}$$

(3) The admitted risk P_{ad} should be determined at the given ratio of incorrectly classified good and bad objects (dissymmetry of recognition), in view of non-equivalence of losses by their wrong classification:

$$E_{gb} = N_{gb} \ / \ N_{bg}. \tag{15.13}$$

Analysis of credit risk and credit activity of a bank. Let the risk P-model be trained and the probabilities of grade-events P_{jr} be known. In order to carry out analysis, we determine contributions of sign-events and grade-events in the object risk and mean risk of a set of objects, as well as the accuracy of the risk LP-model. We shall determine the contributions with the computer program by calculating the differences between the values of the above mentioned characteristics in the optimal mode and those obtained (|) for the zero probabilities of the grade-events [3, 4, 30]. The contribution of the sign (all sign grades) to the risk of the credit i is

$$\triangle P_j = P(i) - P(i) \left|_{P_j=0} \right. ; \ j = 1, 2, \ldots, n. \tag{15.14}$$

The contribution of the feature to the mean risk P_m of the set of credits

$$\triangle P_{jm} = P_{jm} - P_{jm} \left|_{P_j=0} \right. , \ j = 1, 2, \ldots, n. \tag{15.15}$$

The contributions of grades to the mean risk P_m of the set of credits is

$$\triangle P_{jrm} = P_{jm} - P_{jm} \left|_{P_{jr}=0} \right. , \ j = 1, 2, \ldots, n; \ r = 1, 2, \ldots, N_j. \tag{15.16}$$

The contribution of the feature to the criteria function F_{max} is

$$\triangle F_j = F_{\max} - F \left|_{P_j=0} \right. ; \ j = 1, 2, \ldots, n. \tag{15.17}$$

The contribution of a grade to the criteria function F_{max} is

$$\triangle F_{jr} = F_{\max} - F \left|_{P_{jr}=0} \right. , \ j = 1, 2, \ldots, n; \ r = 1, 2, \ldots, N_j. \tag{15.18}$$

We note that the calculation of the contributions of the grade-events to the mean risk and the criteria function ($\triangle P_{jrm}$ and $\triangle F_{jr}$) is incorrect because it is not known how to correct the relative frequencies of other grades $P2_{jr}$ in GIE if one of them is set to zero. Therefore, by analogy with (12.2), instead of the contributions $\triangle F_{jr}$ one should calculate the errors of credit classification for each grade-event:

$$E_{jrg} = (N_{jrg} - N_{jrgg}) \ / \ N_{jrg}; \quad E_{jrb} = (N_{jrb} - N_{jrbb}) \ / \ N_{jrb};$$
$$E_{jrm} = (N_{jr} - N_{jrgg} - N_{jrbb}) \ / \ N_{jr}, \tag{15.19}$$

where the N_{jrg}, N_{jrb}, N_{jr} are the numbers of good, bad, and all credits with grade; N_{jrgg}, N_{jrbb} are, respectively, the numbers of credits with the grade r and correct classification.

15.4 Transparency of assessment method of credit risk

Transparency of the risk scenario. The scenario of the failure credit is associative and is formulated for full set of events as follows: failure occurs if any (one or two),...or all feature-events occur. Let us mark that any scoring method cannot note this scenario with the help of the mathematics. Using this scenario, as presented above, we describe the logic risk function (15.2) and then the probabilistic function of the risk failure (15.4).

Transparency of the criterion function. The formulation of the criteria function is remarkable for simplicity and transparency: the number of corrected recognized good and bad credits must be maximal (15.9). Corrected classification means, that estimations of credits on statistics and on the risk LP-model are coincided. In contrast with the other methods, here do not use building separating hyperplane for classification or least-squares method, or maximum-likelihood method or expert system, which do non-transparency of estimation and analysis of risk.

Simple choice of initiating conditions. Some methods to specify the initiating values of probabilities of grade-events are possible:

- probabilities P_{jr}, $r = 1, 2, \ldots, N_j$; $j = 1, 2, \ldots, n$ are the same and equal to the mean value, easily computed by value of the average risk of credits P_{av} in statistics;
- probabilities $P1_{jr}$, $r = 1, 2, \ldots, N_j$; $j = 1, 2, \ldots, n$ equal to the average value in the group of incompatible events, easily computed by the number of grades in the group;
- probabilities $P1_{jr}$ are computed by frequencies of grade-events in statistic data;
- probabilities $P1_{jr}$, P_{jr} are given from previous session of optimization.

The average risk value in the model P_m should be equal to the average risk value by statistics, and the initiating value of the criterion function F should be equal to half of the number of credits in statistics.

Simple estimation of accuracy and robustness. The estimates of accuracy of the risk LP-models have the optimal properties: efficiency, consistent, and unbiasedness. This follows from the direct (no indirect) criterion function (maximum of the number of correctly recognized credits) and from its accurate determination as integer number.

The estimation of accuracy of determination of probabilities $P1_{jr}$, $r = 1, 2, \ldots, N_j$; $j = 1, 2, \ldots, n$ of grade-events at identification of the risk LP-model is known. They also possess the optimal property. This follows from corrected determination of the largest error $\triangle P1_{min}$ and its possible values on the interval $[\triangle P1_{min}, 0]$. In considered variants of account, this error was no more than 0.25% from value $P1_{jr}$.

Different methods of risk assessment (or one method with different algorithms of training on statistical data) classify the credits to good and bad ones in different ways. A credit can be recognized as bad by one method and as good by another. Estimation of robustness (stability) of the risk P-model on the classification of credits is conducted by the method of comparison of the results of classification by different models or in different variants of optimization. The comparison is conducted on the number of the odds of estimation (good and bad ones).

New tasks. The failure credit risk LP-model allows one to decide new important tasks for analysis and management of the risk (instead of two methods in scoring), in particular:

1. Determination the risk of the credit:
   - estimation of the credit risk;
   - classification of the credit as good or bad;
   - setting price for risk;
   - analysis of the credit risk;
   - reservation.
2. Analysis of the credit activity of the bank:
   - determination of investments signs and grades of credit signs in the average credit risk of the bank;
   - determination of contributions signs and grades of credit signs in the accuracy of classification of credits;
   - decision of reservation the task for risks of credits;
   - optimization of the numbers of signs and grades of signs;
   - fragmentation on intervals: sum and term of credit, age of client, etc.
3. Training (identification) and estimation of quality of the risk LP-model:
   - statistical analysis of the risk model;
   - combinatorial analysis of the risk model;
   - estimation of probabilities of grad-events and the admitted risk;
   - estimation of accuracy of the risk LP-model;
   - estimation of robustness of the risk LP-model;
   - choose of optimal relation of incorrectly qualified good and bad credits (asymmetry recognition).

15.5 Transparency of results of risk assessment and analysis

Transparency of the results is demonstrated with the examples of analysis of the risk model and the credit activity of the bank with expressions (15.14–15.19). The probabilistic risk model is trained by the statistical data [97] on 1000 credits of natural persons, from which 700 were good and 300 bad ones.

Choice of the admitted risk. The admitted risk P_{ad} and the computed number of good and bad credits N_{gc} are determined on the given relation of mistakenly classified good and bad credits E_{bg} from non-equivalence of

Table 15.3. Results on choice of the admitted risk $\boldsymbol{P_{ad}}$

N_{gc}	0	550	580	610	650	700	750	800	1000
F_{max}	300	767	787	809	825	829	831	819	700
N_{gg}	0	508	533	559	587	614	640	659	700
N_{bb}	300	259	254	250	238	215	191	160	0
E_g	1.0	0.274	0.238	0.201	0.161	0.123	0.10	0.058	0
E_b	0.0	0.137	0.153	0.167	0.207	0.283	0.363	0.466	1.0
E_m	0.7	0.233	0.213	0.191	0.175	0.171	0.169	0.181	0.3
P_{ad}		0.3010	0.3012	0.3016	0.3021	0.3031	0.3039	0.3049	
E_{gb}		4.55	3.53	2.65	1.77	1.0	0.51	0.213	

losses at its mistaken identification (15.13). The relation of these numbers is considered as given. For case of credit risks, its value E_{gb} equals from 2 to 10. The needed value E_{gb} is determined after some calculations at different N_{gc}.

As an example of determination of E_{gb}, we shall present results of training the risk P-model of the physical persons on statistics of 1000 credits, from which 700 were good and 300 were bad. For the risk model training, different numbers of good credits were used: N_{gc} = 550, 580, 610, 650, 700, 750, 800 (Table 15.3). We computed risks P_i of each of 1000 credits; maximum of the criterion function F_{max}, numbers of correctly recognized objects N_{gg} and N_{bb}, errors in recognition of credits E_g, E_b, E_m, the admitted credit risk P_{ad}; asymmetric recognition of good and bad credits E_{gb}.

Parameters of the variant $N_{gc} = 610$ are $E_b = 0.167, E_g = 0.201, E_m = 0.191$. Parameters of the variant the variant $N_{gc} = 650$ are $E_b = 0.207$, $E_g = 0.161$, $E_m = 0.175$, respectively. The optimal variant is established by the parameter of asymmetry $E_{gb} = 2.65$. The variants $N_{gc} = 610$ and $N_{gc} = 650$ have symmetrically different mistakes of evaluations E_b and E_g (Fig. 15.5). Variants $N_{gc} = 700$ and N_{gc}=750, though having a greater value of the criterion function, cannot be accepted as optimal ones because of the large error of recognition of the bad credits E_b. We choose variant $N_{gc} = 610$ as optimal.

Cost of risk. The price for the risk (present for credit) is computed on the simple formula (15.8) in dependence on the price for the average risk and difference between the risk of the credit and the admitted risk.

The optimal number of signs for description of the credit. For every sign j after optimal training, the risk model are determined (Table 15.4): the average values of probabilities $P1_{jm}$ and P_{jm}, and decreasing the number of correctly recognized good and bad credits $\triangle F_j$ at eliminating this sign from the risk model. The risk model after this changing is training again.

According to the obtained results, the maximum contribution to the accuracy of the credit recognition is brought with the sign-events Z_1, Z_2, Z_4, Z_5, Z_6, Z_3, Z_{13}. The zero contribution is brought with the sign-events Z_{11}, Z_{12}, Z_{17}, Z_{18}, Z_{19}; excluding these sign-events (last line in Table 15.4) reduces the number of identified credits by two. These sign-events can be used for

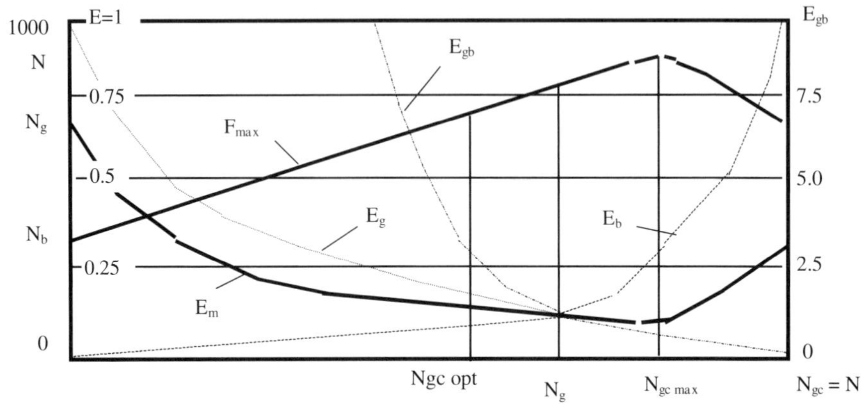

Fig. 15.5. Choice of the admitted risk P_{ad}

Table 15.4. Analysis of contributions of sign-events into accuracy of the credit risk mode

Signs, j	Number of grades, N_j	$P1_{jm}$	P_{jm}	K_j	$\triangle F_j$
1	4	0.272384	0.020226	0.074255	–64
2	10	0.063346	0.012359	0.195102	–27
3	5	0.098475	0.009327	0.094713	–18
4	11	0.090820	0.020927	0.230421	–26
5	10	0.080377	0.017593	0.21888	–20
6	5	0.272148	0.022466	0.082550	–20
7	5	0.206945	0.018549	0.089632	–6
8	4	0.266619	0.017736	0.066521	–6
9	4	0.183897	0.014253	0.077505	–10
10	3	0.318015	0.018295	0.057528	–10
11	4	0.251871	0.018974	0.075331	0
12	4	0.247375	0.017166	0.069392	0
13	5	0.206718	0.018900	0.091428	–16
14	3	0.235637	0.014733	0.062524	–2
15	3	0.261648	0.017591	0.067231	–8
16	4	0.341959	0.021975	0.064261	–2
17	4	0.289853	0.018739	0.064649	0
18	2	0.482499	0.017417	0.036097	0
19	2	0.508613	0.018138	0.035661	0
20	2	0.750896	0.018326	0.024405	–2
11+12+17+18					–4

the change of the logic structure of the risk model with the purpose to increase its accuracy. For this it is necessary to replace the sign-events by their combinations with negations of some variables [3].

Table 15.5. Analysis of contributions of grade-events into accuracy of the credit risk model

$P2_{jr}$	$P2_{0jr}$	$P2_{1jr}$	$P1_{jr}$	P_{jr}	E_{jr}	$E1_{jr}$	$E0_{jr}$	$\triangle F_{jr}-$	$\triangle F_{jr}+$
Term of credit ($j = 2$)									
0.014	0.007	0.007	0.01	0.019	0.214	0.429	0.0	0	0
0.002	0.001	0.001	0.070	0.014	0.5	1.0	0.0	–2	0
0.054	0.032	0.022	0.194	0.038	0.278	0.682	0.0	–20	0
0.017	0.005	0.012	0.159	0.031	0.412	0.5	0.2	–10	0
0.086	0.038	0.048	0.145	0.028	0.256	0.417	0.053	–8	0
0.057	0.019	0.038	0.095	0.019	0.228	0.289	0.105	–16	0
0.224	0.066	0.158	0.067	0.013	0.169	0.196	0.106	–4	0
0.187	0.056	0.131	0.053	0.010	0.203	0.183	0.250	–22	–2
0.359	0.076	0.283	0.016	0.003	0.114	0.081	0.237	0	0
0.0	0.0	0.0	0.1	0.019	0	0	0		
Age of client ($j = 13$)									
0.190	0.080	0.110	0.283	0.027	0.237	0.345	0.087	–18	–14
0.511	0.142	0.369	0.233	0.021	0.186	0.201	0.148	–26	–26
0.248	0.065	0.183	0.093	0.008	0.113	0.082	0.200	–12	–4
0.028	0.007	0.021	0.346	0.032	0.178	0.238	0.0	0	0
0.023	0.006	0.017	0.044	0.004	0.217	0.117	0.5		

Contributions of grade-events into accuracy of the credit risk model. For grade-events (Table 15.5), constructed on intervals of the change of the credit time Z_2 and the client age Z_{13}, for the optimal trained risk LP-model ($F_{max} = 826$; $N_{gc} = 610$) the following parameters are found:

1. Frequencies $P2_{jr}$, $P20_{jr}$, $P21_{jr}$ of grades for all, bad and good credits, respectively;
2. Probabilities of grade-events $P1_{jr}$ and P_{jr};
3. Recognition errors in grades E_{jr}, $E0_{jr}$, $E1_{jr}$ for all bad and good credits, respectively;
4. Changes of the criterion function $dF_{jr}+$ and $dF_{jr}-$ at correction of the grade interval lengths by $\pm$ 25%; the model after such change was not retrained, and the criterion function after one step of calculations was fixed. At that, the lengths of the two next intervals change. Values of changes are placed in the line for the first interval.

On the results of research (Table 15.5), it is possible to make the conclusions:

1. Contributions of sign-events to the credit risk are proportional to values of probabilities P_j, $j = 1, \ldots, n$, equal to grade-event probabilities P_{jr};
2. The change of interval lengths of some grades reduces the criterion function of the risk LP-model up to $\Delta F_{jr} = 20$ units (from 1000 objects);
3. The change of intervals of some grades does not reduce the accuracy of the risk LP-model (ΔF_{jr} =0). It means that at retraining, the criterion function of the risk LP-model can be increased to exceed F_{max}=826.

The optimal number of grade-events N_j in a sign. The accuracy changes of the risk LP-model at the change of the number of grades in a sign was investigated for the first sign (the time of the credit), which in the initial variant had 10 grades. After retraining the risk model, the following results were obtained: in the absence of the sign $F_{max} = 800$; with two grades $F_{max} = 812$; with four grades $F_{max} = 812$; with ten grades $F_{max} = 824$; with hundred grades (in that case there was seventy empty grades) $F_{max} = 828$. Each sign of the credit had an optimum number of grades in certain time.

As a result of the carried out analysis of the credit risk LP-model, it is established that the criterion function can be increased by the structural identification methods up to $F_{max} = 840 \div 845$, it means that the average error in classification of credits can be reduced to $E_m = 0.155 \div 0.15$.

15.6 Comparison of LP method by accuracy and robustness with other methods

Accuracy. For comparison, the different methods used standard statistical data including N = 1000 credits of physical persons, of which $N_g = 700$ are good and $N_b = 300$ are bad [3, 97]. The credit is described by $n = 20$ signs, which have 94 grades in sum.

The risk LP-model (15.2), (15.4) is training without taking into account GIE and with taking into account GIE and with the structural identification (var. 1, 2, and 3 in Table 15.6).

Parameters of the risk LP-model with GIE after the training are $F_{max} = 826$; $N_{gg} = 568$; $N_{bb} = 258$; $N_{gc} = 610$; $E_{gb} = 3$ is asymmetry of recognition; $P_{min} = 0.230$ and $P_{max} = 0.379$ are the minimal and maximal risks among the credits; $P_{ad} = 0.306$; $P_{av} = 0.3$ is the average risk on statistics.

Table 15.6. Accuracy of classification of credits by different methods

Used method	Error of classification of bad objects, E_b	Error of classification of good objects, E_g	The average error, E_m
LDA Resubstitution	0.26	0.279	0.273
LDA Leaving-one-out	0.287	0.291	0.29
QDA Resubstitution (QDA)	0.183	0.283	0.253
QDA Leaving-one-out	0.283	0.34	0.323
CART	0.273	0.289	0.285
Neuron networks 1 (NN)	0.38	0.24	0.282
Neuron networks 2 (NN)	0.24	0.312	0.29
LP-model without GIE (Var. 1)	0.167	0.201	0.191
LP-model with GIE (Var. 2)	0.1433	0.190	0.176
LP-model with GIE and after structure identification (Var. 3)	0.126	0.174	0.155

Table 15.7. Pairwise comparison of stability of different models

With GIE				Without GIE			
Variants	n_g	n_b	n_{gb}	Variants	n_g	n_b	n_{gb}
1 and 2	9	9	18	1 and 2	80	80	160
2 and 3	9	9	18	1 and 3	45	45	90
1 and 3	8	8	16	2 and 3	68	68	136
				1 and 4	60	60	120
				2 and 4	76	76	152
				3 and 4	50	50	100

The comparison of different methods of risk estimation is carried out with accuracy indicators of classification objects E_m, E_g, E_b using standard statistical data. This data was used to evaluate accuracy of nearly 10 different classification method [3, 97] based on linear (LDA) and quadratic (QDA) discriminant analysis, cluster analysis (CARD), and neuron networks (NN). In credit classification, the credit risk LP-model has essentially smaller mistakes $E_m = 0.174$; $E_g = 0.189$; $E_b = 0.143$, than the known techniques, which give $F_{max} = 750 \div 720$; $E_m = 0.25 \div 0.28$.

The results of accuracy comparison of different methods on the same statistical data (Table 15.6) show that the risk LP-model is almost 2 times more accurate than are other classification methods.

Robustness. Different risk models classify the credits to good and bad ones differently. One of the two different risk models may classify a credit as good, whereas another may classify it as bad. A stability (robust) estimate for the risk P-model for classification was carried out, using the data and results from above given examples, by the method of comparison (in pairs) of different variants of the solution in the credit classification. Difference of criterion functions F_{max} reach 10 units. The comparison was based on the number of inconsistencies of the estimates of good n_g, bad n_b, and in unrecognized n_{gb} credits (Table 15.7).

During the training of the risk P-model with GIE, three different solutions are obtained. The stability indicator equals

$$K_{s1} = (18 + 18 + 16)/(1000 \cdot 3) = 0.018.$$

During the training of the risk P-model without GIE, four different solutions are obtained. The stability indicator equals

$$K_{s2} = (160 + 90 + 136 + 120 + 152 + 100) \ / \ (1000 \cdot 6) = 0,128.$$

The ratio of the stability indicator for the risk P-models with GIE and for that without GIE is as follows:

$$K_{s2}/K_{s1} = 0.128/0.018 = 7.1.$$

Thus, the risk LP-model has seven time more robustness in classification of credits than do well-known methods.

The obtained result can be generalized to the instability question for risk models based on neuron networks, where a large number of weights of net links is introduced without any restrictions and GIE. Non-robustness of the methods of the risk estimation based on neuron networks has been noted in some papers [97].

15.7 Investment at LP-model with data of the real bank

We adduce results of investment of the credit risk with the statistic of the real commercial bank PCB below. Of 419 credits, the bank method did not recognize 73 bad credits, while the LP-method only failed to recognize 27 bad credits. We set the most important signs for description of credits of the bank and signs that can eliminate. We also set signs bringing the most and the least contributions in the average value of the credit risk of the bank.

Description of the credit. Credits of natural personals in the commercial bank is described by signs and grades. The sign of the success of credit is Y (2 grades). Credit signs are: Z_1 is the term of the credit (4 grades), Z_2 is the sum of the credit (6), Z_3 is the aim of the credit (3), Z_4 is the credit history in the bank (3), Z_5 is presence of rouble or currency deposit (4), Z_6 is belonging the credit cards of the bank (4), Z_7 is the living condition (3), Z_8 is presence in property of expensive property (3), Z_9 is the age of client (3), Z_{10} is the time of constant living in Saint Petersburg (3), Z_{11} is the official level (4), Z_{12} is the stability of employment or work time in this company (4), Z_{13} is income in the basic place of work (5), Z_{14} is the number of idle members of the family (3).

We use statistics on credits of natural persons of banks. Data for signs and grades are taken from the credit file for every borrower. As this time the balls are replaced by grades. The quantitative data (term, sum of credit) were divided on intervals, where grade number were assigned.

For every credit, the sign Y equals to grade 1 (good)) or grade 0 (bad). In results we obtain the table in which lines are credits, in the first column is the grade of the success of the credit, and in other columns are grades for the sign in every credit. The data structure for credits is presented in Fig. 15.1.

We made the program in *Visual C++*, which allows us to decide all tasks for estimation and analysis of the credit risk of natural persons of PCB bank.

The bank statistics has $N = 419$ credits, from which $N_g = 346$ are good and $N_b = 73$ bad. Results of investment are presented in Table 15.8. In formula (12.15) for computation of increment $\triangle P1_{jr}$ we change: the coefficient $K_1 = 0.02 \div 0.1$ and the number of optimization $N_{opt} = 100 \div 400$. We obtain the following results: the criterion function $F_{max} = 343 \div 348$; the error of recognizing bad credits $E_b = 0.5068 \div 0.5479$; the error of recognizing good

Table 15.8. Results of searching the global extreme

Variant	N_{opt}	K_1	F_{max}	E_b	E_g	E_m	E_{bg}
1	300	0.1	346	0.52	0.1069	0.1789	0.9736
2	300	0.05	348	0.5068	0.0982	0.1694	0.9189
3	100	0.1	343	0.5479	0.1069	0.1837	0.925
4	100	0.05	346	0.5068	0.1098	0.1789	1.027
5	200	0.075	344	0.5753	0.1069	0.1885	0.8809
6	400	0.02	348	0.5205	0.1040	0.1766	0.9473

credits $E_g = 0.0982 \div 0.1098$; the error of recognizing in average $E_m = 0.1694 \div 0.1885$; asymmetry of recognizing good and bad credits $E_{gb} = 0.9189 \div 1.027$.

As optimal parameters can be taken the following training parameters and attributes of the trained credit risk LP-model: $K_1 = 0,02$; $N_{opt} = 400$; $F_{max} = 348$; $E_b = 0.5205$; $E_g = 0.1040$; $E_m = 0.1766$; $E_{gb} = 0.9473$. For this variant, we give histograms of distribution of all credits (black color) and wrongly recognized credits (white color). It is obvious that distributions are not normal ones and have gaunt left "tail" (Fig. 15.6).

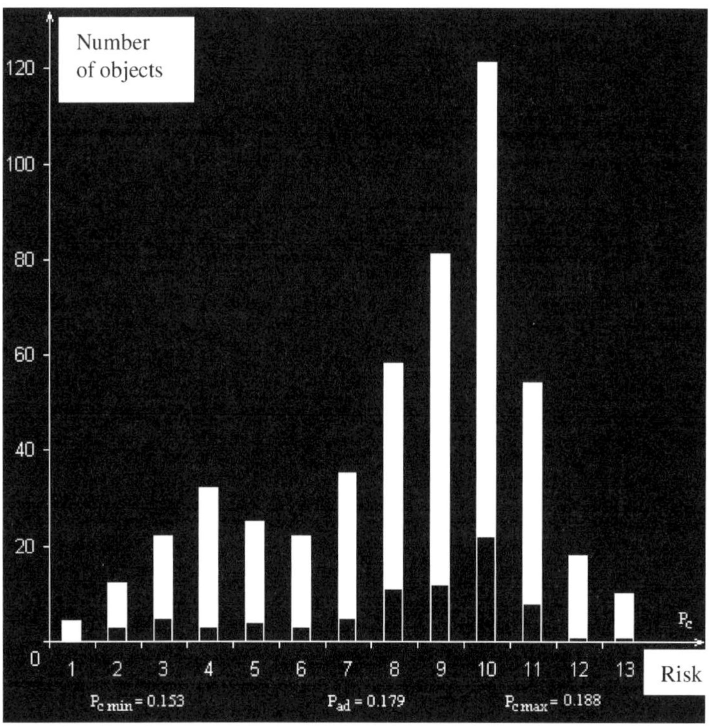

Fig. 15.6. Histograms of distribution of credits on risk

Table 15.9. Investment on choice of the admitted risk P_{ad}

Var.	F_{max}	N_b^{in}	N_g^{in}	E_b	E_g	E_m	E_{gb}	$\triangle P_c$	P_{ad}	P_m	N_{bc}	N_{gc}
1	296	10	114	0.178	0.3179	0.2935	11.40	0.0391	0.177	0.174	179	240
2	314	17	89	0.2191	0.263	0.2553	5.235	0.0305	0.177	0.174	154	265
3	329	21	71	0.2739	0.2109	0.2219	3.381	0.0243	0.177	0.174	129	290
4	339	26	57	0.3698	0.156	0.1933	2.192	0.0267	0.1778	0.1742	104	315
5	345	35	42	0.4794	0.1127	0.1766	1.200	0.0330	0.1789	0.1740	79	340
6	350	46	24	0.6301	0.0635	0.1622	0.522	0.0279	0.1791	0.1740	54	365
7	351	59	11	0.7534	0.0346	0.1599	0.186	0.0287	0.1813	0.1743	29	390
8	348	37	39	0.5342	0.0953	0.1718	1.054	0.0323	0.1780	0.1741	73	346
9	339	27	54	0.3561	0.1589	0.1933	2.000	0.0259	0.1773	0.1744	104	315
10	329	24	68	0.3287	0.1907	0.2147	0.353	0.0266	0.1777	0.1742	119	300

However, the named variant of tooling the credit risk LP-model is not acceptable, because the ratio of the number of correctly recognized good and bad credits E_{gb} (asymmetry of recognition) is less than 1 and non-admitted. Therefore there were made the investments on choice of acceptable computed number good and bad credits N_{gc} and corresponding admitted risk P_{ad}. Results of this investigation is presented in Table 15.9.

Investments are carried out at $N_{opt} = 400$, $K_1 = 0.02$ and changing computed number of good and bad credits $N_{gc} = 240 \div 419$. As this we changed the characteristics of the model the following way: criterion function $F_{max} = 296 \div 351$; the number of non-correctly recognized good credits $N_g^{in} = 11 \div 114$; the number of non-correctly recognized bad credits $N_b^{in} = 10 \div 59$; the error of recognition of bad credits $E_b = 0.178 \div 0.753$; the error of recognition of good credits $E_g = 0.0346 \div 0.3179$; the error of recognition in average $E_m = 0.1599 \div 0.2935$; asymmetry recognition good and bad credits $E_{gb} = 0.186 \div 11.4$; the difference between maximum and minimum risks $\triangle P_c = 0.0243 \div 0.0391$; the admitted risk $P_{ad} = 0.1777 \div 0.1813$; the average risk $P_m = 0.1740 \div 0.1744$.

Investment results are explained in Fig. 15.7 and show that with the increasing of the computed number of the good credits N_{gc}, the number of correctly recognized credits N_{gg} increases and aims to its value on statistics. The number of correctly recognized bad credits N_{bs}, contra, decreases and aims for zero.

Investment results explained in Fig. 15.8 show that with the increasing of the computed number of the good credits N_{gc}, the error of recognition of bad credits E_b increases and aims to 1, the error of recognition of good credits E_g decreases and aims to zero, the error of recognition in average E_m decreases and aims to average value on statistics of the bank, and the asymmetry of recognition good and bad credits E_{gb} decreases.

As the optimal we take the variant . 9: $N_{gc} = 315$; $F_{max} = 341$; $N_g^{in} = 54$; $N_b^{in} = 27$; $E_b = 0.3561$; $E_g = 0.1589$; $E_m = 0.1933$; $E_{gb} = 2.115$; $\triangle P_c = 0.0259$; $P_{ad} = 0.1773$; $P_m = 0.1744$.

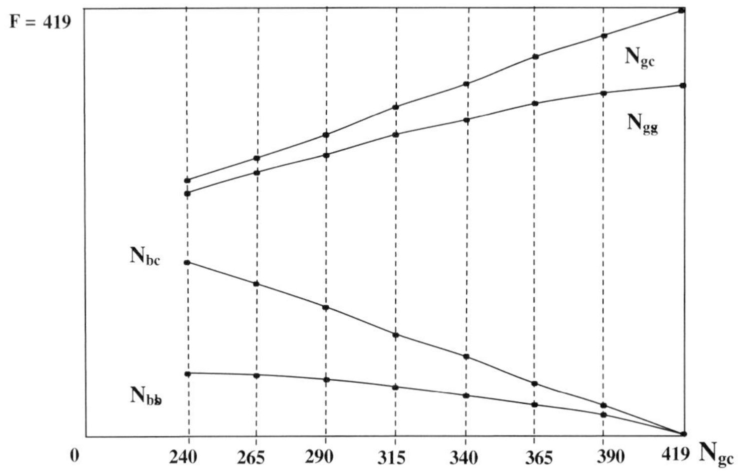

Fig. 15.7. The number of correctly recognized credits in function of the computed number of good credits N_{gc}

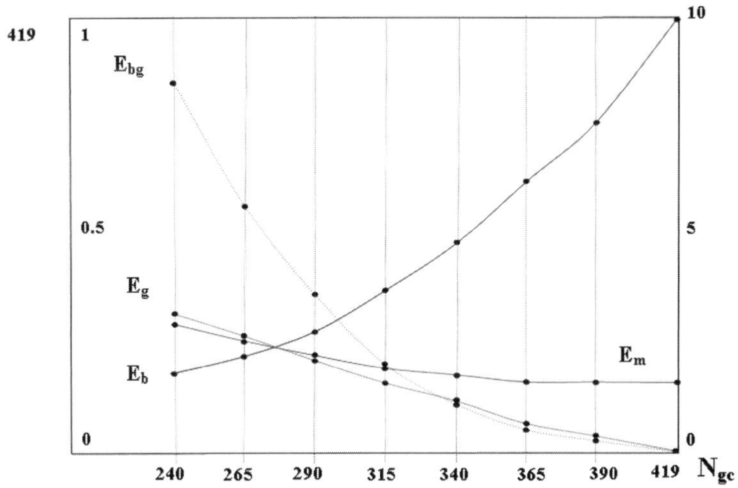

Fig. 15.8. Errors of recognition of credits in function of the computed number of good credits N_{gc}

Reasoning from conducting investments, we can make the following conclusion, that the bank method did not recognize 73 bad credits, but the credit risk LP-model did not recognize 27 bad credits only.

The risk of the credit is analyzed by values of probabilities of grade-events of signs describing credit. This probabilities are found in the results file of estimation of the credit. The more probability of grade-events, the more its contribution in the risk of the credit. Significance of the sign of the credit is determined by two attributes:

(1) the average value of probability of the sign P_{jm} for bank credits by results of identification of the risk LP-model: the more P_{jm}, the more the average bank risk P_m depends from the sign;

2) decreasing value of criterion function $\triangle F_j$ at eliminating the sign j from description of the credit.

Increment $\triangle F_j = F_{max} - F_j$ is computed at given to probability $P_j = 0$ and conducting repeated training the risk LP-model. The more $\triangle F_j$, the more significance the sign j has for classification of credits on good and bad. If changing criterion function $\triangle F_j$ equals or closes to zero, then this sign can eliminate from description of credits, because it does not influence accuracy of classification of credits (accuracy of the risk LP-model).

Results of analysis of sign contributions in the average credit risk and in decreasing accuracy of recognition of bad and good credits $\triangle F_j$ are presented in Table 15.10. Computations are fulfilled for the variant with number good $N_g = 315$ and bad $N_b = 104$ credits; the maximum value of the criterion function equals to $F_{max} = 341$.

The maximum contribution in accuracy of the failure risk LP-model brings the signs 2, 7, 9, 13, 14. The minimum contribution in accuracy of the failure

Table 15.10. Importance of signs of the risk LP-model

Sign, j	Name of sign	Criterion function, F_j	Changing accuracy, $\triangle F_j$	Average risk, P_{jm}
1	Term of credit	341	0	0.013424
2	Sum of credit	334	7	0.013139
3	Aim of credit	340	1	0.013578
4	Credit history in bank	338	3	0.012229
5	Presence of currency/ or rouble deposit in bank	338	3	0.018112
6	Presence plastic cart of bank	340	1	0.015835
7	Living conditions	335	6	0.013888
8	Presence in property expensive property	339	2	0.011828
9	Adg of client	331	10	0.011576
10	The time living in SPb.	340	1	0.012619
11	Official level	338	3	0.013867
12	The work time in company	339	2	0.012948
13	Income on place of work	336	5	0.011901
14	The number idle member family	332	9	0.0151

risk LP-model brings the signs 1, 3, 6, 10; the last signs can be eliminated for description of the credit of the bank.

The maximum contribution in average credit risk of the bank brings the signs 5, 6, 14. The minimum contribution in average credit risk of the bank brings the signs 8, 9, 13. As it is obvious, ratings of signs on contributions in accuracy of the risk model and in the average bank risk do not coincide.

15.8 Conclusions

Thus, in this chapter we get the following new results:

1. The structural and tabular presentations of data for the estimation of credit risks of natural persons;
2. The structural, logical, and probabilistic models of the credit risk;
3. The measure and the price for the credit risk;
4. The analysis method of the credit risk and the bank activity on crediting natural persons;
5. Proofs of the transparency of the LP-model and results;
6. Comparison of the credit risk models, constructed by different methods. The credit risk LP-model has the following essential advantages:
   - almost two times higher accuracy in recognition of good and bad credits than other well-known models of risks;
   - almost seven times higher robustness (stability) in classification than other well-known risk models;
   - the new tasks of analysis and management of risks (14 instead of 2);
   - the absolute transparency of the risk model and results of estimation and analysis of the risk.

16

LP-Model of Credit Risk for Juridical Persons

Your affairs prosper if you have enough money to receive credit in the bank.

Joke

The credit risk LP-model of juridical persons in many respects is similar to the credit risk LP-model of natural persons. We consider the credit risk model on Western and Russian markets. We describe the credit risk LP-model with GIE. Reserve of capital in the case of client bankruptcy is determined in the bank with use of the LP-model of credit risk.

16.1 Credit risk methods to Western market

Let us consider the application for credit from juridical persons by method of *Price Waterhouse* [128]. It has 38 signs, every which has from 2 to 6 grades. The number of all grades equal to 127. Let us pass from the signs of the finance risk, and risk-business to initiating random events of the credit risk. The events, connected with signs, we denote by Z_j. The number of all events are $n = 38$. Events, connected with grades, we denote by Z_{jr}, $j = 1, 2, \ldots, 38$, $r = 1, 2, \ldots, N_j$. Events $Z_1 - Z_{38}$ are initiating ones. Events $Z_{39} - Z_{52}$ are derivative ones. Events lead with some probabilities to errors of the estimation of the credit quality.

In response with the risk LP-theory with GIE [3, 30], we develop the structural risk model of the credit failure, presented in Fig. 16.1. Let us describe signs of the credit of juridical persons and indicate their grades by digits.

(1) The market sectors and spheres of activities of the client: 1 is a financial and investment activity; 2 is a production; 3 is a commercial and purchasing; 4 is a building; 5 is a science and research; 6 is other type of activity.

(2) The branch and the client place in the branch: 1 is the stability positive attributes of increasing the developed branches with the limited access of competitors, branches with the high level of the profit; 2 is developing profitable

E.D. Solojentsev, *Scenario Logic and Probabilistic Management of Risk in Business and Engineering*, DOI 10.1007/978-0-387-77946-1_16,

branches, but having difficulties because of high competitions or the other problems, new branches; 3 is branches with the trend of decreasing business activity, branches with the low norm of the profit.

(3) The size of the client market and its quota: 1 is the dominant or more than 50%; 2 is the big, more than 10% and less than 50%; 3 is the essential, more than 1% and less than 10%; 4 is less than 1%.

(4) The competitiveness and quality of productions (services): 1 is competitive, the excellent quality; 2 is qualitative; 3 is antiquated or non-qualitative.

(5) Clients/distributers, account payable, incomplete production, etc.: 1 is big credit account payable; 2 is credit and debit account payable about equal; 3 is the big debit account payable.

(6) Economic perspective in sense of opposed by economical falling-off of production: 1 is high; 2 is mean; 3 if low.

(7) Geographical position of client and founder of its firm: 1 is foreign companies; 2 is free economical zones; 3 is economically developed regions; 4 is insufficiently developed regions; 5 is back regions of country.

(8) Knowledge, experience, and level of professional management in a company: 1 is very high; 2 is big; 3 is mean; 4 is bad.

(9) Age, stating health and potential duration of management in a company: 1 means duration of management, old persons and with weak health are not; 2 is a lack of duration of management; 3 is managers divided on several groups.

(10) Ability, ductility and realism of managers of firm: 1 is flexible and realistic aims and tasks; 2 is hard-hitting aims and tasks; 3 is non-realistic aims and tasks.

(11) Possibility of founders and department of control: 1 is national structures; 2 is commercial; 3 is foreign companies; 4 is mixed structures.

(12) Department of financial control of client firm: 1 is adequate, on high level; 2 is good with small defects; 3 is bad.

(13) Mutual collaboration, in sense of fairness and honesty of client: 1 is less than 1 year; 2 is from 1 year to 3 years; 3 is from 3 to 7 years; 4 is more than 7 years.

(14) Intensity of relations of the client and its branch with bank: 1 is very high; 2 is high; 3 is mean; 4 is low.

(15) Discipline of client in conduct of account in bank in the past time: 1 is excellent; 2 is satisfying; 3 is weak.

(16) Allocation of credit: 1 is purchase of staple or materials; 2 is purchase of technics or technologies; 3 is repairs and modernization; 4 is capital buildings; 5 is training personnel; 6 is others.

(17) Sum of credit in application: 1 is to 50 rouble thousands ; 2 is from 50 to 500 rouble thousands; 3 is from 500 rouble thousands to 3 rouble millions; 4 is more than 3 rouble millions.

(18) The type of credit: 1 is fixed; 2 is overdraft.

(19) Term of credit: 1 is short-term, less than 1 year; 2 - medium-term, from 1 year to 3 years; 3 is long-term, more than 3 years.

(20) Parts of financial partnership of client in the project: 1 is more than 80%; 2 is from 65 to 80%; 3 is from 50 to 65%; 4 is less than 50%.

(21) Price for credit: 1 is usual; 2 is preferential.

(22) Order of giving credit: 1 is the whole sum at once; 2 is sequential.

(23) Commission fees: 1 is usual; 2 is preferential; 3 is not.

(24) Fees on services: 1 is usual; 2 is preferential; 3 is not.

(25) Commission fees for obligations: 1 is usual; 2 is preferential; 3 is not.

(26) Venture capital that is used: 1 is reliability more than 80%; 2 is reliability from 50 to 80%.

(27) Deposit of guaranty (property): 1 is high-liquid; 2 is mean-liquid; 3 is low-liquid; 4 is non-liquid.

(28) Characteristics and classification of deposit of guaranty, given another creditor: 1 is high-liquid; 2 is mean-liquid; 3 is low-liquid; 4 is non-liquid.

(29) Coefficient of productivity of assets: 1 is high; 2 is mean; 3 is low.

(30) Coefficient of profitability of own capital: 1 is high; 2 is mean; 3 is low.

(31) Coefficient of margin of operating profit: 1 is high; 2 is mean; 3 is low.

(32) Coefficient of per cent cover (on section of profitableness): 1 is high; 2 is mean; 3 is low.

(33) Coefficient of cover (on section of movement of money tools): 1 is high; 2 is mean; 3 is low.

(34) Coefficient of period of payback debt: 1 is high; 2 is mean; 3 is low.

(35) Coefficient of possibility to liquidate short-term obligations: 1 is high; 2 is mean; 3 is low.

(36) Coefficient of service of debt: 1 is high; 2 is mean; 3 is low.

(37) First coefficient of estimation of liquidation of capital: 1 is high; 2 is mean; 3 is low.

(38) Second coefficient of estimation of liquidation of capital: 1 is high; 2 is mean; 3 is low.

In the structural risk model (Fig. 16.1), the events with numbers $1 \div 38$ are initiating and with numbers $39 \div 51$ are derived. The events 52 is the final event, having value 1 or 0 (success or failure of the credit). The risk LP-function Y_{52} for failure is obtained. In the example it without working out in detail on grades is Fig. 16.1:

$$\begin{aligned} Y_{52} = {} & Z_{38} \vee Z_{37} \vee Z_{36} \vee Z_{35} \vee Z_{34} \vee Z_{33} \vee Z_{32} \vee Z_{31} \vee Z_{30} \vee Z_{29} \vee Z_{28} \vee Z_{27} \vee \\ & \vee Z_{26} \vee Z_{25} \vee Z_{24} \vee Z_{23} \vee Z_{22} \vee Z_{21} \vee Z_{20} \vee Z_{19} \vee Z_{18} \vee Z_{17} \vee \\ & \vee Z_{16} \vee Z_{15} \vee Z_{14} \vee Z_{13} \vee Z_{12} \vee Z_{11} \vee Z_{10} Z_{9} \vee Z_{8} \vee Z_{7} \vee \\ & \vee Z_{6} \vee Z_{5} \vee Z_{4} \vee Z_{3} \vee Z_{2} \vee Z_{1}. \end{aligned} \tag{16.1}$$

Events $Z_1 \div Z_{38}$ are grouped for convenience of computation and analysis of events $Z_{39} \div Z_{52}$, which we consider as derivative. This derivative events generalized consider: Z_{39} is branch factor; Z_{40} is purpose, sum and term of credit; Z_{41} is technology of given credit; Z_{42} is guarantee; Z_{43} is outside

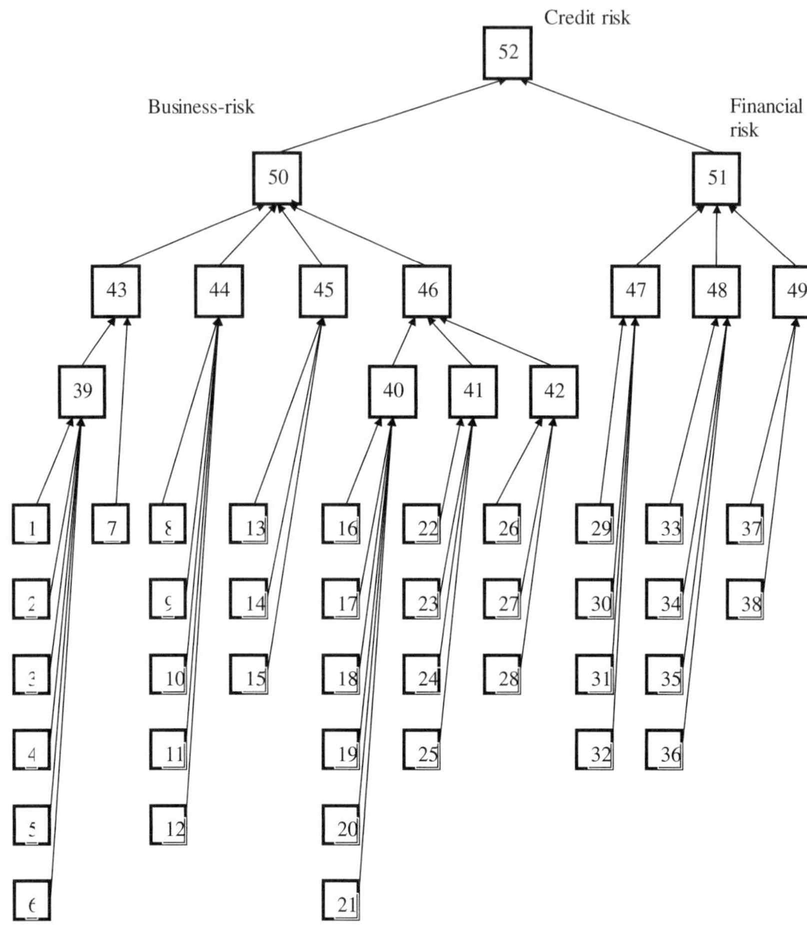

Fig. 16.1. Structural model of credit risk of juridical persons

conditions, in which client acts; Z_{44} is quality of management by client firm; Z_{45} is credit history of client; Z_{46} is attributes of credit application; Z_{47} is a profitability of client firm; Z_{48} is movement of money tools; Z_{49} is liquidity of capital. In one's turn, named events are generalized events: risk-business Z_{50} (outside conditions, quality of management, competentness and honesty of clients, attributes of credit application); financial risk Z_{51} (financial condition and possibility of client). At last, events of business and financial risks determine the credit risk Z_{52}.

Thus, we describe the structural model of credit risk of juridical personals, which, as it is obvious, has the type of "unit" (Fig. 15.2) and the corresponding risk LP-model of type (15.2–15.4).

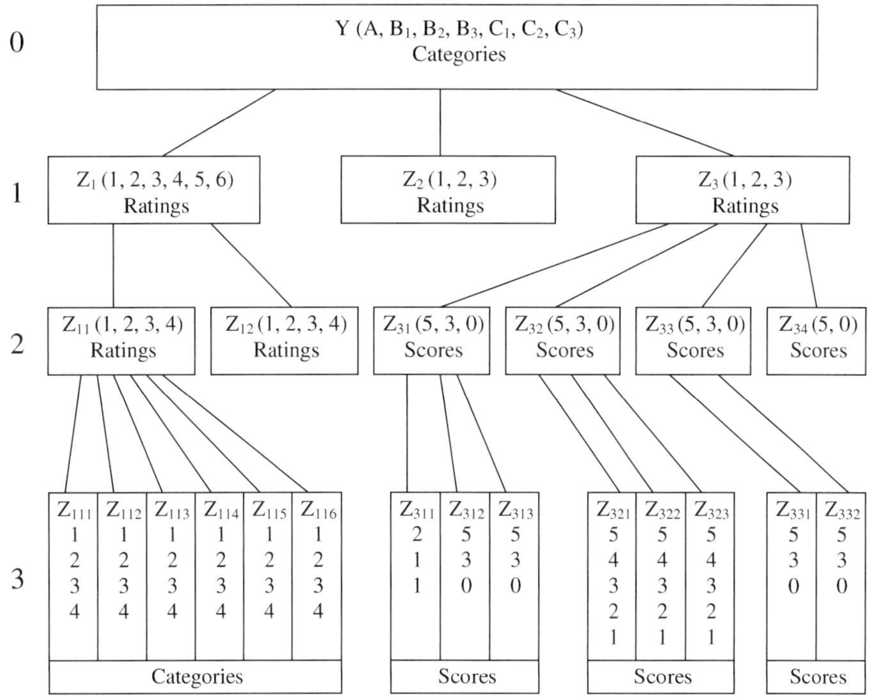

Fig. 16.2. Computation of debtor categories on Russian market

16.2 Credit risk methods according to Russian market

Structure representation of data. Let us build the scheme of computation of categories of debtor on the standard method, recommended by Center Bank of Russia for commercial banks (Fig. 16.2).[1] We mark the identifiers of events in response to Fig. 16.2. The numbers of categories, ratings, and estimations is presented in Table 16.1.

Based on Table 16.1 and Fig. 16.1, let us determine the number of different possible debtors on the level of computed algorithm 0, 1, 2, 3, and the number of grades in deciding independent factors.

The number of different category of debtors (on level 0)

$$T_0 = N_y = 7. \tag{16.2}$$

The number of different debtors (combination) on level 1

$$T_1 = N_1 \cdot N_2 \cdot N_3 = 6 \cdot 3 \cdot 3 = 54. \tag{16.3}$$

The number of different debtors (combination) on level 2

[1] About typing banking risks: The letter CB RF from 23.06.04, N70-T, p. 3.

Table 16.1. Events and their grades

NN.	Identifier	Name of identifier	Number of category, rate or estimation
1	Y	Debtor	$N_y = 7$
2	Z_1	Paying capacity	$N_1 = 6$
3	Z_2	Factor of developing	$N_2 = 3$
4	Z_3	Business risk	$N_3 = 3$
5	Z_{11}	Balance attributes	$N_{11} = 4$
6	Z_{12}	Monetary flow	$N_{12} = 4$
7	Z_{31}	Environment	$N_{31} = 3$
8	Z_{32}	Quality of management	$N_{32} = 3$
9	Z_{33}	Attributes of credit	$N_{33} = 3$
10	Z_{34}	Relation of bank and debtor	$N_{34} = 2$
11	Z_{111}	Finance stability	$N_{111} = 4$
12	Z_{112}	Paying capacity on money	$N_{112} = 4$
13	Z_{113}	Business activity	$N_{113} = 4$
14	Z_{114}	Effectiveness of management of assets	$N_{114} = 4$
15	Z_{115}	Dynamic of money tools	$N_{115} = 4$
16	Z_{116}	Presence and value of assets	$N_{116} = 4$
17	Z_{311}	Branch belonging	$N_{311} = 3$
18	Z_{312}	Competitive ability	$N_{312} = 3$
19	Z_{313}	Reliability of providers and partners	$N_{313} = 3$
20	Z_{321}	Knowledge and qualification	$N_{321} = 5$
21	Z_{322}	Individual quality	$N_{322} = 5$
22	Z_{323}	Effectiveness of management of bank	$N_{323} = 5$
23	Z_{331}	Aim and sum of credit	$N_{331} = 3$
24	Z_{332}	Term and type of credit	$N_{332} = 3$

$$T_2 = N_{11} \cdot N_{12} \cdot N_{31} \cdot N_{32} \cdot N_{33} \cdot N_{34} = 4 \cdot 4 \cdot 3 \cdot 3 \cdot 3 \cdot 2 = 864. \tag{16.4}$$

The number of different debtors (combination) on level 3

$$T_3 = N_{111}N_{112}N_{113}N_{114}N_{115}N_{116}N_{311}N_{312}N_{313}N_{321}N_{322}N_{323}N_{331}N_{332} = \\ = 4 \cdot 4 \cdot 4 \cdot 4 \cdot 4 \cdot 4 \cdot 3 \cdot 3 \cdot 3 \cdot 5 \cdot 5 \cdot 5 \cdot 3 \cdot 3 = 124\,440\,900. \tag{16.5}$$

The number of grades in deciding factors of low level

$$N = N_{111} + N_{112} + N_{113} + N_{114} + N_{115} + N_{116} + N_{12} + N_2 + N_{311} + N_{312} + \\ + N_{313} + N_{321} + N_{322} + N_{323} + N_{331} + N_{332} + N_{34} = 4 + 4 + 4 + \\ 4 + 4 + 4 + 4 + 3 + 3 + 3 + 3 + 3 + 5 + 5 + 5 + 3 + 3 + 2 = 60. \tag{16.6}$$

The number of independent grades in factors of the low level

$$N_{ind} = N - N_{dep} = 60 - 17 = 43, \tag{16.7}$$

where N_{dep} is the number of items in (16.6).

Let us mark the following disadvantages of considered method for estimation of categories of debtors:

1. The method is not automatized and has not corresponding software; estimation of categories of debtors is made by hand computation with use of tables of taking decision.
2. The method has not possibilities of conducting analysis if risk of debtor.
3. The method has interesting description of the problem of estimation of category of debtor, simple mathematic apparatus, and is as the example of the type scoring method for estimation rating with using balls and weights of factors.
4. The method cannot have the high accuracy and robustness in problem of estimation of category debtor from arithmetic adding events.
5. The method has no transparency in estimation of the category of debtors because of incorrect placement of balls, ratings, categories, etc.
6. From 124,440,900 debtors with different levels of credit risk only 7 categories of debtors is recognized.

16.3 LP-model of credit risk for Russian market

Let us develop the LP-model for estimation and analysis of credit risk of juridical persons (debtors), which allows us to eliminate the indicated shortage and to create corresponding program tools.

Let us construct structural, logical, and probabilistic models for credit risk of juridical persons and to decide the task of the reservation of the capital under possible loss on loans. Scheme for computation of category of debtor is transformed in the structural scheme (scenario, model-graph) of credit risk of juridical persons (the risk of failure of the credit). For this, we give all identifiers the status of the logical variables and random events, and edges of a graph of the sense of logical connections OR (Fig. 16.3).

The event Y is the final one and has GIE from seven events.

The events Z_{111}, Z_{112}, Z_{113}, Z_{114}, Z_{115}, Z_{116}, Z_{12}, Z_2, Z_{311}, Z_{312}, Z_{313}, Z_{321}, Z_{322}, Z_{323}, Z_{331}, Z_{332}, Z_{34} are independent initiating events of low level and have GIE, too.

The events $Z_1, Z_3, Z_{11}, Z_{31}, Z_{32}, Z_{33}$ are derivative ones. They also have GIE, after training the risk LP-model on statistical data for analysis of credit risk of juridical persons and credit activity of bank.

The inputting designation give us the transition from arithmetic adding factors (with weights, numbers, ratings, and categories) to logical adding events and to computing its probabilities.

The scenario of credit risk failure of juridical persons is formulated so: *failure occurs, if any one, any two,... or all sign-events occur.*

Proceeding from Fig. 16.3 and using the results of the works [4, 30], the credit risk LP-model of juridical persons in DNF is written:

$$Y = Z_{111} \vee Z_{112} \vee Z_{113} \vee Z_{114} \vee Z_{115} \vee Z_{116} \vee Z_{12} \vee Z_2 \vee Z_{311} \vee \\ \vee Z_{312} \vee Z_{313} \vee Z_{321} \vee Z_{322} \vee Z_{323} \vee Z_{331} \vee Z_{332} \vee Z_{34}. \quad (16.8)$$

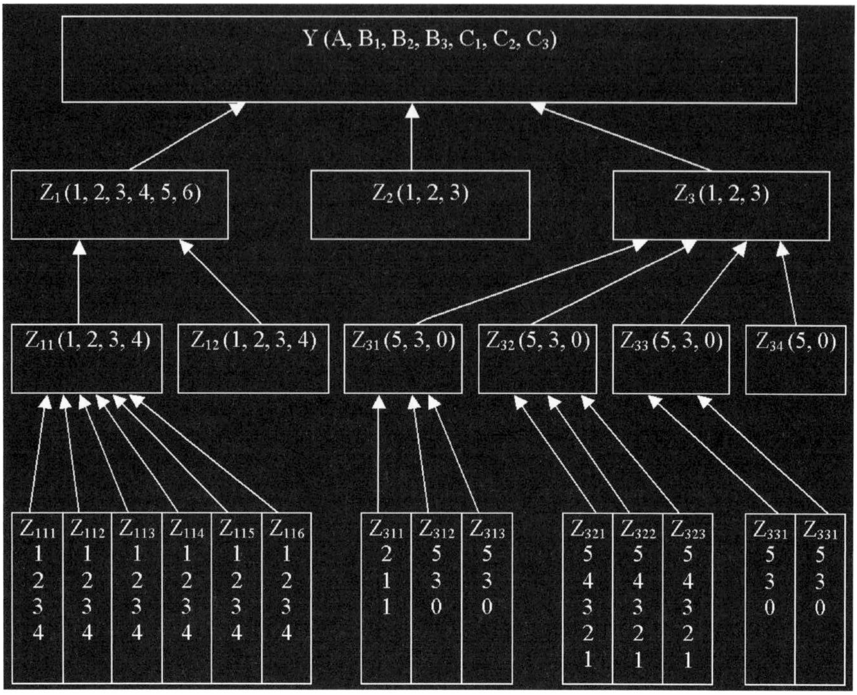

Fig. 16.3. Structural model of the credit risk of a debtor

Let us write the risk LP-function of the credit failure in equivalent form after its orthogonolization

$$Y = Z_{111} \vee Z_{112}\overline{Z_{111}} \vee Z_{113}\overline{Z_{111}}\ \overline{Z_{112}} \vee \ldots \tag{16.9}$$

Let as pass from logical description of credit failure risk to arithmetic description. The P-model (P-polynomial) of the credit

$$P\{Y = 0\} = p_{111} + p_{112}(1 - p_{111}) + p_{113}(1 - p_{111})(1 - p_{112}) + \\ + p_{114}(1 - p_{111})(1 - p_{112})(1 - p_{113}) + \ldots \tag{16.10}$$

Here probabilities $p_{111}, p_{112}, p_{113}, p_{114}, \ldots, p_{331}, p_{332}, p_{34}$ are determined by the method of identification on statistic data of credits of juridical persons of the bank.

The credit risk LP-theory of juridical persons coincides with the credit risk LP-theory of natural persons, including basic equations for GIE, the record of logical and probabilistic risk LP-models, the method of identification of the risk LP-model on statistical data and analysis of the credit and the credit set.

The unique distinction is: debtors must be classified on seven class-categories (see Fig. 16.3) by the risk value in corresponding with Fig. 16.4.

For the credit risk of juridical persons the clients categories can be presented by grades 1, 2, 3, ... for the sign "category of clients". This sign

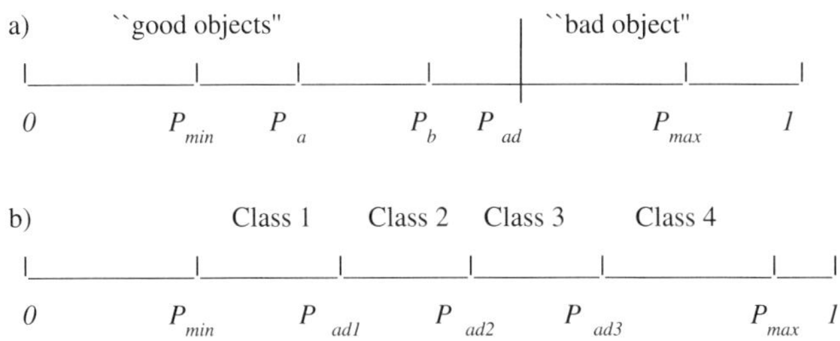

Fig. 16.4. Scheme of classification of debtors by risk

conveniently is placed in the last column of the table and for it is computed the additional attributes of the risk.

Credit changes its states for times, it also must change its category. It should monthly analyze the condition of the credits and its estimations. Sequence of computing operations on training and using LP-model is:

1. To put the credit category for debtor on application data.
2. To take decision on giving out credit, its type, sum, term and pro cent for credit using computed category and risk of the debtor.
3. Monthly to analyze and correct categories of credits on new values of factors of finance risk and business-risk.
4. Monthly to identify the risk LP-model on new statistical data on conditions of credits (debtors) (Table 16.2).

Each month the credit risk LP-model is retrained on basic (for previous months) and supplemented information. For example, in the first month was 300 credits, and the LP-model is trained on 300 credits (credit conditions), in the second month we consider 600 credit conditions (300 for previous month and 300 for current month), in the third one are 900 conditions, etc. After each month i we get the new data and new estimations of credit conditions, which can coincide or no coincide with estimations for previous month. Well-known estimation of correct or non-correct recognized conditions of credits are the base for training the risk LP-model with computed the condition risk for all set of credits on given month.

Thus, the credit risk LP-theory of the juridical persons are also differed by the large transparency of received results. This property of the credit risk LP-theory for example of crediting the natural persons in Chapter 15 is considered in details.

The failure risk LP-model of credits of juridical persons allows to decide the new tasks of analysis and management of the risk, and namely:

1. Determination of attributes of the credit risk: estimation of the credit risk, classification of the credit, determination of the price for the risk, analysis of the credit risk, reservation.

Table 16.2. Factors and conditions of the credit for months

NN	Factors	1	2	3	...	i	...	11	12
1	Z_{111}								
2	Z_{112}								
3	Z_{113}								
4	Z_{114}								
5	Z_{115}								
6	Z_{116}								
7	Z_{12}								
8	Z_{2}								
9	Z_{311}								
10	Z_{312}								
11	Z_{313}								
12	Z_{321}								
13	Z_{322}								
14	Z_{323}								
15	Z_{331}								
16	Z_{332}								
17	Z_{34}								
	Category								
	Risk of condition								

2. Determination of attributes of the risk of the set of credits of the bank: estimation of the admitted risks for classis (categories), determination of distributions of signs and grades of signs to the average credit risk of the bank, determination of distributions of signs and grades of signs to accuracy of classification of credits, decision of reservation task under risks of credits, optimization of the number of signs and grades of signs for description of credits, optimal fragmentation on intervals the sum and the term, age of clients, etc.
3. Ensuring and analysis of quality of the risk model: identification of the risk model on statistical data on credits, statistical and combinatoric analysis of the risk model, estimation of probabilities of grade-events for signs, estimation of the accuracy of the risk model; estimation of the robustness of the risk model, choice of the optimal coefficient of asymmetry of recognition of good and bad credits.

Estimation of the transparency of computation results and analysis of the risk are fulfilled on model data and here are not considered. This estimation coincides with results, stated in Chapter 15. We choose the admitted risk, determine the price for the risk, choose the optimal numbers of signs for description of the credit and analyze contributions of grades in the credit risk and accuracy of the credit risk model, choose the optimal number of grade-events in the sign.

16.4 Software for estimation and analysis of credit risks

Now, after consideration of methodological aspects of an estimation and the analysis of credit risks of natural and juridical persons, we shall state technology of practical realization of these works.

We developed the organization and technology "Bureau of estimation and analysis of credit risks" for rendering services to banks by estimation, analysis, and research of credit risks of physical and legal persons for banks.

Bureau of estimation and analysis of credit risk

The company can provide your bank the service on estimation and analysis of credit risk for natural or juridical persons. Today, only our company uses logical and probabilistic theory of risk with groups of incompatible events, which differs sharply from usual scoring techniques. Our methodology is adequate to risk and surpasses in many parameters existing methodologies and meets "Basel II" Accord requirements of techniques of quantitative estimation of credit risks and Capital Adequacy.

Principal difference of the technique is application of knowledgebase about credits as a system of logical equations (instead of database), use of logical addition of events instead of arithmetical addition of scores or other indicators, adequate credit risk scenario, and correct mathematical identification of credit risk model by statistical data.

The technique and special logical software were developed and were researched during 10 years. Working with banks, we obtained results with twice more accuracy, seven times more robustness, and high transparency in identification of bad and good credits than with other techniques. These results are stated in scientific publications.

Free trial. Bank can use free service during one month for adjustment of technology of data transfer and estimation of accuracy, transparency, and effectiveness of constructed LP-model of credit risks.

Service consists of two parts:

(1) Construction of credit risk model by bank's statistical data and analysis of risk model;

(2) Estimation and analysis of credit risk of borrower and determination of price for the risk.

Information guarding — credit data and results of its estimation and analysis are presented in impersonal form (as a set of numbers).

Technique of risk estimation and software. The logical and probabilistic (LP) risk theory with groups of incompatible events (GIE) and special logical software are used. Every bank is individual because it gives credits to different borrowers in different districts of the city or regions of the country, various branches and business fields, and needs to have its own credit risk models, which are constructed under its own statistics.

Relations between the Bureau and Bank are defined by the economic contract (the Appendix 1). Payment of the order is carried out by transference of money resources of the customer on the current account of the Executor.

Service cost. Service is paid for estimation and analysis of credit risk only. Service cost is 10.0 USD for credit of natural or juridical person. Final price is determined by agreement with Customer. Payment is performed when the Customer will get results of estimation and analysis of one or several credits.

Additional services. In addition, the agreement for following works can be concluded: training of bank staff, optimization of credit's description with choice of optimal numbers of parameters and grades for every parameter, adjustment of intervals for parameters (amount of credit, credit period, age of borrower, etc.), joint research in risk problem.

1. Technology of realization of the order

Order 1. The bank (the credit enterprise) represents statistics given before credits under Form 1. A file with statistics is impersonal: a bank creates it, then archives it, then sends it over e-mail. Names of files should have the following kind:

- Name of archival files: $STNNN - OUT.zip$, where:
  ST is an attribute of a file of statistics,
  NNN is a conditional number of bank (three last figures BIC),
  OUT is the output file.
- Name of an internal file: $NNNN - XXXXXX.txt$, where:
  $NNNN$ is a registration number of bank,
  XXXXXX is a serial number of the message.

Updating statistics and construction of new LP-model of credit risk is carried out periodically in 1–4 quarters under the arrangement with the customer.

Form 1. The depersonalized file $STNNN - OUT.txt$ of statistics of bank credits

N
n
$N_1 \dots N_n$
$Y_1 Z_{11} Z_{21} \dots Z_{n1}$
$Y_2 Z_{12} Z_{22} \dots Z_{n2}$
....
$Y_N Z_{1N} Z_{2N} \dots Z_{Nn}$,
where:
N is a number of credits;
n is a number of attributes;
$N_1 \div N_n$ are numbers of grades in each attribute;
$Y_1 \div Y_N$ is a success of the credit (1 is good, 0 is bad);
$Z_{11} \div Z_{Nn}$ are the number of grade of attributes.

Example:

1000
20
41051110554434453344222
11353312421421311311 1
1125151323121231232 11
12231022422141131121 11
11251413331212312221 2
. . . .
. . . .
. . . .
11151312231413321221 2
14354311421435322111 1
02434531121441311111 1

The data of the customer serve for training LP-model of risk. Numbers of good and bad credits are counted up on a file automatically. Also calculation of number of equally described credits is carried out, and it is established which gradation of attributes is not used for descriptions of credits. It allows one to supervise the data of the Order 1 and results of training LP-model of risk.

For credit risk of legal people, the label "Category of client" is represented by grades 1, 2, 3,. . . .This parameter settles down in the last column of a file and for it additional attributes of risk are calculated.

Order 2. The bank gives the order for an estimation and analysis of the credit risk for one or more borrowers on Form 2. The file with the order is archived also and sent on e-mail. Names of files should have the following kind:

- Name of archival files: $RENNN - OUT.zip$, where:
  RE is an attribute of a file for an estimation of risk (Risk Estimation),
  $NNN$is a conditional number of bank (three last figures BIC),
  OUT is a output file.
- Name of an internal file: $NNNN - XXXXXX.txt$ where:
  $NNNN$ is a registration number of bank,
  XXXXXX is a serial number of the message.

Form 2. The depersonalized file $RENNN - OUT.txt$ of the order on estimate and analysis of credits risks

yy.mm.dd hh:mm:ss
UserID $Z_1 \div Z_n$
where:
yy.mm.dd are year, month and day of order,
hh.mm.ss are hour, minutes and seconds of forming application,
UserID - the identifier of user (are singly formed by the bank),
$Z_1 \div Z_n$ are grade numbers on each sign.

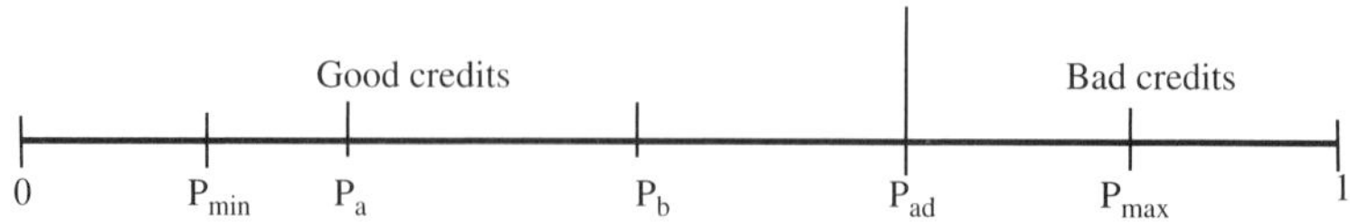

Fig. 16.5. Scheme of credit classification on the risk

Example:

{2006.08.17 01:03:24}
10345678 1 3 5 3 3 1 2 4 2 1 4 2 1 3 1 1 3 1 1 1
10234673 1 2 5 1 5 1 3 2 3 1 2 1 2 3 1 2 3 2 1 1
10543572 2 6 5 4 4 1 5 4 3 1 4 1 2 3 2 1 3 1 1 1
10862346 2 6 3 6 4 1 2 4 3 1 1 4 2 3 1 1 2 1 1 1
. . . .
10642686 1 3 3 1 3 1 2 4 2 1 3 3 1 3 1 1 3 1 2 1
10135467 4 3 5 7 4 2 3 4 2 1 2 1 2 3 2 2 3 1 1 1

Numbers of good and bad credits are counted up on a file automatically. These data serve for option of the program of training risk LP-models. It allows one to supervise the data on credits of bank and results of training the risk LP-model.

For credit risk of legal the label "category of client" is represented as grades $1, 2, 3, \ldots$. This parameter settles down in the last column of a file, and for it additional attributes of risk are calculated.

2. Technology of transfer of results

Results under Order 1. Construction of LP-model credit risk of bank occupies time until 12 o'clock. Results are sent to the customer by e-mail. Names of files have the following kind:

- Name of an archived file: $STNNN - IN.zip$, where:
  ST is an attribute of a file of statistics,
  NNN is conditional number of bank (three last figures BIC),
  IN is an input file.
- Name of an internal file: $NNNN - XXXXXX.txt$, where:
  $NNNN$ is a registration number of bank,
  XXXXXX is a serial number of the message.

The customer is informed, that the LP-model of credit risk is constructed. We send the basic attributes and parameters of quality (accuracy are informed, robustness, asymmetry of recognition) in the Form 3.

Let's make some explanation (Fig 16.5): N_g^{20} is a number of very good credits with the least risk, making 20% from the common number N_{gc} good credits:

$N_{gc} - N_g^{20}$ is the number of good credits;

N_b^{20} is the number of very bad credits with the greatest risk, making 20% from the general common settlement number N_{bc} bad credits;

Form 3. The file *STNNN-IN.txt* with parameters LP-models of credit risk of bank

The identifier	Names of parameters
N	Number of credits in statistics of bank
N_g	Number of good credits in statistics
N_b	Number of bad credits in statistics
P_m	Average risk of credits of bank
P_{ad}	Allowable credit risk
P_{min}	Minimal risk of the credit in the statistical data
P_{max}	Maximal risk of the credit in the statistical data
N_g^{20}	Number of very good credits
$N_{gc} - N_g^{20}$	Number of good credits
N_b^{20}	Number of very bad credits
$N_{bc} - N_b^{20}$	Number of bad credits
E_m	Average mistake in recognition of credits
E_b	Mistake in recognition of bad credits
E_{gb}	Factor of asymmetry in recognition of credits
F_{max}	Number of the recognized credits in statistics of bank
F_{eff}	Reduction of number of non-recognized credits by the LP-technique in comparison with a technique used by bank
F_1	Number reduction of recognized credits at exception of attribute 1
F_2	Number reduction of the recognized credits at exception of attribute 2
...	...
F_n	Number reduction of the recognized credits at exception of attribute n

$N_{bc} - N_b^{20}$ is the number of bad credits.

For legal persons for "categories of clients" (last column in Form 1), the following attributes are calculated in addition: frequencies of the categories in all credits, in bad and in good credits, and also average value of risk of credits for categories. It allows one to estimate adequacy of division of clients on a category.

Results under Order 2. Results of estimation and analysis of the credit risk of one borrower are sent to the customer by e-mail as a file under Form 4 during term not exceeding 8 hours. Names files have the following kind:

- Name of an archival file: $STNNN - IN.zip$, where:
  RE is an attribute of a file of results according to risk of credits (Risk Estimation),
  NNN is a conditional number of bank (three last figures of BIC),
  IN is the input file.
- Name of an internal file: $NNNN - XXXXXX.txt$, where:
  $NNNN$ is a registration number of bank,
  XXXXXX is a serial number of the message.

The information of Forms 3 and 4 allows one to construct the formula for the price (percent) for the credit risk. For example, elementary formula is

$$C_i = C_{ad} + k(P_{risk} - P_{ad}),$$

where: C_i is a cost i-th the credit; C_{ad} is a price for allowable Risk; k is a coefficient.

Form 4. File RENNN-IN.txt with parameters of estimation and analysis of risk

The identifier and the Name of parameter
$UserID$ and the Identifier of the user
Y and the Attribute of quality of the credit: $Y = 0$ is bad; $Y = 1$ is good
P_{risk} and Risk of the credit

Example:

10345678
1
0.199218
...
10234673
0
0.203452

3. A know-how and economic efficiency

Know-how is itself logic and probabilistic theory (LP-theory) of risk with groups of not joint events (GIE) and corresponding software. Protection of intellectual properties is carried out on a basis: (1) Certificates of the Russian Federation about official registration of computer program No. 2006610004 "Logic-probabilistic estimation and the analysis of credit risks"; (2) Publications in scientific Russian both foreign magazines and books.

Competitiveness. Now LP-techniques and LP-software for an estimation and the analysis of credit risks on the market are absent. In the market are available scoring techniques and program products for an estimation of risk on the basis of linear and square-law discriminant analysis, neural networks and data mining, having essentially worse characteristics on accuracy, robustness, transparency, and solving the limited number of tasks of the analysis and managements of credit risk.

Advantages in comparison with existing techniques:

- Twice bigger accuracy in an estimation of credit risk;
- Seven times bigger robustness (stability of an estimation of risk);
- Absolute transparency in an estimation and the analysis of risk of the credit and most models of risk;
- Opportunity to operate risk, changing asymmetry of recognition of good and bad credits, number of parameters, and the gradation describing the credit.

Approbation. Logic and probabilistic technique of an estimation and analysis of credit risks and corresponding software are approved on the data of the Western bank (1000 credits) and two Russian banks (500 credits of physical and juridical persons).

4. Essence of the LP-theory of credit risk

Description of credits. In each bank, the credit is described in parameters (attributes), each of which has gradation. On to practice the number of attributes can be from 10 up to 20 and number of gradation in attributes from 2 up to 11. For example, credits of physical persons in one CB of the Russian Federation were described by the following attributes (parameters) and their grades (Table 16.3).

Table 16.3. The description of attributes and grades of the credit

Number of attributes	Name of attributes	Number of grade	Grades of attribute
1	Term of the credit	1	Up to 6th month and less
		2	From 6th month and up to 1.5 years
		3	From 1.5 years up to 5th years
		4	From 5th till 15 years
2	Sum of the credit	1	Up to 45000 rouble
		2	From 45000 up to 100000 rouble
		3	From 100000 up to 200000 rouble
		4	From 200000 up to 300000 rouble
		5	From 300000 up to 500000 rouble
		6	From 500000 rouble and more
3	Purpose of the credit	1	Express trains - credits
		2	Consumer
		3	On purchase of habitation
4	Credit history in bank	1	Diligent credit history
		2	Comprehensible credit history
		3	Also did not use credits
5	Possession of plastic cards of bank	1	Also there is no cardmap
		2	VISA Electron etc.
		3	VISA Classic (Eurocard/Mastercard Mass)
		4	VISA Gold (Eurocard/Mastercard Gold)
6	Living conditions	1	Presence in the property of house
		2	Live in a municipal apartment, rent
		3	Other variants
7	Presence in the property	1	There is no such property
		2	Automobile has been not earlier of 3 years
		3	Market securities for sum not less than 1000
8	Age of the borrower	1	18 - 25 years
		2	26 - 50 years
		3	50 - 75 years
9	Official level	1	Top manager, the head of firm
		2	Manager of an average link
		3	Qualified professional
		4	Expert
10	Stability of work in company	1	Up to 2 years
		2	From 2 till 4 years
		3	From 4 till 6 years
		4	From above 6
11	Income in the basic place of work	1	Up to 10000 rouble in year
		2	From 10000 up to 15000 rouble in year
		3	From 15000 up to 30000 rouble in year
		4	From 30000 up to 50000 rouble in year
		5	From 50000 rouble in year and more
12	Amount of idle members of family	1	There are none
		2	Less than 2
		3	2 and more

Table 16.4. Credits and their signs

Number of credit both i	Sign 1 Z_1	...	Sign j Z_j	...	Sign n Z_n	Sign of efficiency credit, Y
1						
...	...	...	...	...	...	...
i			Z_{jr}			
...	...	...	...	...	...	...
N						

Attribute of the success of the credit is Y (2 grades). Attributes of the credit: Z_1 is a term of the credit (4 grades), Z_2 is a sum of the credit (6), Z_3 is a purpose of the credit (3), Z_4 is a credit history in bank (3), Z_5 is possession of plastic cards of bank (4), Z_6 is living conditions (3), Z_7 is a presence in the property of expensive property (3), Z_8 is an age of the borrower (3), Z_9 is an official level (4), Z_{10} is a stability of employment (the period of work in specified the companies) (4), Z_{11} is the income in a place of work (5), Z_{12} is a number of idle members of family (3).

The quantitative data (term, the sum of the credit) are broken into intervals by which numbers of gradation (attributes 1, 2, 8, 10, 11) are given.

Formal tabular representation of bank statistics on credits. The data under credits of bank can be presented as the table (Table 16.4) containing in lines credits $i = 1, 2, \ldots, N$. In column of tables are attributes (parameters) of the credit $Z_1, \ldots, Z_j, \ldots, Z_n$. In turn, attributes have grades $Z_{jr}, r = 1, 2, \ldots, N_j; j = 1, 2, \ldots, n$. Grades are in cells of the table. In the first column of Tables 16.4 is parameter of efficiency of the credit Y.

Every credit has an attribute (parameter) of success of the credit Y, which has two gradations: gradation of 1 (good) or gradation 0 (bad). As a result, it is received the file - table, in which lines are credits, in the first column is the grade of success of the credit, and in the other columns are grades for attributes of credits.

Grades are considered as random variables or grade-events. Generally, grades are linearly disordered and it is impossible to tell, that grade 3 is worse or better than grade 4 for final event, which also has grades. Set of credits divides on two classes - grades: the class 1, the credit is returned; a class 0, the credit is not returned.

Grade-events for each attribute form group of incompatible events (GIE). The greatest possible number of combinations (different credits) is equal

$$N_{max} = N_1 \cdot N_2 \cdot \ldots \cdot N_j \cdot \ldots \cdot N_n,$$

where $N_1, N_2, \ldots, N_j, \ldots, N_n$ are numbers of grades in parameters.

The number of credits in statistics of bank should be not less than $20n$, where n is a number of attributes for the description of credits.

Basic equations and criterion function of optimization. We shall consider attributes of the credit and their grades as a casual events: sign-events and grade-events. These events with certain probability bring to the failure of the credit.

The script of the failure of the credit is associative and is formulated for all sets of possible events: the failure occurs, if there is any one, any two, ... or all initiating events. We shall notice that any well-known scoring techniques for credit risk cannot use such script. The logic variable Z_j is equal to 1 with probability p_j, if the attribute j has led to failure, and is equal to 0 with probability $q_j = 1 - p_j$ in the opposite case. A logic variable Z_{jr}, corresponding with gradation r an attribute j, is equal to 1 with probability p_{jr} and it is equal to 0 with probability $q_{jr} = 1 - p_{jr}$. The vector $Z(i) = (Z_1, \ldots, Z_j, \ldots, Z_n)$ describes object i from Table 16.4. At the task object i instead of logic variables $Z_1, \ldots, Z_j, \ldots, Z_n$, it is necessary to substitute variables Z_{jr} for grades of attributes of this object i.

We write down the failure risk L-function for the credit

$$Y = Z_1 \vee Z_2 \vee \ldots \vee Z_n.$$

We write down the failure risk L-function for the credit in the equivalent orthogonalization form as follows

$$Y = Z_1 \vee Z_2\,\overline{Z}_1 \vee Z_3\overline{Z}_2\,\overline{Z}_1 \vee \ldots.$$

Now we come from the logic description of failure risk to the arithmetic one. P-model (P-polynomial) of failure risk of credit is

$$P = p_1 + p_2 \cdot q_1 + p_3 \cdot q_1 \cdot q_2 + \ldots.$$

"Arithmetic" of risk L-model is such that for final event, the value of risk belongs to the range [0,1] at any values of probabilities of initiating events.

Criterion function (CF) for identification of the failure risk P-model on statistical data is formulated so: the number of correctly classified credits must be maximum:

$$F = N_{bb} + N_{gg} \Rightarrow \max_{Pjr},$$

where N_{gg}, N_{bb} are the numbers of objects classified as good and bad using both the statistics and the risk P-model (both estimates should coincide).

Transparency of LP-models of credit risk and results estimations and analysis of risk is provided with an opportunity of calculation of contributions of signs and grades to risk of the credit, in average risk of all sets of credits and to accuracy of credit risk LP-model. Under the agreement with the bank, these contributions can be submitted in addition.

5. Intellectual protection

The method and the special logic software were developed and investigated for 10 years. There are the certificate of the Russian Federation on official registration of computer program No 2006610004 "Logic and probabilistic estimation and analysis of credit risks," 2005, and also a lot of publications for the period 1995 – 2006. We name some past publications only:

1. Solojentsev E. D. Scenario logic and probabilistic management of risk in business and engineering. Sec. Edition, SPb.: Business-press, 2006, 560 p.

2. Solojentsev E. D. Scenario Logic and Probabilistic Management of Risk in Business and Engineering. Springer: 2004. 391 p.

3. Solojentsev E. D., Stepanova N. V., Karasev V. V. Transparency of methods of estimation of credit risks and ratings. SPb.: Publishing house St.-Petersburg University, 2005, 196 p.

6. The contact information

Executor is the Bureau of Estimation and Analysis of Credit Risks. It is a juridical person, a small enterprise with charter and legal address, giving the paid service consisting of an estimation and the analysis of credit risk of the borrower.

Ltd INO-TEL, Russia, St. Petersburg,
Tel.: 7(812) 321-47-66;
Tel. and Fax: 7(812)590-00-35; risk@sapr.ipme.ru

16.5 Conclusions

As a result of developing the credit risk LP-model of juridical persons are obtained the following:

1. The basic concepts of construction of the credit risk LP-model for juridical persons are stated.
2. The credit risk model on Western market and on Russian market is considered.
3. The credit risk LP-model of juridical persons are suggested.
4. The credit risk LP-model of juridical persons has essential advantages on accuracy, stability, and transparency of the model and results in comparison with other methods. Transparency is reached by possibility of computation of contributions of initiating events and its grades to the risk of the credit and the set of credits.

17

Scenario Logic and Probabilistic Risk Models of Bribes

Everyone has his own price, and the price he named was too close to that of mine!
Abraham Lincoln on a bribe

We consider the following risk LP-models of bribes [153]:

1. Institutions by parameters of their functioning;
2. Officials by parameters of their behavior;
3. Institutions and officials by parameters of service;
4. In complex by logical adding 1, 2, 3.

Bribe risk LP-models, developed on statistical data, are intended for:

- departments of "Economic crimes" of towns,
- services of internal security and checking of companies and banks,
- development of norms and standards on service parameters.

17.1 Problems of bribes and corruption

The scenario logic and probabilistic (LP) bribe risk models are offered with the purpose of revealing, estimating, and analyzing bribes on the basis of the statistical data. Problems of bribes and corruption are of great computing complexity and are solved only by means of special logical software.

Problems with bribes and corruption have been actual at all times and in all countries. The website www.vzyatka.ru informs us about troubles in Russia with bribes and corruption.

Books and articles on corruption and on bribes [65, 150], on social statistics [117, 149], have thorough substantial descriptions and analysis, as well as a great number of various examples, comments on the law and on the criminal codex, but they do not contain any mathematical models of bribes.

For solution of social and organization problems (including problems of detecting frauds, bribes, and corruption), it needs, according to John von Neumann and Norbert Wiener [3, 30], the mathematical apparatus of logic, discrete mathematics, and combination theory, more suitable than differential equations.

E.D. Solojentsev, *Scenario Logic and Probabilistic Management of Risk in Business and Engineering*, DOI 10.1007/978-0-387-77946-1_17,

Such adequate mathematical apparatus is being developed and it is called "Logic and probabilistic (LP) theory of risk with groups of incompatible events" [3, 30]. It has been tested for the estimation and for the analysis of credit risks, security portfolio risk, the risk of the loss of quality, and the risk of non-success of the management of the company. The LP-models of risk have a high quality. For example, the credit risk LP-models have shown accuracy almost two times higher, robustness almost seven times greater than other methods and also an absolute transparency in classifying than other methods.

In the current work, an attempt has been made to use the LP-approach and the LP-calculus [3] for the solution of the actual problem — for the estimation and analysis of the probability of bribes and corruption.

17.2 Axioms of the bribe theory

Corruption is regarded as the basic kind of the so -called shadow economy. More often corruption implies the reception of bribes and illegal monetary incomes by state bureaucrats who extort them from citizens for the sake of personal enrichment. That is a brazen violation of public morals and the norms of law.

For the construction of the system and the technique of the struggle against bribes and corruption, the following axioms have been accepted [3, 30]:

- Under the pressure of circumstances, everyone may swindle if valuables are not guarded well enough and if it is possible to conceal the trickery for some time and when the control over the validity of the decisions taken is insufficient.
- Without a quantitative estimation and without the analysis of the probability of bribes, it is impossible to struggle against the swindle, bribes, and corruption.
- Each commercial bank or company is capable of a swindle or corruption if there is no transparency in their business and no control over their activities.
- Behind the non-transparency of the techniques of the estimation of credit risks and ratings of the banks and of the borrowers, there may be bribes and swindles.
- Complexity of organizational structure of an institution or company can be a sign of swindle and corruption.

Let's explain by an example only the first axiom [65]. The most honest president of the USA, Abraham Lincoln, had once thrown out of his office a man who had offered him a big bribe. Being asked what had irritated him so much that he was beyond himself, Lincoln answered: "Everyone has his own price, and the price he named was too close to that of mine!"

Concepts of the probability of bribes and corruption are close to those of reliability and safety in engineering and they are also close to the notion of risk in economy, in business, and in banks. Most frequently bribes take place when people receive licenses (in education, tourism, medicine, construction, etc.), sanctions (GAI, customs, etc.), in education (certificates, diplomas, examinations, etc.), registration (bodies of the Ministry of Internal Affairs, embassy, bodies of local authorities, etc.).

The scenarios and technique of a bribe are various for the ministry, for the mayoralty, for institutions, for companies, for banks, for officials, for doctors, for teachers, etc. The bribe implies two objects: the briber and the bribe-taker, either of whom has his benefit. The briber solves his problem faster, more qualitatively, receives privileges, bypasses the law, etc. The bribe-taker has monetary or material benefit, etc. We use the following terms: probability of corruption and of a bribe, probability of success and non-success, probability of the absence or of the presence of a bribe, probability of a good or bad project (of an object, of an official, of an institution). We consider those terms from the point of view of the size of the probability of a bribe.

For a quantitative estimation and for the analysis of the bribe probability we use the logic and probabilistic non-success risk LP-theory (LP-theory) with groups of incompatible events (GIE) [3, 4, 30], and some bribe LP-models are constructed on the basis of the statistical data. The paper is one of the first mathematical publications on the probability of bribes and does not claim to consider all the aspects of this complex problem and to develop all the scenarios of bribes. Here we just give the description and the construction of the bribe model, try to give the estimation and the analysis of the probability of a bribe, and hardly ever touch upon the social, legal, and organizational problems of bribes.

17.3 The LP-theory of bribes with groups of incompatible events

Events and probabilities. An event of a bribe is described by signs and their grades, which happen to be random variables and are regarded as the logic variable of casual sign-events and grade-events having certain probabilities. The sign-events are connected by the logic connections OR, AND, NOT and can have cycles. The grade-events for a sign make a group of incompatible events (GIE) [3, 30].

The bribe LP-theory completely coincides with the risk LP-theory with GIE, stated in Chapter 11.

Identifying the bribe LP-model on the basis of statistical data. The problem of the identifying the bribes LP-model is solved by the algorithmic iterative methods [3, 30]. Here we use the connection between probabilities P_{jr}, $\mathrm{P1}_{\mathrm{jr}}$, and $\mathrm{P2}_{\mathrm{jr}}$ in response to Fig. 17.1. The following scheme of solving the problem is proposed here. Let the probabilities for the grades $P_{jr}, r = 1, 2, \ldots, N_j, j = 1, 2, \ldots, n$ be known as a first approximation and the risks $P_i, i = 1, \ldots, N$ for the objects in statistics be also calculated, each of which might be accompanied by bribes. In the statistics of good projects,

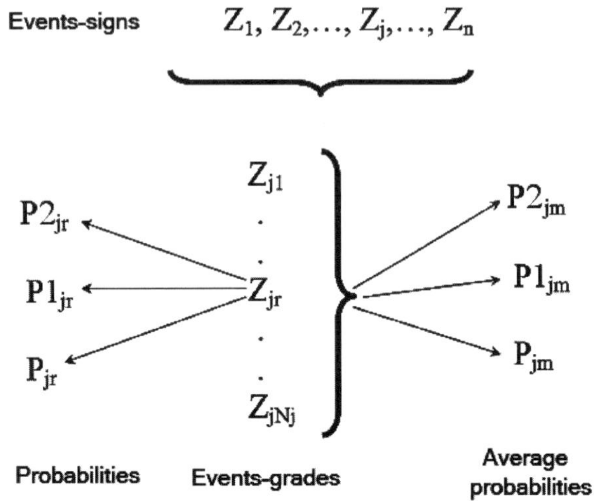

Fig. 17.1. Probabilies in group of incompatible events

the symbol N_g is used and in the statistics of bad projects the symbol N_b is used. We will determine the admitted risk P_{ad} (Fig. 17.2) so that the number of projects accepted by us without bribes (the good projects) N_{gc} had the lesser risk than the admitted one and, accordingly, the number of the projects with bribes (the bad projects) $N_{bc} = N - N_{gc}$ had the risk that exceeded the admitted one. At the step of optimization, we shall change the probabilities $P_{jr}, r = 1, 2, \ldots, N_j; j = 1, 2, \ldots, n$ in such a way that the number of the recognizable projects might be increased. The variables P_{ad} and N_{gc} are connected unequivocally. In the algorithm of the problem, it is more convenient to give N_{gc} and to determine P_{ad}.

The condition $P_i > P_{ad}$ specifies the following types of projects: N_{gg} - denotes the projects that are good according to both - their models and the statistics; N_{gb} denotes the projects that are according to their models and bad by the statistics; N_{bg} denotes the projects that are bad by the model and good by the statistics; N_{bb} denotes the projects bad by both the models and the statistics. The risks of the projects N_{gg}, N_{gb}, N_{bg}, N_{bb} move relative to P_{ad} at the change of P_{jr}. At the transition of some projects to the right from P_{ad} on value of the risk, some number of projects passes to the left. The change P_{jr} that translates the projects N_{gb} and N_{bg} through P_{ad} towards each other will be the optimal one.

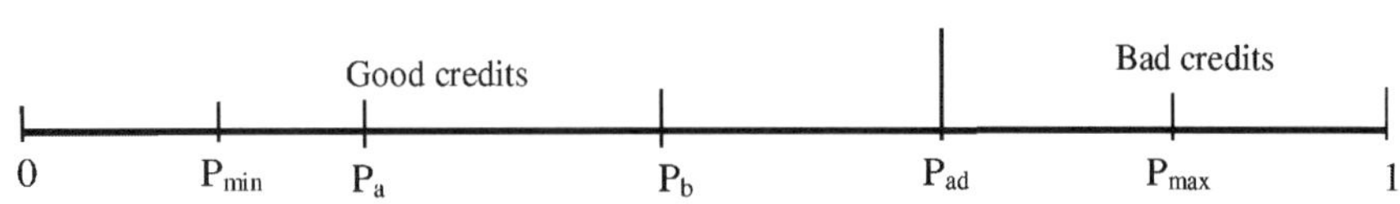

Fig. 17.2. The circuit of classification of objects

The problem of the identification of the LP-model of the bribe completely coincides with the problem of the identification of the risk LP-model with GIE, stated in Chapter 12.

The analysis of bribe probability. Let the bribe P-model and the probabilities P_{jr} be defined and known. We shall determine the contributions of sign-events and grade-events into the probability of the bribe for the project and for a set of projects, and also into the accuracy of the bribe LP-model. For this purpose, we shall calculate the differences between the values of the characteristics for the optimum model and if the probabilities are endowed with the corresponding grade-events of zero values [3, 30].

The analysis of the contributions to the probability of the bribe grades, of signs, of the project, and of the project set allows us to optimize the bribe model for increase of its accuracy.

The problem of the bribe LP-analysis completely coincides with the problem of the risk LP-analysis with GIE, stated in Chapter 13.

17.4 The bribe LP-model at institutions

The institution is making decisions on some projects (on the cases or affairs of the citizens). There are a lot of projects. The projects are either successful (good) or non-successful (bad). The reasons for the non-success projects are the unjustified sanctions, given out as a result of bribes.

Elements of the scenario and of the bribe LP-model are the functional departments $Z_1, \dots Z_n$, each of which has N_j officials who make decisions.

Generally, the object with the elements $Z_1, \dots, Z_j, \dots, Z_n$ is complex as it includes connections OR, AND, NOT, and repeated elements and cycles. Officials in the j-department $Z_{j1}, \dots, Z_{jr}, \dots, Z_{jNj}$ are GIE. The official, making a decision, signs the corresponding document. The construction of the bribe LP-model consists in the calculation of the probabilities $P_{jr}, j = 1, 2, \dots, n; r = 1, 2, \dots, N_j$ with which officials take bribes on the basis of the statistics from N successful and non-successful projects.

We shall consider the bribe LP-model, say, of a bank. The statistics about the success of the credits is used. The reasons of the non-success of the credits are explained by bribes.

Let the bank have five functional groups of the officials who take decisions on giving out the credits. The logic variables Z_1, Z_2, Z_3, Z_4, Z_5 correspond with these functional groups. These groups have accordingly N_1, N_2, N_3, N_4, N_5 officials taking decisions. The number of the officials in groups coincides with the number of the grades in GIE.

The given credits are either successful (grade 1) or non-successful (grade 0). There are documents on the given credits where the officials making decisions fix their signatures.

The greatest number of any possible combinations of the client's passing through the institution and the bribes is equal to

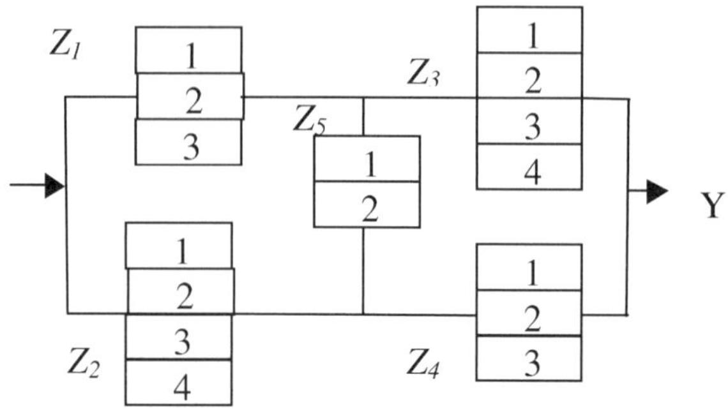

Fig. 17.3. The structural model of the bribes of the "bridge" type

$$N_{max} = N_1 \cdot N_2 \cdot N_3 \cdot N_4 \cdot N_5. \tag{17.1}$$

The logic function of the bribes in the perfect disjunctive normal form (PDNF) has N_{max} of the logical terms, and we may write

$$Y = Z_1 Z_2 Z_3 Z_4 Z_5 \vee \overline{Z_1} Z_2 Z_3 Z_4 Z_5 \vee \cdots \vee \overline{Z_1 Z_2 Z_3 Z_4 Z_5}. \tag{17.2}$$

Every logic variable from Z_1, Z_2, Z_3, Z_4, Z_5 or its denial (the line over a variable) goes into any conjunct. All the conjuncts are orthogonal in pairs, that is PDNF is the orthogonal form of the logic function. At calculation of the probability of the event Y, we put in (17.2) the probabilities P_1, P_2, P_3, P_4, P_5 instead of the events Z_1, Z_2, Z_3, Z_4, Z_5 and the sign "OR" is replaced by "+."

PDNF is the cumbrous recording of the logic function. In reality the logic bribe model may be recorded more simply if taking into account the structure of bank departments and their connection. It can be of any kind. To be specific, we will assume that the structure of the risk model is presented by the "bridge"(Fig. 17.3).

The officials from Z_1 and Z_2 check the maintenance of the credits, and the officials from Z_3 and Z_4 take the decision on the size and on the terms of the credit. The top officials (chiefs) from Z_5 control the process. The client visits one of the top officials who either advises the client or takes a bribe and directs the client to the officials from the groups Z_1, Z_2,Z_3 or Z_4 who take bribes, too. The number of the officials in each functional group corresponds with the number of the grades in the sign. There are four different ways the client may take and four trajectories of bribes (Fig. 17.3).

The logic model (L-model) of bribes in disjunctive normal form (DNF) (the records of the logic functions without the brackets) on the basis of the shortest way of functioning is

$$Y = Z_1 Z_3 \vee Z_2 Z_4 \vee Z_1 Z_5 Z_4 \vee Z_2 Z_5 Z_3. \tag{17.3}$$

The probabilistic model (P-model, P-polynomial) of bribes is obtained after the orthogonalization of the logic function (17.3)

$$P_i = p_2\, p_4 + p_1\, p_3 + q_1\, p_2\, p_3\, q_4\, p_5 + p_1\, q_2\, q_3\, p_4\, p_5 - p_1\, p_2\, p_3\, p_4. \tag{17.4}$$

Table 17.1. Average probabilities for the sign-events

Signs j	Factor K_j	Probabilities P_{jm}	Probabilities $P1_{jm}$	Number of grade N_j
1	1.916036	0.478113	0.249540	4
2	4.586733	0.348310	0.075949	10
3	2.233272	0.298833	0.133823	5
4	3.341342	0.388857	0.116348	11
5	3.177692	0.291868	0.091775	10

Example 1. For training the bribe P-model, the statistics from 1000 credits is used (700 are good and 300 are bad). The average bribe risk is equal to $P_{av} = 300/1000 = 0.3$. Five sign-events have from 4 to 11 grades; all in all there are 40 grades.

As the result of training, the probabilities P_{jr} and $P1_{jr}$ for all grade-events have been obtained, and the following parameters of the bribe LP-model have been calculated: the criterion function is equal to $F_{max} = 720$ and the admitted risk is equal to $P_{ad} = 0,3094$.

Some results of computing research are given in Tables 17.1 and 17.2. The probabilities $P2_{jr}$ and $P1_{jr}$ of the grades, though they make a total 1 in GIE, can differ essentially (Table 17.2). The probabilities of bribes (probabilities P_{jr}) differ more than 10 times.

The sign-events 1 and 4 have the maximum average probabilities P_{jm}. The same events bring the maximum contributions to the average risk P_m. The average probabilities P_{jm} for sign-events differ nearly in two times.

Table 17.2. Probabilities of grade-events and their frequencies

Grade number	Probabilities P_{1r}	Probabilities P_{1r}	Frequencies $P2_{1r}$
Sign Z_1			
1	1.000000	0.522300	0.274
2	0.596084	0.311103	0.269
3	0.248278	0.129579	0.063
4	0.070927	0.037017	0.394
Sign Z_2			
1	0	0	0
2	0.687703	0.149933	0.014
3	0.227359	0.0495688	0.002
4	1.000000	0.218209	0.054
5	0.510577	0.111316	0.017
6	0.704722	0.153643	0.086
7	0.570149	0.124304	0.057
8	0.448856	0.097859	0.224
9	0.434821	0.094799	0.187
10	0.001675	0.000365	0.359

17.5 The bribe LP-model on the basis of officials' behavior

A bribe is not a crime that is made a parade. There is no question about "corpus delict" at a robbery of a bank, which is witnessed by the employees or by the clients. A bribe differs from any other kinds of crime by the difficulty of its revealing. However, bribes have a mass character and there are many data on bribes both in judicial law-courts and in the controlling units.

For each type of bribes, it is possible to find signs [30, 65] that are associated with a similar crime. Each of such signs has at least 2 grades. The bribe P-model can be identified on the statistical data.

The investigation of the bribe can be carried out only in the case when there are serious reasons to believe that the bribe had actually taken place in the past. The value of this "seriousness" can be estimated quantitatively on the probability of the bribe, and the final decision is taken by an office head.

Special signs testify to the bribes taken by the officials (doctors, teachers). There are following signs of the person's unusual behavior and organization:

- Age;
- Duration of the period of work at an institution or in company;
- Purchase of a house, of an apartment, of a summer residence, of a car, etc., at the price not adequate corresponding with the level of the wages;
- Debts;
- Financial inquiries;
- Predisposition to gambling;
- The way of life beyond the habitual frameworks;
- Unusual behavior;
- Presence of complaints;
- Vague or criminal past;
- Dishonest or unethical behavior in the office;
- Absence of the division of duties;
- Absence of independent checks;
- Absence of the proper authority;
- Absence of the necessary documents and records;
- The neglect of the existing rules;
- The inadequate system of document circulation, etc.

Elements of the scenario and of the bribe LP-model, listed above, are presented by the signs $Z_1, \ldots, Z_j, \ldots, Z_n$, each of which has several grades. The grades for the j-sign of $Z_{j1}, \ldots, Z_{jr}, \ldots, Z_{jNj}$ form GIE. The bribe scenario of the official is described as follows: a bribe can take place if any one sign-event or any two sign-events or … all sign-events take place. The scenario of a bribe is given in Fig. 17.4 in the form of a structural graph.

The construction of the bribe LP-model consists in the calculation of the probabilities $P_{jr}, j = 1, 2, \ldots, n; r = 1, 2, \ldots, N_j$ (with which the official takes bribes) on the statistics of the bribe facts established by the courts.

The bribe L-function (the bribe LP-model)in DNF is [3, 30]

$$Y = Z_1 \vee Z_2 \vee \ldots \vee Z_n. \tag{17.5}$$

The bribe L-function in the equivalent orthogonal form (ODNF) after orthogonalization (17.5) is

$$Y = Z_1 \vee Z_2\overline{Z}_1 \vee Z_3\overline{Z}_2\overline{Z}_1 \vee \ldots \tag{17.6}$$

The bribe P-function (model, polynomial) is

$$P = p_1 + p_2(1 - p_1) + p_3(1 - p_1)(1 - p_2) + \ldots \tag{17.7}$$

"Arithmetics" in the bribe P-model is such that for the final event, the bribe probability value is within the limits of [0,1] at any values of the probabilities of initiating events. For every grade-event in GIE, we use the three probabilities $P2_{jr}, P1_{jr}, P_{jr}$, introduced before (Fig. 17.1). The maximum number of different bribes is equal to

$$N_{max} = N_1 \cdot N_2 \cdot \ldots \cdot N_j \cdot \ldots \cdot N_n, \tag{17.8}$$

where $N_1, \ldots, N_j, \ldots, N_n$ are the numbers of the grades in the signs. If the number of signs is equal to $n = 20$ and each sign has $N_j = 5$ grades, the number of different bribes (conjuncts in a perfect orthogonal disjunctive normal form — PODNF) equals the astronomical number $N_{max} = 5^{20}$, which explains the difficulties of the struggle with bribes and corruption. (PODNF comprises various conjuncts, each of which comprises all the variables $Z_1, Z_2, \ldots, Z_n$ or their denials). The conjunctions are connected by the logical operation *OR*.

The bribe LP-model (17.5–17.7) describes all kinds of bribes and is the most complete and accurate one. In some cases, however, it is not necessary to take into account all possible bribes. For example, it is known from the statistic data that there were bribes when one or two events occurred from $Z_1, Z_2, \ldots, Z_n$. Then, to simplify the model, you should use the bribe model for a limited number of bribes [30].

If we have a logical bribe model of four elements

$$Y = Z_1 \vee Z_2 \vee Z_3 \vee Z_4. \tag{17.9}$$

then for a limited number of bribes, when either one or two events occur, the bribe model will be recorded as

$$\begin{aligned} Y = {} & Z_1\overline{Z_2}\ \overline{Z_3}\ \overline{Z_4} \vee Z_2\overline{Z_1}\ \overline{Z_3}\ \overline{Z_4} \vee Z_3\overline{Z_1}\ \overline{Z_2}\ \overline{Z_4} \vee Z_4\overline{Z_1}\ \overline{Z_2}\ \overline{Z_3} \vee Z_1Z_2\overline{Z_3}\ \overline{Z_4} \vee \\ & \vee Z_1Z_3\overline{Z_2}\ \overline{Z_4} \vee Z_1Z_4\overline{Z_2}\ \overline{Z_3} \vee Z_2Z_3\overline{Z_1}\ \overline{Z_4} \vee Z_2Z_4\ \overline{Z_1}\ \overline{Z_3} \vee Z_3Z_4\overline{Z_1}\ \overline{Z_2}. \end{aligned} \tag{17.10}$$

In the bribe L-model, all the logical summands are orthogonal in pairs, which allows the bribe P-model (P-polynomial) to be written directly:

Table 17.3. Probabilities and errors of recognition for the grade-events

$P2_{jr}$	$P2_{0jr}$	$P2_{1jr}$	$P1_{jr}$	P_{jr}	E_{jr}	$E1_{jr}$	$E0_{jr}$
Sign Z_2							
0	0	0	0	0	0	0	0
0.014	0.007	0.007	0.01	0.019	0.214	0.429	0.0
0.002	0.001	0.001	0.070	0.014	0,5	1.0	0.0
0.054	0.032	0.022	0.194	0.038	0.278	0.682	0.0
0.017	0.005	0.012	0.159	0.031	0.412	0.5	0.2
0.086	0.038	0.048	0.145	0.028	0.256	0.417	0.053
0.057	0.019	0.038	0.095	0.019	0.228	0.289	0.105
0.224	0.066	0.158	0.067	0.013	0.169	0.196	0.106
0.187	0.056	0.131	0.053	0.010	0.203	0.183	0.250
0.359	0.076	0.283	0.016	0.003	0.114	0.081	0.237
Sign Z_{13}							
0.190	0.080	0.110	0.283	0.027	0.237	0.345	0.087
0.511	0.142	0.369	0.233	0.021	0.186	0.201	0.148
0.248	0.065	0.183	0.093	0.008	0.113	0.082	0.200
0.028	0.007	0.021	0.346	0.032	0.178	0.238	0.0
0.023	0.006	0.017	0.044	0.004	0.217	0.117	0.5

$$P\{Y\} = p_1q_2q_3q_4 + p_2q_1q_3q_4 + p_3q_1q_2q_4 + p_4q_1q_2q_3 + p_1p_2q_3q_4 + \\ + p_1p_3q_2q_4 + p_1p_4q_2q_3 + p_2p_3q_1q_4 + p_2p_4q_1q_3 + p_3p_4q_1q_2. \tag{17.11}$$

Example 2. The author did not have any factual data about the bribes established by courts on criminal cases. The modeling data were used as the statistical data. From 1000 officials, suspected of bribes, against whom suits were brought, only 300 were condemned, and 700 were considered to be innocent. Thus, the average risk of bribes is equal to $P_{av} = 300/1000 = 0.3$. The suspected officials are described by $n = 20$ signs whose total sum is 94 grades.

The identification of the bribe P-model (17.7) consists in defining the probabilities $P_{jr}, r = 1, 2, \dots, N_j; j = 1, 2, \dots, n$ for grade-events. The bribe probability for every suspected official is calculated on the optimization step and is compared with the admitted risk P_{ad}. The suspected official is either a bad or good one. The criterion function is formulated as follows: the number of the correctly classified suspected officials should be as great as possible.

Contributions of the grade-events into the accuracy of the bribe LP-model will be considered by us by the example of sign-events (Table 17.3) of the signs Z_2 and Z_{13} for the optimal identified bribe LP-model ($F_{max} = 826$). The grade frequencies for all $P2_{jr}$, for the bad $P2_{0jr}$ and for the good $P2_{1jr}$, the probabilities of the grad-events $P1_{jr}$ and P_{jr}; for the mistakes of recognition on grades for all E_{jr}, for the bad $E0_{jr}$ and for the good $E1_{jr}$ officials who are under suspicion, are provided in Table 17.3.

The contribution of the sign-event into the the probability of a bribe by an official is proportional to the probability $P_j, j = 1, 2, \dots, n$, which equals

Table 17.4. Analysis of the contributions of the sign-events into the accuracy of the bribe model

Signs, j	Number of grades, N_j	$P1_{jm}$	P_{jm}	K_j	$\triangle F_j$
1	4	0.272384	0.020226	0.074255	–64
2	10	0.063346	0.012359	0.195102	–27
3	5	0.098475	0.009327	0.094713	–18
4	11	0.090820	0.020927	0.230421	–26
5	10	0.080377	0.017593	0.21888	–20
6	5	0.272148	0.022466	0.082550	–20
7	5	0.206945	0.018549	0.089632	–6
8	4	0.266619	0.017736	0.066521	–6
9	4	0.183897	0.014253	0.077505	–10
10	3	0.318015	0.018295	0.057528	–10
11	4	0.251871	0.018974	0.075331	0
12	4	0.247375	0.017166	0.069392	0
13	5	0.206718	0.018900	0.091428	–16
14	3	0.235637	0.014733	0.062524	–2
15	3	0.261648	0.017591	0.067231	–8
16	4	0.341959	0.021975	0.064261	–2
17	4	0.289853	0.018739	0.064649	0
18	2	0.482499	0.017417	0.036097	0
19	2	0.508613	0.018138	0.035661	0
20	2	0.750896	0.018326	0.024405	–2

the probability of the grade-event P_{jr}. The probabilities P_{jr} of a sign-grades differ more than 10 times. The grade errors E_{jr} in the classification of bribes differ almost 5 times.

The LP-analysis of the bribe model is carried out with use of (13.3 –13.8). For each sign j (Table 17.4): the average values of probabilities $P1_{jm}$ and P_{jm} were determined and also the decrease of the number of the identified good and bad suspected officials $\triangle F_j$. When this sign was excluded from the bribe model, the bribe LP-model was retrained. The decrease of the number of the suspected officials that could be recognized is determined in relation to the optimal trained bribe model with all signs.

The maximum contribution into the accuracy of the suspected official is brought with the sign-events: $Z_1, Z_2, Z_4, Z_5, Z_6, Z_3, Z_13$. The zero contribution is brought by the sign-events $Z_{11}, Z_{12}, Z_{17}, Z_{18}, Z_{19}$; excluding sign-events $11, 12, 17$, and 18 reduces the number of the identified suspected only by 4.

The accuracy of the bribe LP-model changes with the change of the number of grades in a sign. The sign Z_2, which in the initial variant had 10 grades, was investigated. After retraining the bribe model, the following results were obtained: in the absence of the sign Fmax $= 800$; with two grades $F_{max} = 808$; with four grades $F_{max} = 812$; with ten grades $F_{max} = 824$; with hundred grades (in that case there were seventy empty grades) $F_{max} = 828$.

We built a graph of bribe probabilities for 1000 suspected officials before and after sorting according to the value of the probability. Approximately 15% of the suspected officials had small bribe probabilities and are good, and 15% of suspected officials had high bribe probabilities and are very bad. It shows that is necessary to classify the suspected officials according to the bribe probability value into four classes.

17.6 The bribe LP-model on the basis of analysis of service parameters

Let's estimate the bribe probability using the statistics of the service parameters. These parameters can be, for example, the time it takes the officials to solve the problem or it takes the dentist to make a denture (from the beginning till the end of the process). Such statistics should contain the service precedent number, sufficient for construction of the discrete or analytical distribution function.

Let us have the statistics (service times) for N clients $Y_i, i = 1, 2, \ldots, N$. If we construct a normal distribution law for the parameter Y_i with the average value and the dispersion, it will lead to an essential decrease of the bribe estimation accuracy.

The service parameter can have either continuous or discrete values. In both cases, with the purpose of increasing the bribe model adequacy and using the apparatus of LP-calculus, we shall build the discrete distribution on the chosen intervals of splitting the parameter values. We give the grade number for each interval. The grades make group of incompatible events (GIE). The probabilities of grade-events are determined by the formula

$$P_j = N_j \ / \ N. \qquad (17.12)$$

where: N_j is the number of parameter values in statistics with the given grades; N is the number of parameter values in the statistics.

The service parameter has the average value Y_m and admitted value Y_{ad} (Fig. 17.4). The probability $P\{Y < Y_{ad}\}$ will be named by us the bribe risk. The scenario of the bribe is formulated as follows: if the service parameter is greater (smaller) than the admitted value, then we suspect that there must have been a bribe.

Thus, for the service parameter at the given *Risk*, we can compute the following: the admitted value Y_{ad} , the number of the values of the parameter in the "tail" of the distribution N_{ad}, the entropy of the probabilities of the parameter in "tail" of the distribution H_{ad}.

In numerous publications in the field of *VaR (Value-at-Risk)* theory, the authors investigated tails of distribution. For these purpose, various distributions and conditional probabilities are suggested, which have no rigorous justification. In the bribe LP-theory, it is not to be done because we use any distribution laws given by a discrete line.

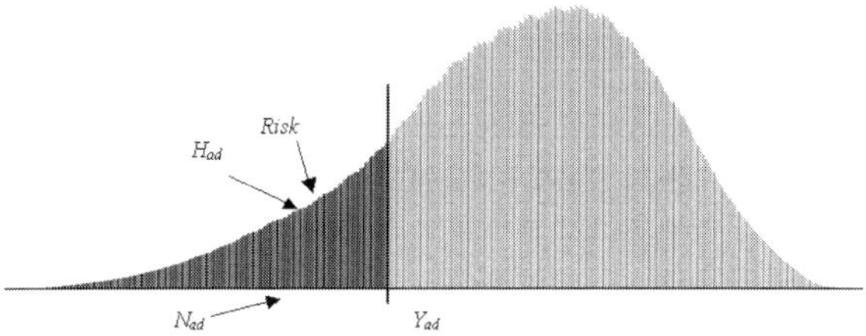

Fig. 17.4. The discrete distributions of service parameter

Example 3. $N = 700$ clients have been served. The parameter Y_1 determines the duration of the service by days and has $N_1 = 30$ days-grades. The probabilities $P1_r$ are calculated, $r = 1, 2, \ldots, 30$ by (17.28). The admitted parameter value is $Y_{1ad} = 10$ and the risk value is $Risk1 = 0,2$. The suspicion of the bribe is caused if $Y_1 < Y_{1ad}$.

Let there be one more, the service parameter Y_2 having $N_2 = 20$ grades, the admitted parameter value Y_{2ad}, and the risk $Risk2$. The logic variables corresponded with the service parameters. The logic variables can be dependent, but is not initially, for they are contained in the certain logic formula that determines the dependency between them.

For the case of two service parameters Y_1 and Y_2, we have $N = N_1 \cdot N_2 = 30 \cdot 20 = 600$ combinations of service. The L-functions for two different service combinations $Y_1 Y_2$ are orthogonal (the product of the logic functions of different combinations is equal to zero) as these combinations contain different grades for Y_1 and Y_2, belonging to GIE.

The property of orthogonality of different service combinations allows us to pass from the L-functions to the algebraic expressions for probabilities, that is the L-variables are to be replaced by the probabilities and the signs "or" are to be replaced by "plus."

It is easy to calculate the number of combinations satisfying the condition

$$P\{(Y_1 \leq Y_{1ad}) \vee (Y_2 \leq Y_{2ad})\} \qquad (17.13)$$

and calculate the bribe probability for this condition.

Example 4. The service parameter at acceptance in kindergarten

At acceptance in kindergarten, one shows the following documents, which we set with the following serial numbers:

1. The application from parents of the acceptance in the kindergarten,
2. The birth certificate,
3. The passport of some parents,
4. The document, which verifies the benefit,
5. The medical card.

Statistical data of the acceptance of children in the kindergarten are present in Table 17.5. In all was accepted 50 children with the different time of queuing: from 1 day to 400 days. Statistical data are quite enough to build the probability density function for the random service parameter — the waiting time Y. We shall also use the information of the produced documents (their full set are 1, 2, 3, 4, 5) and especially; of the document, which verifies the benefit (the document 4). We pick out the following estimations of the parameters of the statistics of the acceptance of children:

- $Y_{min} = 1$ is the minimal waiting time;
- $Y_{max} = 400$ is the maximal waiting time;
- $Y_{av} = 60.2$ is the average waiting time;
- $\sigma = 9.9$ is the least square value, if the waiting parameter distribution Y is the normal one;
- $N_{ben} = 7$ is the number of accepted in the kindergarten with benefit;
- $Y_{ben\ min} = 21$ is the minimal waiting time of children with benefit;
- $Y_{bem\ max} = 156$ is the maximal waiting time of children with benefit;
- $Y_{ben\ av} = 62$ is the average waiting time of children with benefit.

From the given values of the parameters of the statistics, we see the low of the probability density function for the service parameter Y is not normal. For choice of the probability density low for the service parameter Y, the considered period {1,400} is divided into intervals in 15 days (column 2 in Table 17.5) and the number of children $N_j; j = 1, 2, 3, \ldots,$ in the intervals is calculated (column 3). The probability of hitting in the interval is equal (column 4):

$$P_j = N_j \ / \ N. \tag{17.14}$$

The sum probabilities for the first five intervals equals to nearly 1. The probability density low may be selected as Weibullized distribution for the service parameter Y because it is usually used in the reliability problems with the greatest probability density in the distribution begin.

The interval of the acceptance waiting time of children with benefit {21,156} is not placed in the beginning of the interval {1,400}, but is removed in the right namely from bribes: some children are accepted in the kindergarten for bribes for the short time than children with benefits.

Let us denote the admitted value of the service parameter as Y_{ad} (column 5 in Table 17.5) and shall consider the children with the service parameter

$$Y < Y_{ad} \tag{17.15}$$

are accreted in the kindergarten for the bribe.

Table 17.5. Statistics of acceptance in the kindergarten

Index of acceptance	Waiting time of acceptance in the kindergarten, in days	Index of produced documents in the kindergarten
1	3	1
2	92	1, 2, 3, 4, 5
3	365	1, 2, 3, 5
4	1	-
5	10	1, 2, 3, 5
6	52	1, 2, 3
7	12	1
8	45	1, 2, 3, 4, 5
9	400	1, 2, 3, 5
10	80	1, 2
11	5	-
12	25	1, 2, 3, 5
13	40	1, 2, 3, 5
14	60	1, 2, 3, 5
15	12	1
16	18	1, 3
17	35	1, 2, 3, 5
18	92	1, 2, 3
19	21	1, 2, 3, 4, 5
20	30	1, 2, 3, 5
21	62	1, 2, 3, 5
22	152	1, 2, 3, 5
23	4	1, 2, 3, 5
24	25	1, 3
25	65	1, 2, 3, 4, 5
26	42	1, 2, 3, 5
27	252	1, 2, 3, 5
28	49	1, 2, 3
29	2	1
30	1	-
31	10	-
32	23	1, 2, 3
33	45	1, 2, 3, 5
34	123	1, 2, 3
35	56	1, 2,3
36	3	3
37	34	1, 2
38	112	1, 2, 3, 5
39	38	1, 2, 3, 4, 5
40	28	1, 2, 3, 5
41	72	1, 2, 3
42	62	4
43	12	1

continues

Table 17.5. **Continuation**

Index of acceptance	Waiting time of acceptance in the kindergarten, in days	Index of produced documents in the kindergarten
44	43	1, 2, 3
45	72	1, 2, 3
46	31	1, 3
47	23	1
48	3	2
49	156	1, 2, 3, 4
50	11	1, 2

The number of children, accepted for bribe N_{ad}, is calculated at the condition (17.16) (column 6). Thus, the risk of the bribe (suspicion in bribe) equal to (column 7):

$$Risk = N_{ad} \ / \ N. \tag{17.16}$$

Thus, Table 17.6 shows the dependence between the parameters of risk $N_i, P_i, Y_{ad}, N_{ad}, Risk$ at the acceptance in the kindergarten.

17.7 Conclusion

Problems of identifying the bribe LP-models, the estimation, and the analysis of bribe probabilities on the basis of the risk LP-theory with groups of incompatible events, as follows from expression (17.6), have an extremely high computing complexity and can be solved only by means of modern computers and special logic *software*. A complex of *software* for the solution of all those problems of training, modeling, and of analysis of bribes has been elaborated. You may get the detailed information on these software in [3, 30], on *www.ipme.ru/ipme/labs/iisad/sapr.htm*, and also by *E-mail*: *risk@sapr.ipme.ru*.

The bribe LP-models and the corresponding *software* are intended for the internal security and checking services of companies and banks and the department of "Economic crimes" of towns with the purpose of detecting bribes according to the statistical data. They can be used for development of norms and standards on service parameters, too.

Basic results of the current work are the following:

1. It is offered to use the risk LP-theory with GIE for the development of the bribe LP-models with the purpose of revealing, estimating, and analyzing their probabilities on the evidence of the statistical data.
2. The construction of the risk LP-model comprises the following:

Table 17.6. Dependence between the parameters of risk $N_i, P_i, Y_{ad}, N_{ad}, Risk$

Index	Interval of waiting time, in days	Number of accepted in intervals, N_j	Probability of acceptance in intervals, P_j	The admitted value of of the service parameter, Y_{ad}	Number of accepted with bribe, N_{ad}	The risk of bribes $Risk$
1	2	3	4	5	6	7
1	1 – 15	14	0.28	15	14	0.28
2	16 – 30	8	0.16	30	22	0.44
3	31 – 45	9	0.18	45	31	0.62
4	46 – 60	3	0.06	60	34	0.68
5	61 – 75	6	0.12	75	40	0.8

- The presentation of the bribe risk L-model in PDNF for the estimation of the possible number of the bribe combinations and the computer complexity of algorithms;
- Recording the bribe L-model in DNF recording the scenario of the bribe in the form of remarks or graphs, or in the form of the shortest ways, or the limited set of events.
- The transformation of the bribe risk L-model from DNF into ODNF;
- Recording the bribe P-model according to ODNF;
- Identifying the B-model of the bribe according to the statistical data taking into account GIE;
- Analysis of the B-model of the bribe with the calculation of the contribution of the signs and grades into the possibility of a bribe, the average probability of bribes, and the exactness of the risk LB-model.

3. We considered the scenarios and the risk LP-models of bribes:
   - institutions by parameters of their functioning,
   - officials by parameters of their behavior,
   - institutions and officials by parameters of the service.
4. Examples are given of the estimation and analysis of probabilities of bribes on the basis of identifying the bribe LP-model on the statistical data.
5. The developed bribe risk LP-models can be used both individually and all together.
6. Software has been developed for identifying risk LP-model and for the estimation and analysis of the probabilities of bribes.

18

LP-Model of Security Portfolio Risk

In the world there are more important things than the finest discovery, it is a knowledge of a method by which they have been made.

G. Leibniz

The logic and probabilistic theory of security portfolio risk (*Logic & Probabilistic Value-at-Risk*), based on arbitrary discrete distribution of security yields and not using the normal law of distribution, is proposed. The logic and probabilistic functions are suggested, problems of selection of an optimal portfolio are formulated, and methods of risk analysis and forecasting portfolio risk by the *LP–VaR* theory are described.

Investments are the basis of the market economy in developed countries. The security portfolio theory is the most widespread modern theory of investments. It makes it possible to optimize, simulate, analyze the risk and operate by the security portfolio risk. It solves the problems of forecasting and optimization of yield and risk.

In Markowitz's theory and VaR-approach (Value-at-Risk), "models of averages and dispersions" are used [17, 18, 132–134]. For each security in a portfolio the yield, as the mean of distribution, and the risk, as the mean square deviation and measure of uncertainty of yield, are taken into account. Such concepts as diversification, curves of indifference of the investor, available and efficient sets of portfolios, are used. The normal distribution laws of yield both for each security, and for total portfolio, are used.

The problem of portfolio optimization was formulated and solved in 1952 by H. Markowitz, who is one of the founders of the modern portfolio theory. For those results he later received the Nobel Prize in economics. Further, the theory of the optimum portfolio was developed as the theory of *VaR* in papers of many scientists.

E.D. Solojentsev, *Scenario Logic and Probabilistic Management of Risk in Business and Engineering*, DOI 10.1007/978-0-387-77946-1_18,

18.1 Selection of the optimum portfolio by VaR

The yield of security j for one period can be calculated using the formula [17, 18, 133]

$$Z_j = (C_{1j} - C_{0j}) \ / \ C_{0j}, \tag{18.1}$$

where C_{0j} is the security yield at the moment $t = 0$; C_{1j} is the security yield at the moment $t = 1$. At first, let us consider the traditional method of the portfolio selection on *VaR*. Yields of security portfolio $Z_1, \ldots, Z_j, \ldots, Z_n$ are random values with the normal distribution laws, defined by their mean values $Z_{1m}, \ldots, Z_{jm}, \ldots, Z_{nm}$ and dispersions. The portfolio yield Y, as the function of random variables $Z_1, \ldots, Z_j, \ldots, Z_n$, is also the random variable, determined by the mean value Y_m and dispersion (Fig. 18.1). A yield distribution of the portfolio Y is described by many-dimensional function of a normal distribution with covariance matrix of the random yields $Z_1, \ldots, Z_j, \ldots, Z_n$.

The problem of the portfolio selection is to find optimal values x_j, $j = 1, 2, \ldots, n$ of the capital, invested in securities. The *VaR*-method of the portfolio selection uses the criterion of affordable losses. It is the typical case of the investor who tries to avoid risk.

$$Y_m = \sum_{j=1}^{n} x_j Z_{jm}. \tag{18.2}$$

On the parameters x_j, $j = 1, 2, \ldots, n$ the restriction is imposed: its sum equals to 1. The formula for standard deviation of yield of portfolio security is as follows:

$$\sigma_y = [\sum_{i=1}^{N} \sum_{j=1}^{N} Z_i \cdot Z_j \cdot \sigma_{ij}]^{1/2},$$

where σ_{ij} is the covariance of yield of security i and j. The optimal portfolio is determined from the condition of the maximum of the admitted yield of the portfolio at the given risk: [17, 18, 133]:

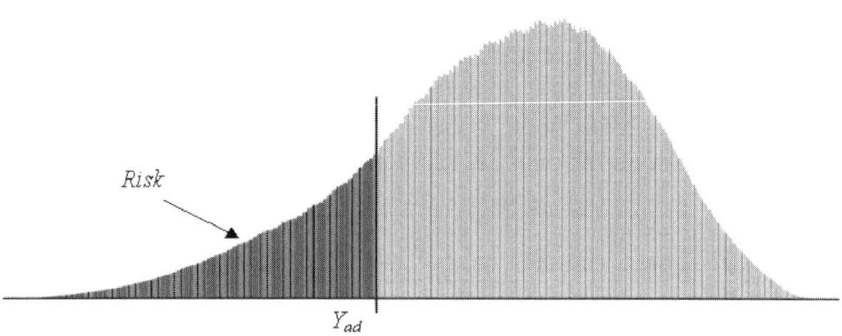

Fig. 18.1. Distribution of the yield of portfolio security

Table 18.1. Initial data of securities

Number of security	Yearly rate of yield, Z_{jm}, %	Standard deviation of yield, σ_j, %
1	12	1
2	13	2
3	12	1
4	14	1

Table 18.2. Optimum structures of portfolios by *VaR*

Parts of securities	Optimization by *VaR*	
	Risk = 0.05 %	*Risk* = 0.01 %
x_1	0.0	0.075
x_2	0.087	0.103
x_3	0.0	0.077
x_4	0.913	0.745

$$Y_{ad} = Y_m - VaR \rightarrow \max_x, \tag{18.3}$$

where Y_m, Y_{ad} are the expected and admitted yields of the security portfolio respectively, Risk is a risk, VaR is the possible losses, depending on *Risk*.

Initial data for the problem of the portfolio structure optimization are the list of securities, which can get to the portfolio, their expected yields and standard deviations from the mean values for each security.

In Table 18.1, the parameters of four securities [134, 135] are given. For simplicity, it is assumed that yields of securities are independent; then the factors of correlation $\rho_{jk} = 0$ for $j \neq k$, and $\rho_{jj} = 1$. Results of calculations are given in Tables 18.2 and 18.3.

The obtained results demonstrate the properties of diversification of the portfolio. The expected yield of the portfolio is placed between maximal and minimal yields of securities. The portfolio variation is less than the least variation of securities. The readiness of the investor to risk is determined in the

Table 18.3. Parameters of optimum portfolios by *VaR*

Optimization	Expected yield of portfolio, Y_m, %	Standard deviation yield, σ_p, %	Admitted yield portfolio, *Risk* = 0.05, %	Admitted yield portfolio, *Risk* = 0.01, %
By *VaR*, *Risk* = 0.05	13.91	0.93	12.38	—
By *VaR*, *Risk* = 0.01	13.59	0.78	—	11.77

terms of *VaR*, namely, by *Risk*. The less *Risk*, the less the tendency of the investor to risk.

In the selection theory of the optimal portfolio by *VaR*, the law of yield distribution of each security was assumed to be normal, defined by the mean value and the dispersion; interdependence of security yields is taken into account by covariance matrix. Such assumptions result in essential shortcomings:

- The horizon of investment limitations of few weeks, because the law of distribution cannot be considered as the normal law;
- Non-transparency of analysis and forecasting portfolio non-success risk by behavior of "a tail" of distribution.

18.2 Selection of the optimal security portfolio by *LP-VaR*

In numerous works in the field of *VaR* theory of security portfolio, the authors investigated "fat" tails of yield distribution of the portfolio [3, 30, 125, 126, 151]. For this purpose, various distributions and condition probabilities for "tails" are suggested, which have no rigorous justification.

All these shortcomings force one to look for other approaches and methods for the selection, analysis, and management of portfolio risk, in particular, on a basis of the logic and probabilistic approach, which does not use analytical distributive laws. The credit risk LP-theory, using "standard" data, is shown to be two times more accurate and seven times more robust in the classification of credits than other well-known methods.

Initial data. The initial one for computation are the statistical data of the security yields, presented as tabular. The table includes itself the security yields $1, \ldots, j, \ldots, n$ in certain moments of time, the interval between them can be equal to day, month, week, etc. States of portfolio is paginated $1, \ldots, N$. Yields of security portfolio $Z_1, \ldots, Z_j, \ldots, Z_n$ are the system of random values, which determines the yield of the portfolio Y.

Quantization. For passing to discrete distributions, we separate changing security yield to intervals Z_{jr} (Fig. 18.2). The number of intervals in separating the yield of different securities N_j in common case can be various. Numerated intervals for all of them are considered as the grades $r = 1, 2, 3, \ldots, N$ of the states of the share. Thus, intervals correspond to the random grade-events, creating GIE. Statistic data on securities will be presented in Table 18.4. Grades of securities are placed in cells of this table.

The sum of probabilities of grade-events in every GIE equals 1. Every grade-event is characterized by the yield (the average on the interval) and by the probability of appearance in Table 18.4

$$P_{jr} = N_{jr} \ / \ N, \tag{18.4}$$

where N_{jr} is the number of finding oneself in the interval r.

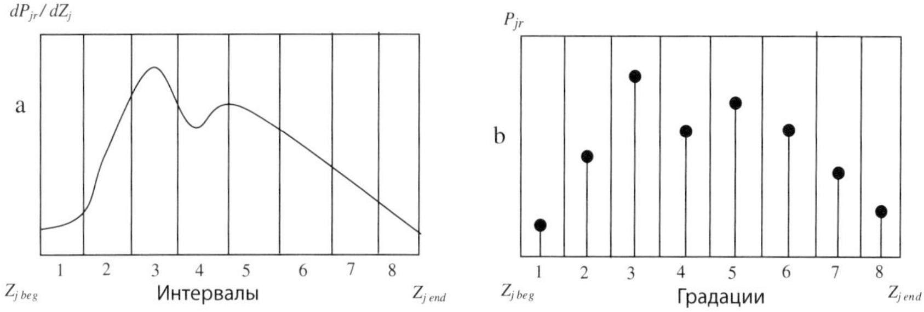

Fig. 18.2. Replacement of continuous distribution of yield by discrete one

Logic. Let us describe the logical function for the possible states of the portfolio in the perfect disjunctive form

$$Y = Y_1 \vee Y_2 \vee \ldots \vee Y_i \vee \ldots \vee Y_N, \tag{18.5}$$

where each state of the portfolio from N ones is defined by L-function

$$Y_k = Z_1 \wedge \ldots \wedge Z_j \wedge \ldots \wedge Z_n, \tag{18.6}$$

including all logical variables (securities of portfolio).

Every logical variable in (18.6) has many values equaled to the number of grades or intervals, on which its yield is separated. The logical functions for two different states of portfolio, for example

$$\begin{aligned} Y_i &= Z_1 \wedge \ldots \wedge Z_{jr} \wedge \ldots \wedge Z_n; \\ Y_{i+1} &= Z_1 \wedge \ldots \wedge Z_{j\ r+1} \wedge \ldots \wedge Z_n, \end{aligned} \tag{18.7}$$

is orthogonal, because Z_{jr} and Z_{jr+1} belong to one GIE

$$Z_{jr} \wedge Z_{j\ r+1} = 0. \tag{18.8}$$

The property of orthogonalization of terms in (18.5) allow us to pass from logical functions to algebraic expression for computing probabilities of discrete distribution of portfolio yield on intervals, analyze the portfolio risk by

Table 18.4. States and grades of security yields

States i	Security 1, yield, Z_1	... ...	Security j, yield, Z_j	... ...	Security n, yield, Z_n	Portfolio, yield, Y
1						
...	...	...	...	...	...	...
i			$Z_{j\ r}$			
...	...	...	...	...	...	...
N						

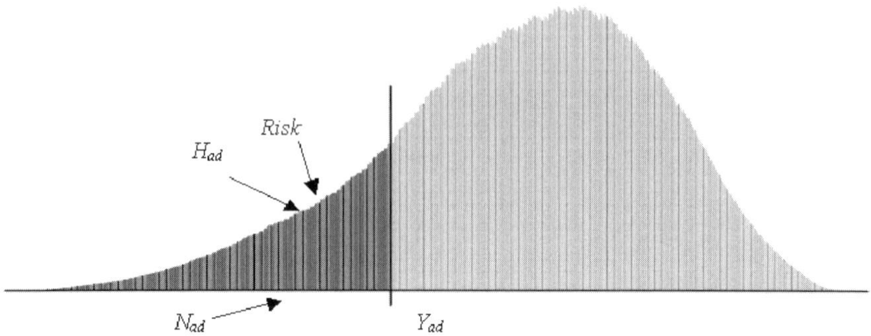

Fig. 18.3. Discrete distribution of portfolio security: Y_{ad} *is the admitted yield, Risk is a risk,* N_{ad} *is the number of states of portfolio in "the tail"* H_{ad} *is the entropy of probabilities in "the tail,"*

contributions of security grades of securities of portfolio, compute the condition probabilities, and also overcome exponential computing difficulty of algorithm.

Proceed from discrete distribution of security yields $j = 1, 2, \ldots, n$, the number of the possible states of portfolio or the number of logical terms in (18.5) is equal to

$$N_{max} = N_1 \cdot N_2 \cdot \ldots \cdot N_j \cdot \ldots \cdot N_n. \tag{18.9}$$

where $N_1, N_2, \ldots, N_j, \ldots, N_n$ are the numbers of grades in yields.

Selection of optimum portfolio by *LP-VaR*. As a criterion function for the solution of the problem of optimization, one of the following criteria is used (Fig. 18.3):

(1) Maximization of minimally admitted yield of the portfolio, at given risk:

$$Y_{ad} \to \max_x; \quad Risk = \text{const}, \tag{18.10}$$

(2) Minimization of risk at the given admitted yield:

$$Risk \to \min_x; \quad Y_{ad} = \text{const}. \tag{18.11}$$

After computing values x_j at the step iterative optimization, they are normalized in order to fulfill the condition

$$\sum_{j=1}^{n} x_j = 1; \; j = 1, 2, \ldots, n.$$

Determination of the admitted yield. Calculation of of the admitted yield of the portfolio Y_{ad} for the given risk *Risk* is a complex algorithmic problem. In Section 11.6, we consider different methods of its solution: the methods of interpolation, sorting, and bisection. At algorithmic realization of the mentioned methods of calculation of the admitted yield Y_{ad}, we should take into account the following property of integer numbers, investigated by G. Weyl [3, 7]. The yield distribution of the portfolio in each interval, which it is broken, can have one value only, if $x_j = 1$ and other x_j are equal to zero, otherwise the yield can have many different values. From the known Weyl's theorem about the remainder at division of integers, it follows that we will have different values of the portfolio yield with probability 1 in the interval. Variety of the values can be wide. The number of different values of the portfolio yield in one interval depends on multiplicity of $x_1, x_2, \ldots, x_n$ (we can turn them into integers and find out the maximum common divisor). For example, if $x_1 = x_2 = x_3 = x_4 = 0.25$, then in one yield interval there are only four different yields ($100/25 = 4$ where 100 is the sum of x). If the values of the parameters are $x_1 = 0.1; x_2 = 0.2; x_3 = 0.3; x_4 = 0.4$, then in one yield interval there are only ten different yields ($100/10 = 10$). If $x_1 = 0.11$; $x_2 = 0.17$; $x_3 = 0.19$; $x_4 = 0.4$, then in the yield interval there are 100 different yields ($100/1 = 100$). Number of different yields increases as the number of digits behind the decimal grows. At the algorithmic method of resolving the optimization problem for the portfolio, we can accordingly choose steps of increment of $x_1, x_2, \ldots, x_n$.

Risk analysis. To analysis of the portfolio risk by behavior of "tail" of distribution (Fig. 18.3) are devoted a lot of works [120 – 125], but its results are not transparency. Let us formulate the task of analysis of portfolio risk in *LP-VAR*. Let the optimal portfolio be constructed and the stock parts $x_1, \ldots, x_j, \ldots, x_n$, invested in each security $1, 2, \ldots, n$, be known. The portfolio analysis is fulfilled by *LP–VaR* using the algorithmic method on a computer by calculating contributions of grade-events into "the tail."

The contributions of grade-events of securities to the portfolio admitted yield Y_{ad} are equal to

$$D_{jr} = N_{jr}/N_{ad}, \quad j = 1, 2, \ldots, n; \quad r = 1, 2, \ldots, N_j, \tag{18.12}$$

where N_{ad} and N_{jr} are the numbers of all conditions of the portfolio and of conditions of the portfolio with grade r of security j, satisfying the condition

$$Y < Y_{ad}. \tag{18.13}$$

The contributions of grade-events to *Risk*:

$$C_{jr} = P_{jr}/Risk, \quad j = 1, 2, \ldots, n; \quad r = 1, 2, \ldots, N_j, \tag{18.14}$$

where P_{jr} is the summary probability of conditions of the portfolio with grade-event r of security j, satisfying the condition (18.3).

Grades or their groups, having the greatest contributions, are the best indicators, showing opportunity of financial loss for a client. These contributions indicate grades of securities, which are the most dangerous in the portfolio and which one should take into account at forecasting non-success. These contributions are the base for management of portfolio by replacement of some securities on other ones or changing parts $x_1, x_2, \ldots, x_n$ of the capital invested to portfolio.

Selection of optimal structure of portfolio leads to increasing the number of the dangerous states of the portfolio in "the tail." Therefore the criterion function of the optimization task of the structure of the portfolio at the given *Risk* can also be given as follows:

$$N_{ad} \rightarrow \max_{x}. \tag{18.15}$$

Entropy of yield probabilities in "the tail" of distribution. The level of heterogeneity or variety of probabilities of the set of states-events in "the tail" depends on the number of states and its probabilities. For measuring variety of probabilities in the set we use entropy, calculating from expression

$$H_{ad} = -\sum_{i=1}^{N} P_i \cdot lnP_i, \tag{18.16}$$

where H_{ad} is the entropy of the set, P_i is a probability of appearing events.

The entropy by (18.16) has the following properties:

1. One equal to zero, when appearing one event is a certain event.
2. One has maximum, when appearing this event is equiprobable attribute.
3. One increases at increasing the number of events in the set.
4. One has the property of additivity.

The computing research shows the *Risk* and the entropy H_{ad} are connected by linear function [126].

18.3 Portfolio with independent yields of stocks

The probability of the grade-event r in the security-events j is equal to

$$P_{jr} = P\{Z_{jr} = 1\}. \tag{18.17}$$

The mean yields of security j, as a random j, is as follows

$$Z_{jm} = \sum_{j=1}^{N_j} Z_{jr} \cdot P_{jr}. \tag{18.18}$$

The yield of any condition i of the portfolio is equal to the sum of corresponding weighted yields included in it:

$$Y_i = x_1 \cdot Z_{1\ J1} + \ldots + x_j \cdot Z_{j\ Jj} + \ldots + x_n \cdot Z_{n\ Jn}, \tag{18.19}$$

where $x_1, \ldots, x_n$ are the stock parts, invested in each stock.

The probability of any condition i of the portfolio is equal to

$$P_i = P_{1\ J1} \cdot P_{2\ J2} \cdot \ldots \cdot P_{j\ Jj} \cdot \ldots \cdot P_{n\ Jn};\ i = 1, 2, \ldots, N. \tag{18.20}$$

In (18.19 and 18.20), indexes $J1, J2, \ldots, Jn$ belong to following sets of values

$$J1 \in \{1, N_1\}; \quad J2 \in \{1, N_2\}; \quad \ldots; \quad Jn \in \{1, N_n\}.$$

From orthogonalization of terms in (18.5), it is followed

$$\sum_{i=1}^{N} P_i = 1. \tag{18.21}$$

The mean value of the yield of the security portfolio is equal to

$$Y_m = \sum_{i=1}^{N} Y_i \cdot P_i. \tag{18.22}$$

Let us build the discrete distribution of yield of the portfolio. For that, all interval of changing portfolio yield we break up on intervals and fulfil summation of probabilities of portfolio states P_i on intervals of the portfolio yield $r = 1, 2, \ldots, N_y$.

Computing difficulty. Passing from (18.9), (18.19), and (18.20), it arises the question about the computed difficulty of the algorithm of the portfolio risk LP-method. The algorithm has the exponential difficulty, which depends on the number of securities in the portfolio n and on the number of intervals N_j, on which security yields are broken.

Seemingly it must put us into shock, as values of the parameters N_j and n are big enough: $n = 10 \div 100$; $N_j = 30 \div 100$. However the real approach to the task takes away all apprehension. The task is not in absolute search of portfolio states (18.9), but in building the discrete distribution of portfolio yield. The limited number of random states of the portfolio we can obtain by modeling Monte Carlo method [136]. Of course, in this case it does not fulfill (18.21) and probabilities of obtained states should normalize to get in sum 1.

Comparison *VaR and LP-VaR*. Let us solve the problem of selection of the optimum portfolio by *LP-VaR*, by using the data on four securities with the normal laws of distribution of yields (Table 18.1).

We replace the analytical normal distributions of yields of securities by discrete distributions with the interval 0.5%, making the corresponding computing by the formula for normal low at $x_1 = x_2 = x_3 = x_4 = 0.25$; $\triangle Z = 0.5\ \%$; $Y_{ad} = 11.35\ \%$; $Risk = 0.05$.

The problem of selection of the portfolio optimal structure with the criterion function (18.10) is solved using different steps of quantization $\Delta Y = 0.5\ \%$

Table 18.5. Results of investigation by *VaR* and *LP-VaR* with the normal distributions

Vari-ants	x_1	x_2	x_3	x_4	Step ΔY, %	*Risk*	Y_m, %	Y_{ad}, %	Commen-tary
1–1	0.0	0.087	0.0	0.913	–	0.05	13.91	12.38	*VaR*
1–2	0.0	0.084	0.0	0.916	0.5	0.05	13.92	12.08	*LP-VaR*
2–1	0.075	0.103	0.077	0.745	–	0.01	13.59	11.77	*VaR*
2–2	0.081	0.083	0.080	0.757	0.5	0.01	13.59	11.46	*LP-VaR*

for *Risk* = 0.05 (Var. 1–1, 1–2, 1–3), and for *Risk* = 0.01 (Var. 2–1, 2–2, 2-3). The results, given in Table 18.5 for the parameters x_1, x_2, x_3, x_4, and the mean Y_m and admitted Y_{ad} yields of the portfolio, confirm close coincidence of solutions obtained using the discrete distributions and analytical normal distributions yields of securities and the portfolio.

Decreasing the step of yield digitization $\triangle Z$ for securities and portfolio with 0.5 to 0.1% increases the accuracy of parameters x, Y_{ad}, Y_m. Investigations determine we should choose the digitization step, equal for securities and portfolio, to avoid the systematic error from features of yield distribution inside intervals.

Let us solve the problem of selection of the optimal portfolio by the method of *LP–VaR*, when the classical *VaR* cannot manage it. We use arbitrary non-normal discrete distributions of yields for four securities (Fig. 18.4). The discrete distribution of the portfolio yield is constructed at $x_1 = x_2 = x_3 = x_4 = 0.25$. The diagrams are shown by points of different type for the middle points of intervals. For convenience, the points are connected by lines. It is notable that the distribution of the portfolio yield is visually similar to the normal one. It is quite natural, for the distribution is constructed on vast quantity of different orthogonal conditions of the portfolio (18.5).

The problem of selection of the optimum structure of the portfolio with the criterion function (18.10) is solved with steps of quantization $\Delta Z = 1\%$ both for *Risk* = 0.05 (Var. 1–1) and *Risk* = 0.01 (Var. 2–1). The solutions given in Table 18.6 for the parameters x_1, x_2, x_3, x_4, and the mean Y_m and admitted Y_{ad} yields of the portfolio confirm the efficiency of the proposed LP-theory of selection of the optimum portfolio with use of discrete non-normal distributions of security yields

In Table 18.7, we bring contributions C_{jr} and D_{jr} for the grade-events, which hit upon "the tail" of yield distribution for the optimal portfolio with

Table 18.6. Results of investigation by *LP-VaR* with average distributions

Vari-ants	x_1	x_2	x_3	x_4	N_{opt}	Step ΔY, %	*Risk*	Y_m, %	Y_{ad}, %
1–1	0.649	0.018	0.333	0.0	80	1.0	0.95	17.535	12.621
2–1	0.589	0.116	0.294	0.0	90	1.0	0.99	17.206	10.358

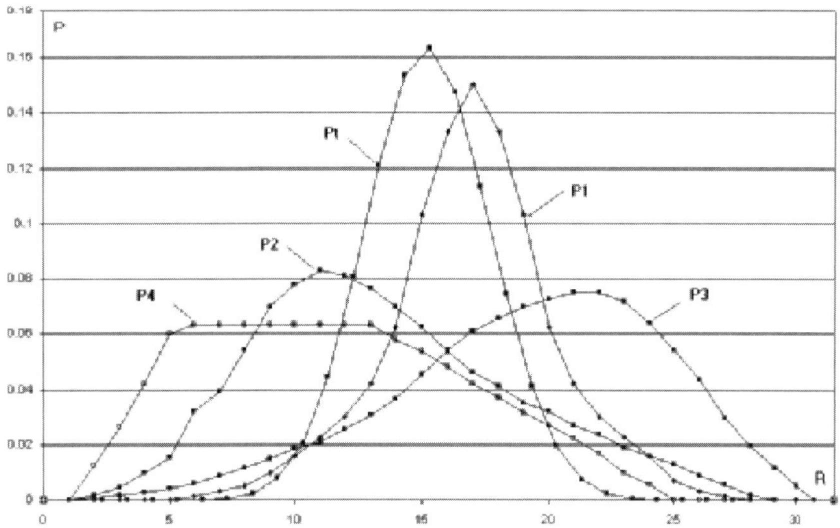

Fig. 18.4. Discrete non-normal distribution low of the security yields 1, 2, 3, 4 and the portfolio at $x_1 = x_2 = x_3 = x_4 = 0.25$; $\triangle Z = 1\%$; $Y_{ad} = 10.74\%$; $Risk = 0.5$

$x_1 = 0.075$; $x_2 = 0.103$; $x_3 = 0.077$; $x_4 = 0.745$ at $Risk = 0.01$ with normal yield distribution (Table 18.1). Contributions of grade-events C_{jr} and D_{jr} in the tail change differently. The contribution of grade-events D_{jr} into the yield Y_{ad} is a monotone decreasing with growth yields of grades or decreasing, and then equal to zero for remaining grades. The contributions of grade-events C_{jr} in Risk have points of extreme inside intervals. The most dangerous are contributions C_{jr} and D_{jr} of the first three grades of the yield of the fourth security.

Let us make the risk estimation of the current state of the portfolio. Using the current security yields $Z_1, \ldots, Z_j, \ldots, Z_n$, we compute the portfolio yield Y by formula (18.19) and from Table 18.4 determine $Risk$ on value Y by the interpolation method.

18.4 Portfolio with dependent yields of stocks

Let us have obtained the real data (Table 18.4). We correlate each asset j the logic variable Z_j and each interval of yield r of the asset j with logic variable Z_{jr} and with random grade-events respectively. For each asset, these grade-events arrange GIE.

For calculation of probability of each portfolio condition, it is necessary to consider n-dimensional system of random variables or n-dimensional random variable. For simplicity, we shall consider the case of a portfolio with two assets. Joint distribution of yields of two financial tools is presented in

Table 18.7. Contributions of grade-events of securities to "the tail"

r	$Z_1 = Z_3, \%$	$C_{1r} \approx C_{3r}$	$D_{1r} \approx D_{3r}$	$Z_2, \%$	C_{2r}	D_{2r}	$Z_4, \%$	C_{4r}	D_{4r}
1	8.5	0.0010	0.0845	6.0	0.0016	0.0565	10.5	0.044	0.361
2	9.0	0.0045	0.0819	6.5	0.0036	0.058	11.0	0.214	0.295
3	9.5	0.0158	0.0793	7.0	0.0058	0.0559	11.5	0.513	0.202
4	10.0	0.0432	0.0767	7.5	0.0102	0.0541	12.0	0.221	0.108
5	10.5	0.0913	0.0741	8.0	0.0181	0.0527	12.5	0.008	0.031
6	11.0	0.1502	0.0715	8.5	0.0287	0.051	13.0	0.0000	0.002
7	11.5	0.1918	0.0688	9.0	0.0453	0.049	13.5	0	0
8	12.0	0.1927	0.0665	9.5	0.0639	0.047	14.0	0	0
9	12.5	0.1533	0.0641	10.0	0.0745	0.045	14.5	0	0
10	13.0	0.092	0.0618	10.5	0.0886	0.0437	15.0	0	0
11	13.5	0.0429	0.0593	11.0	0.0869	0.0414	15.5	0	0
12	14.0	0.0155	0.0566	11.5	0.0954	0.0398	16.0	0	0
13	14.5	0.00436	0.0541	12.0	0.1017	0.0391	16.5	0	0
14	15.0	0.0009	0.0515	12.5	0.0955	0.0383	17.0	0	0
15	15.5	0.0002	0.0491	13.0	0.0882	0.0361	17.5	0	0
16				13.5	0.0664	0.0348			
17				14.0	0.0462	0.0326			
18				14.5	0.0322	0.0305			
19				15.0	0.0186	0.0292			
20				15.5	0.0125	0.0269			
21				16.0	0.0078	0.0256			
22				16.5	0.0043	0.0239			
23				17.0	0.0023	0.0217			
24				17.5	0.0009	0.0203			
25				18.0	0.0004	0.0180			
26				18.5	0.0002	0.0161			
27				19.0	0.0001	0.0147			
28				19.5	0.0001	0.0125			
29				20.0	0.0000	0.0097			

Table 18.8. Each sell of this table is the portfolio state. The probability sum of states in this table is equal to 1.

Each condition of the portfolio corresponds with a cell in this table. As it was said above, these conditions form GIE and the sum of their probabilities is equal to 1.

Probability (or frequency) of i-th condition of the portfolio:

$$P_{1r1-2r2} = N_{1r1-2r2} \;/\; N, \tag{18.23}$$

where $N_{1r1-2r2}$ is the number of occurrences of the set of event-gradations Z_{1r1} and Z_{1r2}.

For a portfolio with three assets, Table 18.8 will have the form of a parallelepiped in three-dimensional space. For "n" assets it will take the form

Table 18.8. Discrete distribution of yields for two securities

Z_1 / Z_2	Z_{21}	...	Z_{2-r2}	...	Z_{2-N2}
Z_{11}	P_{11-21}	...	P_{1r1-21}	...	P_{1N1-21}
...	...	...	...	...	...
$Z_{1\eta}$	P_{11-2r2}	...	$P_{1r1-2r2}$	...	$P_{1N1-2r2}$
...	...	...	...	...	...
Z_{1N1}	P_{11-2N2}	...	$P_{1r1-2N2}$	...	$P_{1N1-2N2}$

of n-dimensional parallelepiped. As probabilities of portfolio conditions, we shall accept frequencies of their occurrence. Thus, the probability of an i-th condition of the portfolio is defined as follows:

$$P_{1r1-2r2-\ldots-nrn} = N_{1r1-2r2} \;/\; N, \tag{18.24}$$

where $P_{1r1-2r2-\ldots-nrn}$ is the number of conditions of assets in a cell $1r1 - 2r2 - nrn$.

When we use formula (18.24), only some possible portfolio conditions their probabilities are distinct from null. All other conditions, which probabilities are equal to null, can be treated as impossible. Thus, there is a limited set of conditions of the portfolio. It is possible to calculate yields and probabilities of all conditions appearing in Table 18.8. Having sorted them, we shall obtain distribution of portfolio yield (Table 18.9).

Joint distribution comprises the information on dependence between asset yields, which has both advantages and disadvantages. The distribution gives the information not about all conditions of the portfolio, but of those that appear in the statistical data table. For distribution of a portfolio to be authentic it is necessary to collect information for sufficiently big period of time, which is not always possible. This allows us to assume that the technique does not suit, for modeling distribution of portfolio yield for big horizons of investment. The decision of the named problems with using dependency of the security yield from external factor and connection of **copula** is described below.

18.5 Portfolio with stock yields depending on external factor

Let us accept the following assumption: yields of securities do not depend, each on all others, but all yields depend on an external factor, for example,

Table 18.9. Distribution of the portfolio yield Y

Y	Y_1	Y_2	...	Y_i	...	Y_N
P	P_1	P_2	...	P_i	...	P_N

Table 18.10. Distribution of security yield concerning the factor f

Z_j / f	f_1	...	f_{rf}	...	f_{Nf}
Z_{j1}	P_{j1f1}	...	P_{j1frf}	...	P_{j1fn1}
...	...	...	...	...	...
Z_{jrj}	P_{jrff1}	...	$P_{1r1-2r2}$	...	$P_{1N1-2r2}$
...	...	...	...	...	...
Z_{1N1}	P_{11-2N2}	...	$P_{1r1-2N2}$	...	$P_{1N1-2N2}$

an index of the world market [138]. To carry out calculations, information is necessary not only about the prices of assets, but also about change of the external factor in parallel with the prices of assets. Such factor can be, for example, any index of the stock market, or yield of any well-known valuable stock. Then it is possible to trace its yield in percentage in parallel with yield of asset. An interval of change of the external factor can be broken into grades. Then, the initial data are presented as a table of the statistical data similar to Table 18.4. The last column of the table contains grades of the selected factor.

Using this table, a new table is constructed, similar to Table 18.4, where each cell contains grade-events. Let us construct conditional distributions of asset yields concerning the factor f. For this purpose, it is necessary to construct conditional distribution for each grade-event. For an asset j such distributions are shown in Table 18.10.

The sum of probabilities in each column is equal to 1:

$$\sum_{rf=1}^{N_f} P_{jr_j\ fr_f} = 1;\ r_j = 1, 2, \ldots, N_j;\ r_f = 1, 2, \ldots, N_f. \tag{18.25}$$

Thus, probability $P_{jr_j\ fr_f}$ is the frequency of occurrence of the grade Z_{jr_j} of asset j in the statistical data, provided that the factor f has the grade f_{r_f}

$$P_{jr_jfr_f} = P(Z_j = Z_{jr} \mid f = f_{r_f}),\ r_f = 1, 2, \ldots, N_f. \tag{18.26}$$

From statistics we determine probabilities of conditions of the factor f and write these values in Table 18.11. Probability P_{r_f} is a frequency of the grade f_{r_f} of the factor f in the statistic data:

$$P_{rf} = N_{rf}\ /\ N, \tag{18.27}$$

where N_{rf} is the number of grade occurrences f_{r_f} of the factor f; N is the number of considered conditions. The grade-events $f_1, \ldots, f_{N_f}$ arrange GIE.

18.6 Comparison of portfolio modeling methods by LP-VaR

Let us compare three above mentioned ways of modeling a security portfolio. To be specific we shall model the portfolio with two assets. For these purposes

Table 18.11. Probabilistic distribution of the factor f

F	f_1	f_2	...	f_{r_f}	...	f_{N_f}
P_f	P_1	P_2	...	P_{r_f}	...	P_{N_f}

we shall choose common stocks of the Russian companies: Russian Joint Stock Company "United Energy System of Russia" (UESR) and LUKOIL. As the factor we shall choose change of the index Russian Trade System (RTS), which depicts the general direction of movement of the market. The statistics on daily yields from October 2002 to September 2004 is used. Yield is calculated through the prices on closing by using the following simple formula, which does not take into account dividend payments:

$$Z_t = 100 \cdot (C_t - C_{t-1}) \;/\; C_{t-1}, \tag{18.28}$$

where Z_t is the yield of the current period (in this case, for a day), C_t is the price on closing the current period, C_{t-1} is the price on closing the previous period.

As a result of the analysis of distribution of histograms for the analyzed stocks and the factor, we ascertain that their distribution low differs from the normal one. Such conclusion can be made on the basis of Kolmogorov-Smirnov's test for check of the form of distribution. Conditional and unconditional distributions of assets related to the factor are presented in Tables 18.12 and 18.13. Distribution of the factor is given in Table 18.14.

Values of yields of portfolio conditions are identical in all three cases. Therefore we shall exclude them from consideration. Values of probabilities are different. For example, let us compute the probability of one state of portfolio for the case when all dependencies are not taken into account (case 1) and for the case with taking into account dependency from the factor (case 3). Let this state be the combination of grades Z_{11} of asset UESR and Z_{23} of asset LUKOIL.

The probability without taking into account dependency from the factor is equal: $P_{p1} = P_{11} \cdot P_{23} = 0.01027 \cdot 0.57084 = 0.00586$.

Table 18.12. Probabilistic distributions of securities of UESR (conditional and unconditional)

UESR Z_1	$f_1 \vee f_2 \vee \ldots \vee f_7$ (unconditional)	f_1	f_2	f_3	f_4	f_5	f_6	f_7
Z_{11}	0.0103	0.5	0.1667	0	0.0093	0	0	0
Z_{12}	0.0328	0.5	0.1667	0.0968	0.0217	0.0109	0	0
Z_{13}	0.1745	0	0	0.2742	0.1739	0.1196	0	1
Z_{14}	0.5092	0	0.5	0.4677	0.5373	0.4674	0	0
Z_{15}	0.2033	0	0.1667	0.11290	0.2019	0.2717	0.5	0
Z_{16}	0.0534	0	0	0.0484	0.0404	0.1087	0	0
Z_{17}	0.0164	0	0	0	0.0155	0.0217	0.5	0

Table 18.13. Probabilistic distributions of securities of LUCOUL (conditional and unconditional)

LUCOUL, Z_2	$f_1 \vee f_2 \vee \ldots \vee f_6 \vee f_7$ (unconditional)	f_1	f_2	f_3	f_4	f_5	f_6	f_7
Z_{21}	0.0164	0.5	0.3333	0.0161	0.0124	0	0	0
Z_{22}	0.1889	0.5	0.1666	0.4839	0.1677	0.0652	0	0
Z_{23}	0.5708	0	0.1666	0.3871	0.6615	0.4239	0	1
Z_{24}	0.2053	0	0.3333	0.0806	0.1491	0.4783	0.5	0
Z_{25}	0.0144	0	0	0.0161	0.0093	0.0326	0	0
Z_{26}	0.0021	0	0	0	0	0	0.5	0
Z_{27}	0.0021	0	0	0.0161	0	0	0	0

Table 18.14. Probabilistic distribution of the factor f

F	f_2	f_2	f_2	f_2	f_2	f_2	f_2
P	0.00411	0.0123	0.1273	0.66119	0.18891	0.00411	0.00205

The probability with taking into account the dependency from the factor is:

$P_{p1} = (P_{11f_1} \cdot P_{23f_1})P_{f1} + (P_{11f_2} \cdot P_{23f_2})P_{f2} + \ldots + (P_{11f_7} \cdot P_{23f_7})P_{f_7} = (0.5 \cdot 0) \cdot 0.0041 + (0.1666 \cdot 0.1667) \cdot 0.0123 + (0 \cdot 0.3871) \cdot 0.1273 + (0.0093 \cdot 0.6615) \cdot 0.6612 + (0 \cdot 0.4239) \cdot 0.1889 + (0 \cdot 0) \cdot 0.0041 + (0 \cdot 0) \cdot 0.0020 = 0.0069.$

Similarly we computed all other portfolio states. The number of all states is equal to $N = 7{\cdot}7 = 49$. Probabilities for the case with taking into account total dependence are computed trivially, without using the logical function. Probabilities of portfolio states for three cases are presented in Tables 18.15, 18.16, and 18.17.

From Tables 18.15, 18.16, and 18.17 it is seen that probabilities for all three cases are different. And at the full account of dependence, probability of a significant part of portfolio conditions are equal to zero (Table 18.6). From 49 possible portfolio states, the zero value take 22 states, or 45% from all possible states. It is explained the size of statistics for computing probabilities is not enough with taking into account of dependency between assets yields.

Table 18.15. Probabilities of the all portfolio states without taking into account the dependency

Z_1 / Z_2	Z_{21}	Z_{22}	Z_{23}	Z_{24}	Z_{25}	Z_{26}	Z_{27}
Z_{11}	0.000168	0.001939	0.005861	0.002108	0.000147	2.108E-05	2.108E-05
Z_{12}	0.000539	0.006206	0.018754	0.006746	0.000472	6.746E-05	6.746E-05
Z_{13}	0.002867	0.032972	0.099633	0.035839	0.002508	0.000358	0.000358
Z_{14}	0.008365	0.096201	0.290695	0.104567	0.007319	0.001045	0.001045
Z_{15}	0.003339	0.038403	0.116044	0.041742	0.002922	0.000417	0.000417
Z_{16}	0.000877	0.010085	0.030476	0.010963	0.000767	0.000109	0.000109
Z_{17}	0.000269	0.003103	0.009377	0.003373	0.000236	3.373E-05	3.373E-05

Table 18.16. Probabilities of the all portfolio states with taking into account the dependency

Z_1/Z_2	Z_{21}	Z_{22}	Z_{23}	Z_{24}	Z_{25}	Z_{26}	Z_{27}
Z_{11}	0.00821	0.00205	0	0	0	0	0
Z_{12}	0.00205	0.02258	0.00821	0	0	0	0
Z_{13}	0.00205	0.06981	0.09035	0,01232	0	0	0
Z_{14}	0.00410	0.08418	0.33675	0.07803	0.00616	0	0
Z_{15}	0	0.00616	0.11704	0.07597	0.00411	0	0
Z_{16}	0	0.00205	0.01437	0.03080	0.00411	0	0.00205
Z_{17}	0	0.00205	0.00411	0.00821	0	0.00205	0

Table 18.17. Probabilities of all portfolio states with taking into account from the factor

Z_1/Z_2	Z_{21}	Z_{22}	Z_{23}	Z_{24}	Z_{25}	Z_{26}	Z_{27}
Z_{11}	0.00178	0.00240	0.00441	0.00160	5.74E-05	0	0
Z_{12}	0.00208	0.00987	0.01549	0.00480	0.0004	0	0.00020
Z_{13}	0.00199	0.03764	0.10120	0.03076	0.00237	0	0.00056
Z_{14}	0.00742	0.09517	0.29649	0.10203	0.00715	0	0.00096
Z_{15}	0.00257	0.03302	0.11595	0.04731	0.00315	0.0010	0.00023
Z_{16}	0.00043	0.00879	0.02874	0.01429	0.00102	0	9.94E-05
Z_{17}	0.00012	0.00199	0.00853	0.00452	0.00023	0.0010	0

For the case without account of dependence of asset yields (Table 18.15), zero probabilities are absent in principle, but there are small ones. This case badly reflects the reality.

In Table 18.7 only 15 from 22 probabilities of portfolio state yield are reconstructed, which take zero values in Table 18.16. It is reached thanks to the use of orthogonal logical function, which helps to reconstruct even those states of the portfolio, that were not met in statistics. Dependence between assets yield is taken indirectly over dependence of asset yield from the factor.

This variant reflects reality rather well, if, for example, we take as a factor index of assets; yield values depend on selected assets. In this variant about 70% probabilities of portfolio states is reconstructed, which in the case taking into account dependence between yields of assets have zero values.

18.7 Examples of portfolio optimization by LP-VaR

For research, a sample of values of daily percentage yields was generated with stocks of Russian Joint Stock Company United Energy System of Russia, LUKOIL, and the Savings Bank, from October 2002 to September 2004. Yield was computed without taking into account dividend payments. Results of calculation of descriptive statistical characteristics and matrix of correlations are presented in Tables 18.18 and 18.19.

Table 18.18. Statistics of securities

Stocks	Average	Minimum	Maximum	Standard deviation	Dispersion
UESR	0.2560	–10.1085	9.5338	2.8483	8.1130
LUKOIL	0.1294	–7.3192	11.5702	2.1234	4.5088
Sberbank	0.1847	–8.8496	9.2796	2.0004	4.0017

Table 18.19. Matrix of correlations

Stocks	1	2	3
UESR	1.0000	0.5664	0.4657
LUKOIL	0.5664	1.0000	0.5923
Sberbank	0.4657	0.5923	1.0000

Table 18.20. Results at independent yields

	x_1	x_2	x_3	Y_{ad}, %	N_{ad}	*Risk*
Before	0.333	0.333	0.333	–2.695	263	0.2
After	0.118	0.398	0.482	–2.532	271	0.2

Using the obtained statistical characteristics of the analyzed stocks, it is possible to conclude that the mean value for all assets is positive, it changes in the range from 0.129 to 0.256. It can be explained by mainly positive price dynamics of the Russian stocks during this period on the background of sufficiently high volatility of the prices. Positive (negative) yield for one trading day for the analyzed stocks achieved 11.57 (-10.1) percent. Such market conditions are characterized by high values of standard deviations of yield, which change in the range from 2 to 2.84 for the analyzed stocks.

Proceeding from the analysis of correlations, it is possible to make a conclusion on sufficiently strong positive connection of the stock yields. It is explained by narrowness of the Russian stock market, which is basically concentrated on the stocks of oil-and-gas branch and electric power industry, what determines dynamics of the market as a whole.

Further in our investigation, the portfolio was formed of the analyzed stocks. Conditions of portfolio yield were modeled for the case of independent and dependent asset yields, and also for the case of dependence of asset yields on the external factor. Optimization of the portfolio was carried out with the purpose of definition of optimum stocks shares in the portfolio, with the purpose to maximize the admitted yield at the given risk value. Then for the case of dependence of asset yields on the external factor, the analysis of portfolio risk was carried out. We also calculated contributions of the event-gradations to the "tail" of yield distribution for the risk and the admitted yield.

Results of optimization by the criteria $Y_{ad} \rightarrow \max$ at $Risk = 0.2$ are presented: for the independent assets yields in Table 18.20; for the dependent

Table 18.21. Results at dependent yields

	x_1	x_2	x_3	Y_{ad}, %	N_{ad}	*Risk*
Before	0.333	0.333	0.333	−0.882	522	0.2
After	0.000	0.000	1	−0.796	513	0.2

Table 18.22. Results at yields dependent from the factor

	x_1	x_2	x_3	Y_{ad}, %	N_{ad}	*Risk*
Before	0.333	0.333	0.333	−1.472	441	0.2
After	0.112	0.486	0.400	−1.074	451	0.2

assets yields in Table 18.21; for the dependency of assets yields from external factor (index RTS) in Table 18.22.

Results of optimization for the three cases illustrate distinction in methods of modeling conditions of portfolio yield. In case of modeling independent asset yields, we have non-zero probabilities of all possible conditions of portfolio yield; this explains the easy process of optimization on all conditions of yield. In the case of dependent asset yields, probabilities of portfolio conditions are determined from statistics that not necessarily contain all possible conditions of yields; as consequence there is a number of zero probabilities of conditions. As a result, the process of optimization becomes more difficult. A solution to the given problem is seen by the authors in construction of a logical function describing all portfolio conditions. In the third case the logic function sets dependence of asset yields on the external factor. Advantage of the scheme is by the fact that both dependence between asset yields, and dependence of yield of asset on external factors can be refined as logic function. It allows us to approach in a more flexible way to the problem of dependence of the initial data, and also to solve the problem with limited size of statistics for management of risk of a portfolio of securities.

Analysis of the portfolio risk by *LP-VAR*. Let us determine contributions of gradations to the risk and to the admitted portfolio yield for the case of dependence of yields of stocks of UESR, LUKOIL, and the Savings Bank on the external factor (the index of RTS). Gradations or their groups, having the greatest contributions are the indicators showing possibility of losses higher than the admitted value. These contributions specify gradations of those financial tools that are most dangerous in the portfolio and to which special attention should be paid at forecasting unsuccess and management of structure of the portfolio. Calculations are presented in Table 18.23 for the case of three assets in the optimum portfolio: $x1 = 0.112$, $x2 = 0.486$, $x3 = 0.400$, $Risk = 0.2$.

The most dangerous contributions C_{jr} to the risk of the portfolio are grade-events numbered by 3, 5, 6, and 7 of the first set of stocks (UESR), grade-event 5 of the second set of stocks (LUKOIL), and grade-events 4, 5 of the third set of stocks (Savings Bank).

Table 18.23. Distributions of grade-events of securities in "tail" of distribution of the portfolio yield

j	Y_{1m}, %	C_{1r}	D_{1r}	Y_{2m}, %	C_{2r}	D_{2r}	Y_{3m}, %	C_{3r}	D_{3r}
1	–10.1	0.020	0.109	0.00	0	0	–9.67	0.020	0.261
2	–8.22	0.039	0.109	–8.85	0.020	0.239	–8.45	0.020	0.239
3	–6.95	0.176	0.087	–6.23	0.059	0.239	–5.89	0.059	0.239
4	–4.93	0.098	0.087	–4.64	0.098	0.239	–4.34	0.390	0.109
5	–2.71	0.255	0.087	–2.68	0.824	0.239	–2.31	0.491	0.130
6	–0.84	0.235	0.087	–0.79	0	0.043	–0.56	0.020	0.022
7	0.78	0.137	0.087	0.82	0	0	0.91	0	0
8	2.86	0.020	0.087	2.76	0	0	2.59	0	0
9	4.94	0.020	0.087	4.81	0	0	4.78	0	0
10	6.70	0	0.087	7.01	0	0	6.89	0	0
11	8.66	0	0.087	9.03	0	0	9.30	0	0

The most dangerous contributions D_{jr} to the portfolio yield are grade-events 1, 2 of the first set of stocks, grade-events 2, 3, 4, 5 of the second set of stocks, and grade-events 1, 2, 3 of the third set of stocks. In total, "tail" of distribution of yield of a portfolio has $N_{ad} = 271$ "dangerous" states of a portfolio.

Thus, knowing the event-gradations of stocks that bring the greatest contribution to the risk and yield of our portfolio, it is possible to operate effectively both the risk and the yield by changing dynamically shares of corresponding assets in the portfolio when new statistics appear in next periods.

In results of optimization of the portfolio by the entropy with initial parts of assets in the portfolio $x_1 = x_2 = 0.5$, entropy $H_{ad} = 0.640$, risk $Risk = 0.16$ at given yield $Y_{ad} = -1.3$ % are obtained: the criterion function $Y_{ad} = 0.469$, the optimal parts $x_1 = 0.08$; $x_2 = 0.92$ and $Risk = 0.1047$. Thus, the entropy of yield probabilities in "tail" of distribution can be used as the attribute of the portfolio for aim of its optimization.

18.8 Efficiency of portfolio management by LP-VaR

LP-VaR technique before the beginning of a new trading day optimizes an investment portfolio, taking into account statistics for the previous period. We introduce algorithm of investment portfolio management *LP-VaR* technique:

Table 18.24. Results of optimization of the portfolio by the entropy H_{ad}

State of portfolio	Admitted yield, Y_{ad}=const	Value of criterion function, H_{ad}	Value of risk, $Risk$	Part of asset 1, x_1	Part of asset 2 x_2
Before optim.	−1.3%	0.640	0.16	0.5	0.5
After optim.	−1.3%	0.469	0.10	0.08	0.92

1. Modeling conditions of the investment portfolio on the basis of statistics on daily yields of securities for the analyzed period;
2. Optimization of the portfolio for a day according to the chosen criterion of optimization;
3. The account of the day T in statistics and modeling conditions of the portfolio with the updated statistics;
4. Optimization of the portfolio for the day $(T+1)$ according to criterion of optimization (and so on, for all period of management of the investment portfolio).

For estimation of the portfolio management efficiency by *LP-VaR* technique, we analyzed the data on daily yields of three Russian stocks: UESR, LUKOIL, and the Savings Bank. The data from October 2002 to September 2004 were used for calculation of conditions probabilities of portfolio yields, for the case of dependence of asset yields in the portfolio on the external factor (the index of RTS). On the basis of the data from September 2004 to October 2004 (Fig. 18.5), the cost of the portfolios including the stocks in equal shares and the assets, whose shares in a portfolio change every day according to the suggested technique *LP-VaR*, was estimated and compared.

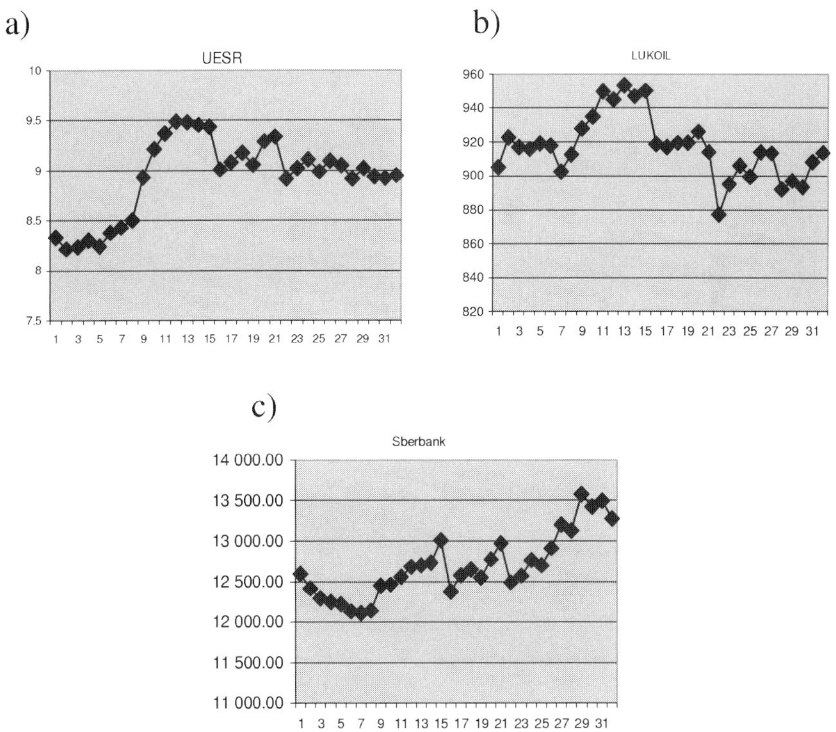

Fig. 18.5. Dynamics of prices from 21 September 2004 to 21 October 2004 for securities: (*a*) is UESR, (*b*) is LUKOIL, (*c*) is Sberbank

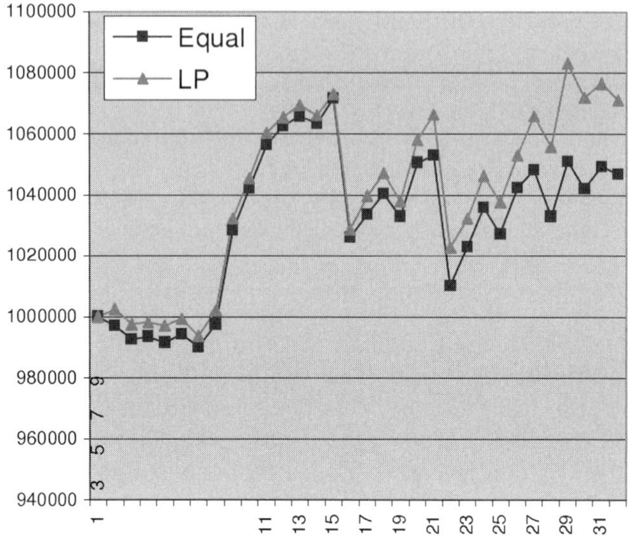

Fig. 18.6. Dynamics of security parts in the process of management by *LP-VaR*

Thus, in the considered example an estimation of efficiency of the technique was made on the data that was not presented in statistics at the stage of modeling conditions of the portfolio,. Optimum shares in the portfolio were calculated by *LP-VaR* proceeding from the problem of optimization (18.10). In management of the portfolio, we kept its risk at constant level $Risk = 0.05$.

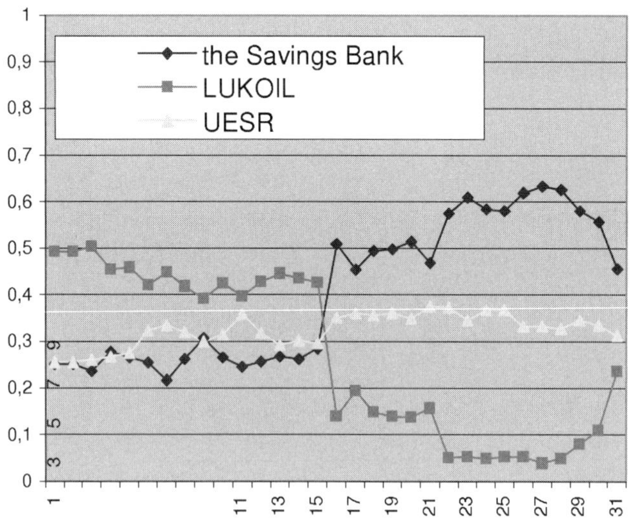

Fig. 18.7. Comparative graph of changing the portfolio cost

Table 18.25. Management of portfolio from 21.09.2004 to 21.10.2004

Variant	CA in begin period, rub.	CA in end period, rub.	Transaction for period, rub.	Commission (0,08 %), rub.	Clear profit, rub.	Yield for period, %	Profit / max. loss
Equal parts	1 000 000	1 046 793	2 046 794	1637	45 156	4.51	0.99
LP-VaR	1 000 000	1 071 228	4 789 290	3831	67 397	6.74	1.51

For the test period, the dynamics of the assets parts in the portfolio is presented in Fig. 18.6. The assets parts in the portfolio on *LP-VaR* changed in a complex way, therefore it is intuitively clear that dynamic adaptation of stock shares in the portfolio can provide better result than fixation of equal stock shares in the portfolio for all test period.

Figure 18.7 confirms efficiency of application of *LP-VaR* technique for the test period. The portfolio asset value in management by *LP-VaR* exceeded portfolio asset value including stocks in equal shares. Result of comparison of the portfolio, including equal stock shares, with the portfolio, including variable stock shares and managed by *LP-VaR* technique, are shown in Table 18.25.

In process of estimation of efficiency of *LP-VaR* technique, we took into account transaction costs, that is payment for fulfillment of operations on the securities market, which is very important for substantiation of results of asset management of the investment portfolio (Table 18.25). The ratio "Profit/Max loss" is calculated as the ratio of net profit by investments to the maximal reduction of asset for the test period and characterizes riskiness of investments. The result of *LP-VaR* management has the better ratio "Profit/Max loss" than passive asset management (equal stock shares during asset management) and higher yield at the fixed risk of investments $Risk = 0.05$.

18.9 Portfolio risk with dependent yields of stocks on basis of copula functions

Technique of application of copulas in modeling security portfolio by logical-and-probabilistic theory with groups of incompatible events is suggested. Results of investigations with two types of copulas are presented. Efficiency of combination of logical-and-probabilistic theory and copulas is proved [59 - 62].

The objectives of presented analysis are the following:

- give an account of LP theory of risk with GIE;
- show advantages of LP risk theory application in problems of risk analysis and forecasting;
- give an account of copulas;

- perform calculus investigations by the example of Ali-Michael-Hack and Clayton copulas;
- determine efficiency of combination of copulas with LP theory of security portfolio risk.

Copulas. Let n be an integer $n \geq 2$. The n-dimensional copula is n-dimensional cumulative distribution function, denoted by $C_n(u_1, \ldots, u_n)$, whose support is the n-dimensional hypercube $[0,1]^n$ and whose univariate marginal distributions are uniform on $[0,1]$. The properties of copula are

- $C_n(u_1, \ldots, u_n) = 0$ if $u_j = 0$ for any $j \leq n$;
- $C_n(1, 1, \ldots, u_j, \ldots, 1, 1) = u_j$, for all $j \leq n$;
- $C_n(u_1, \ldots, u_n)$ is n-increasing function (this is extension of increasing function concept for one-dimensional case).

Actually these properties mean that C_n is a positive probability measure. The class of described functions is very important because it gives possibility to define the dependence structure between the margins of a multivariate distribution. Indeed, let us think about N random variables $(z_1, \ldots, z_N)$ with multivariate distribution F and univariate margins $(F_1, \ldots, F_N)$. Then we have the canonical decomposition

$$F(z_1, z_2, \ldots, z_n) = C(F_1(z_1), F_2(z_2), \ldots, F_n(z_n)). \tag{18.29}$$

Thus having defined copula and knowing marginal distributions, it is possible to calculate joint multivariate probability distribution. There are several families of copulas and several types in each family. Defining type and parameters of a copula is separate problem. Expert estimation can help to define a type of copula. Determining parameters of particular copula may involve statistical data processing. V. Krymsky, F. Akhmedjanov, and K. Balaba to solve the problem suggest a method based on maximum entropy principle [139].

Modeling joint assets' yield distribution using copulas

Let discrete assets' yield distributions be defined. The range of change of yield for each asset j was split into N_j intervals. Particular value of yield Z_{jr} corresponds with each interval r. The probability P_{jr} was calculated for each yield. Thus yield distribution of asset j is defined by discrete set (Table 18.26).

In Table 18.27, numeric example of discrete distributions for two assets A and B is given. The example will be used in further calculations.

Let's move from discrete distributions to accumulated discrete distributions (Table 18.28) using the below formula:

Table 18.26. Discrete yield distribution of asset j

Z_{jr}	Z_{j1}	Z_{j2}	$\ldots$	Z_{je}	$\ldots$	Z_{jNj}
P_{jr}	P_{j1}	P_{j2}	$\ldots$	P_{jr}	$\ldots$	P_{jNj}

Table 18.27. Accumulated discrete distributions for assets ***A*** and ***B***

Yield of asset A, %,	P_{Z_A}	Yield of asset B, %	P_{Z_B}
–3	0.001	–3	0.004
–2	0.009	–2	0.058
–1	0.109	–1	0.300
0	0.409	0	0.699
1	0.859	1	0.941
2	0.999	2	0.995
3	1.00	3	1.00

Table 18.28. Discrete yield distribution of assets ***A*** and ***B***

Yield of asset A, %	P_{Z_A}	Yield of assets B, %	P_{Z_B}
–3	0.001	–3	0.004
–2	0.008	–2	0.054
–1	0.100	–1	0.242
0	0.300	0	0.399
1	0.450	1	0.242
2	0.140	2	0.054
3	0.001	3	0.004
	Sum 1.00		Sum 1.00

Table 18.29. Joint yield distribution for assets ***A*** and ***B***

	–3	–2	–1	0	1	2	3
–3	4.43E-06	3.55E-05	0.000443	0.00133	0.001994	0.00062	4.43E-06
–2	5.4E-05	0.000432	0.005399	0.016197	0.024296	0.007559	5.4E-05
–1	0.000242	0.001936	0.024197	0.072591	0.108887	0.033876	0.000242
0	0.000399	0.003192	0.039894	0.119683	0.179524	0.055852	0.000399
1	0.000242	0.001936	0.024197	0.072591	0.108887	0.033876	0.000242
2	5.4E-05	0.000432	0.005399	0.016197	0.024296	0.007559	5.4E-05
3	4.43E-06	3.55E-05	0.000443	0.00133	0.001994	0.00062	4.43E-06

$$F_r = \sum_{l=1}^{r} P_l. \tag{18.30}$$

Having calculated discrete distributions of security yields, it is possible to compute joint distribution without taking dependence into account:

$$P = \prod_{j=1}^{n} P_{r_j}. \tag{18.31}$$

In Table 18.29, computational results for assets A and B are given. On Fig. 18.8, the graph of joint yield distribution for the case is shown.

As it was mentioned above, copulas allow, to get joint distributions having marginal distributions using formula (18.29). If joint accumulated distribution

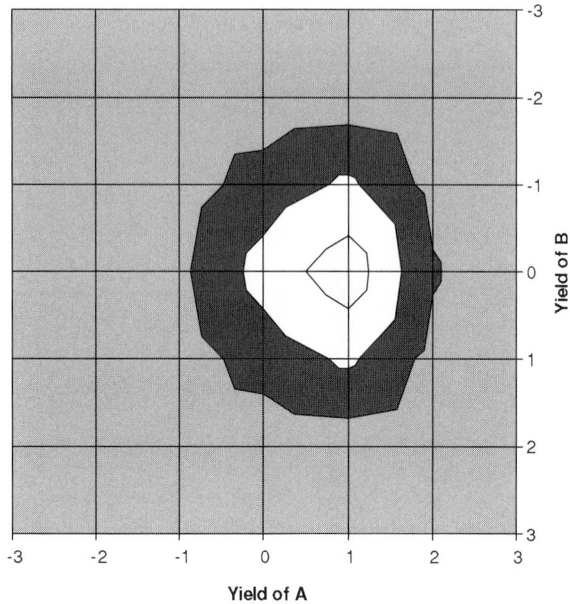

Fig. 18.8. Joint yield distribution for assets A and B without taking dependence into account

function is known, it is possible to move to discrete yield distribution, which contains information about probabilities for each state of portfolio. The yield of each state is calculated by formula (18.30). As it was mentioned above, having these data it is possible to get discrete portfolio yield distribution, which gives ability to perform risk analysis and calculate admissible yield for given Risk value. Thus copulas help to solve the problem of modeling the dependency between assets, namely to refuse multivariate normal distributions, which present dependence in the form of correlated matrix.

Examples of yield distribution modeling of the security portfolio by means of various copulas

Ali-Mikhail-Haq copula. *Ali-Mikhail-Haq copula* increases dependence of bottom "tail" for positive value of parameter Q. In other words, it extends left "tail" of joint yield distribution of two assets. Copula function for a two-dimensional case has the following form:

$$C(u,v) = \frac{uv}{1 - Q(1-u)(1-v)}. \tag{18.32}$$

where

Q is a parameter in an interval [-1,1], describing a degree of dependence;

u is a value of distribution function of the first random variable;

v is a value of distribution function of the second random variable.

Density of the copula is represented on Fig. 18.9 (a derivative on both arguments).

Let's calculate joint yield distribution of two assets at $Q = 0.8$. The distribution graph is presented on Fig. 18.10.

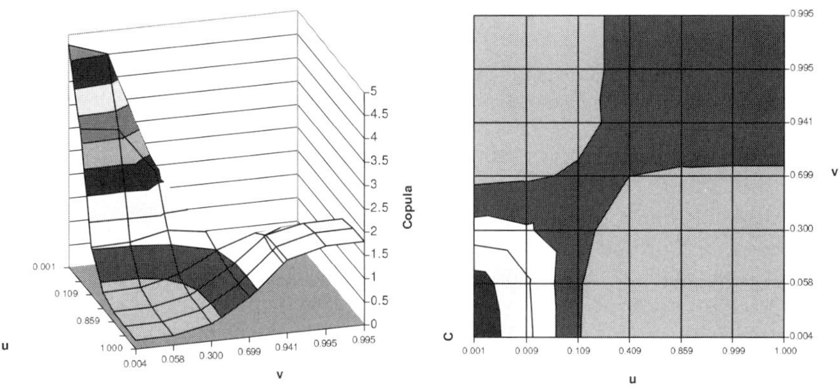

Fig. 18.9. The density of Ali-Mikhail-Haq *copula*

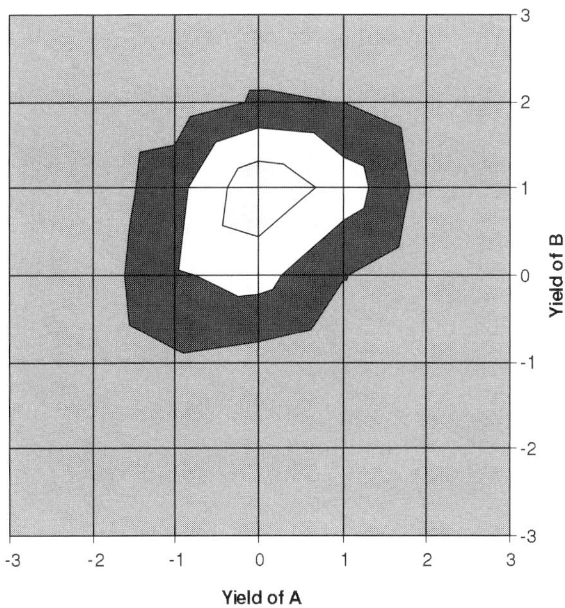

Fig. 18.10. Joint yield distribution for securities A and B with taking dependence into account (*Ali-Mikhail-Haq copula*, $Q = 0.8$)

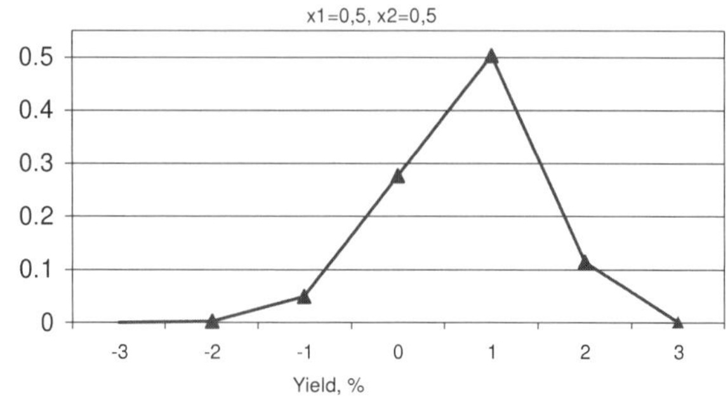

Fig. 18.11. Yield distribution of portfolio containing two assets A and B with taking dependence into account *(Ali-Mikhail-Haq copula,* $Q = 0.8$)

Let's count up the yield distribution of security portfolio having set the shares of its components. Graph of yield distributions for equal shares of assets is shown on Fig. 18.11.

From Fig. 18.11, it is visible that left "tail" of yield distribution of security portfolio is extended. Changing shares of assets, it is possible to change the distribution view and perform optimization by one of the suggested criteria.

Clayton copula. Clayton copula sets strong dependence in right "tail" of joint distribution. This copula function has the following form:

$$C(u,v) = u + v - 1 + ((1-u)^{1/a} + (1-v)^{1/a} - 1)^{-a}, a > 0. \tag{18.33}$$

Density of the copula is represented on Fig. 18.12.

Let's execute calculation at $a = 0.6$. The distribution graph is presented on Fig. 18.13. Graph of yield distributions for equal shares of assets is shown on Fig. 18.13.

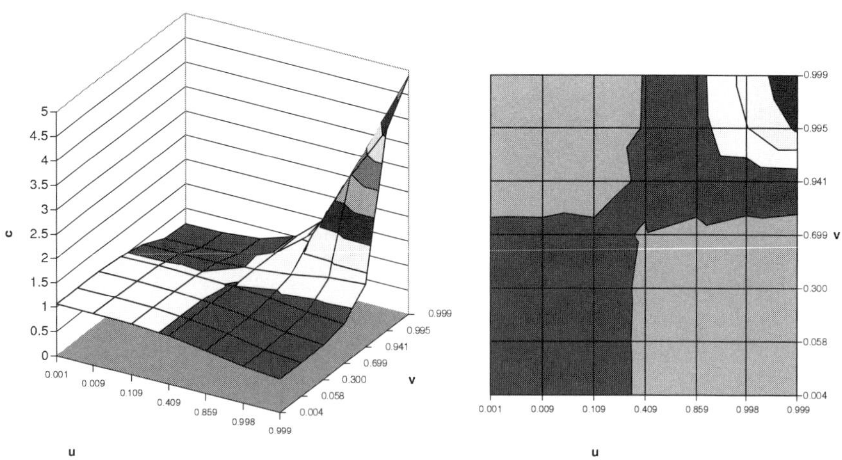

Fig. 18.12. Clayton copula density

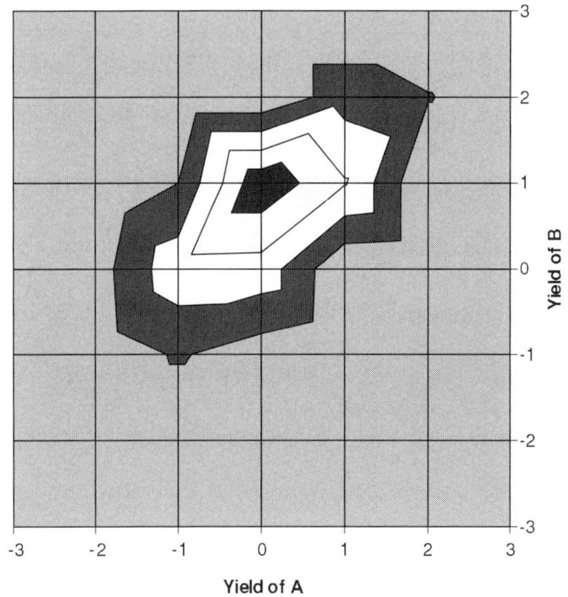

Fig. 18.13. Joint yield distribution of assets A and B with taking dependence into account (Clayton copula, $a = 0.6$)

To compare two figures (see Fig. 18.13 and Fig. 18.14), it is seen that the last distribution has shorter left "tail" and longer right one. Thus copula sets structure of dependence between risk factors. Applying various copulas and varying their parameters, it is possible to model various kinds of dependence.

Copulas allow solving the challenge of assets interdependence accounting. Unlike multivariate normal distributions setting dependence in kind of a correlation matrix, copulas allow one to consider more complex kinds of dependence. To choose copula type and its parameters correctly, it is possible to increase reliability of modeling results. Hence, opportunities of the analysis

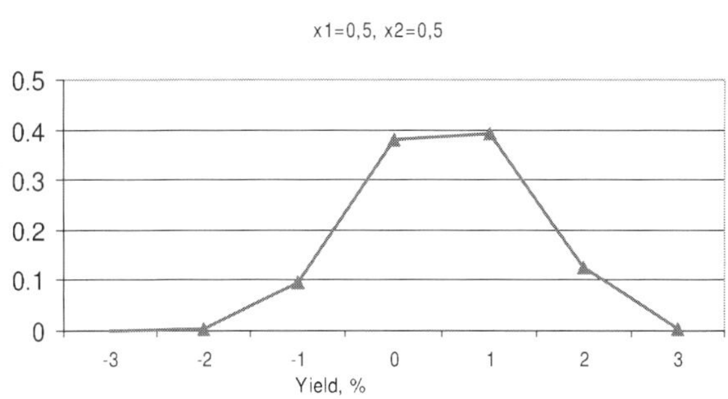

Fig. 18.14. Yield distribution of portfolio containing two assets A and B with taking dependence into account (Clayton copula, $a = 0.6$)

and forecasting the risk are extended. Optimum copula parameters can be determined, using a method of probabilities entropy maximization [139].

From results of the research, it is possible to draw the following conclusions:

1. LP-theory of security portfolio risk with GIE uses statistical data initially received in discrete view for construction of discrete yield distributions of assets that allows:

- to enter concepts of events, gradation, groups of incompatible events, and to use LP-calculus for modeling and for portfolio risk analysis;
- to refuse construction of analytical yield distributions of assets and portfolio;
- to apply arbitrary yield distributions of assets in discrete representation;
- to keep more information at processing statistical data;
- to increase accuracy and transparency of modeling and forecasting risk.

2. LP-theory of security portfolio risk solves the problem of analysis of risk and portfolio yield:

- it is possible to examine in detail the most important part of yield distribution — left "tail";
- to calculate contributions of all events gradation of each asset in risk and portfolio yield;
- to reveal the most dangerous assets and gradations of assets and to make the decision on portfolio re-structuring.

3. Copulas allow one to solve the challenge of assets interdependence accounting in LP-theory of security portfolio:

- to refuse the multivariate normal distributions setting dependence in kind of correlation matrix;
- to set more complex kinds of dependence and to increase reliability of results of portfolio yield and risk modeling;
- to expand opportunities of analysis and forecasting the risk.

4. Results of researches for *Ali-Mikhail-Haq copula and Clayton copula* have shown that they differently consider structure of interdependence of assets that is reflected in final portfolio distribution and, hence, in yield and risk parameters.

5. Copula approach is effectively combined with LP-theory of security portfolio. The combination allows one to increase reliability of results of calculations and to expand opportunities of risk analysis due to LP-approach on the one hand and to improve opportunities of assets interdependence modeling due to copulas on the other hand.

18.10 Conclusions

The basic results of the current chapter are

1. Choice of the security portfolio by *VaR* has the following defects: (1) assuming the normal low of distribution of assets yields and portfolio, which decreases the horizon of forecasting the portfolio risk, (2) absence of transparency in analysis and forecasting the risk.
2. The risk LP-theory of security portfolio by *LP-VaR* is stated, using the arbitrary discrete distributions and connections of assets yields, which secures transparency of forecasting the risk and allows one to solve the new tasks of portfolio risk analysis.
3. L-models and P-models of the portfolio risk for the cases of independency and independency of assets yields are presented, allowing one to build the portfolio yield distribution.
4. LP-problem of choice of the optimal portfolio with different criteria is formulated: maximum of the admitted yield, maximum of the number of portfolio states in "tail" of distribution, minimum of the risk, minimum of the entropy of probabilities in "tail."
5. LP-methods for analysis and forecasting the portfolio risk by contributions of assets yield grades in the admitted yield and risk of portfolio are suggested.
6. The same accuracy of attributes of the optimal portfolio by *VaR* (at describing assets yield distributions by the normal low) and by *LP-VaR* (at describing assets yield distributions by discrete low) is proved.
7. Possibility of replacement in *LP-VaR* of the complete set of portfolio states by the limited set is put, obtained by the Monte Carlo method from the number of possible states of the portfolio.
8. Taking into account dependency assets yield from external factors allows to reconstruct about 70% of probabilities of portfolio states.
9. For taking into account dependency of assets yields in *LP-VaR*, the function of connection *copula* can be used.
10. Computing investment on real assets, the expediency of regularities of the theory and technology of *LP-VaR* and their transparency at selection and analysis of security portfolio risk for arbitrary distributions of their yields are proved.

19

Risk LP-Models of Quality and Efficiency

Quality of a spoilage constantly grows.
Evgeny Kosheev

The maintenance of efficiency, quality, and accuracy is one of the main and complex problems of economics and modern production [30, 143, 144]. At the same time, this problem is not completely understood in the scientific plane and is not solved satisfactorily in the applied sense. First, we shall consider the classical modern approach to estimation of quality, and then we shall turn to the logic and probabilistic theory of efficiency, quality, and accuracy.

19.1 General problem of quality management in business

For achievement of effective production along with the quality of production itself, it is also necessary to consider the whole system of economic relations, to develop and investigate the management processes of all the activity of the enterprise, and to pay attention to many aspects of the activity, such as finances, resources, personnel, etc. [30, 144]. Now in use are American, European, Japanese, Russian, and other standards for estimation and management of quality. In their essence, these standards are identical and differ only in terminology and numerical estimations of importance of separate criteria.

Management systems by quality are used for:

1. Self-estimation of activity by companies in the framework of standards of national awards in the field of quality;

2. Management of developing advantage in business.

The model criteria are divided into two categories: opportunities and results (Fig. 19.1). Half of the criteria of the quality system defines capabilities of the company, and the other half of the criteria defines results of its activity. We denote these criteria by identifiers with indexes. The category of the capabilities Z_{10} is determined by estimates of the following criteria: role of leaders

E.D. Solojentsev, *Scenario Logic and Probabilistic Management of Risk in Business and Engineering*, DOI 10.1007/978-0-387-77946-1_19,

in the work organization Z_1, using people potential Z_2, planning in the area of quality Z_3, using resources Z_4, technological processes of manufacturing, advertising, service Z_5. The category results Z_{11} is determined by estimates of criteria of satisfaction of different people: employee Z_6, customers Z_7, society as the whole Z_8 and the financiers (business results) Z_9. Elements of the system are interconnected. There is the strong connection between “Business results” Z_9, “Quality” Y, and “Satisfaction of the customers” Z_7.

Indexes for the numerical estimation of the criteria, included in quality, are introduced. For example, the Association of Customers and Satisfaction Index (ACSI) takes into account the product image, expectations of customers, the perceived quality as product characteristics, the perceived quality as service characteristics, the product perceived value and further adherence to the repeated purchase. There are statistical proofs of dependence ACSI and economic parameters. For example, this is reaction of the stock exchange and the income by shares after publication of ACSI for various types of productions and branches.

In many countries, quality estimation systems are known as national premiums and are used with the purpose of stimulation of improvement of quality and competitiveness of production (services) and encouragement of organizations, which use advanced and effective methods of quality management. For example in Russia, premiums by the Government of the Russian Federation in the field of the quality were founded in 1996.

We note some characteristics of the premiums system. Centers of quality control carry out consultation and estimation of the documentation applied for competition on the premium. The information, submitted by the companies, is strictly confidential and is not subject to disclosure. The estimation of quality of management systems of the companies is done by a group of independent, highly professional specialists who are specially prepared for this work. Usually not more than 12 premiums a year are given. Periodicity of self-estimation of quality of management systems is equal to one year.

The maximal numerical estimations of criteria in points for the Russian Federation Government Award model in the quality field are shown in Fig. 19.1. The sum of all maximal estimations is equal to 1000. The given distribution of estimates is recommended to be used for any enterprise and organization irrespectively of the kind of production and activity, of the sizes and forms of ownership.

The self-estimation of the company quality system is carried out by “a special group” of the company. They give the estimation for each criterion Z_1–Z_9 in points and calculate the ratio of this estimation to the greatest possible value. The system of participation in competitions makes it possible to estimate the achieved level of criteria in per cent of the greatest possible value. Thus, the flaw of the company in each criterion and the activity direction is objectively visible. Achievements of the company can be determined by comparison of numerical values of criteria Z_1–Z_9 on years.

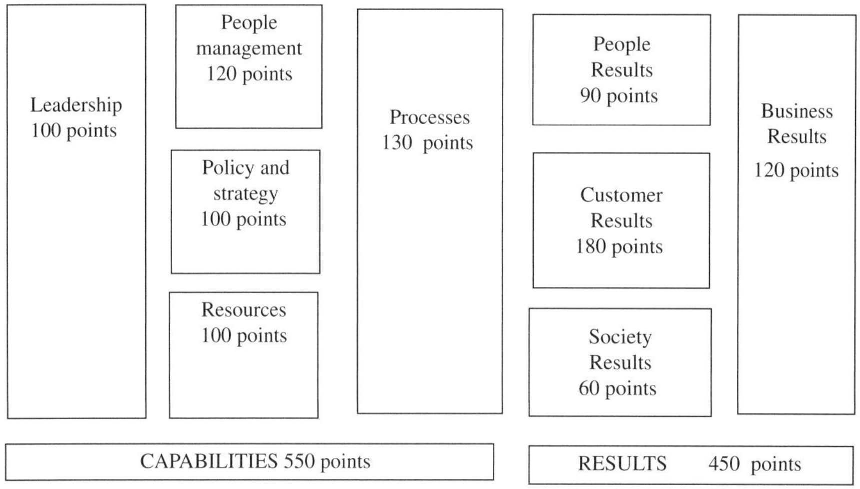

Fig. 19.1. Model of premium in the area of quality

Risk of loss of quality and market. Let us construct structural, logic, and probabilistic models of risk of loss of quality and market. We denote random events, corresponding with the quality criteria), by logic variables Z_1–Z_9, property "Quality" by logic variable Y and derivative events of "Capabilities" and "Results" by Z_{10} and Z_{11}. We shall construct the structural model of quality loss risk (Fig. 19.2) or the non-success model with logic connections OR.

"Quality" is the parameter that is difficult to measure. Introduction of the maximal points to criteria and estimations of criteria for the specific company does not give a rigorous solution of all problems of risk. It may happen that it is more convenient to use as the final event it is the easily measured criterion "Business results." Earlier we already marked that the criteria "Quality" and "Business results" are closely connected.

We write down the risk L-function in DNF for the graph in Fig. 19.2, a:

$$Y = Z_1 \vee Z_2 \vee X_3 \vee Z_4 \vee Z_5 \vee Z_6 \vee Z_7 \vee Z_8 \vee Z_9. \tag{19.1}$$

The corresponding risk L-function in ODNF can be written as follows:

$$Y = Z_1 \vee Z_2\overline{Z}_1 \vee Z_3\overline{Z}_2\overline{Z}_1 \vee \ldots \tag{19.2}$$

The corresponding risk P-function can be written down

$$P\{Y = 1\} = p_1 + p_2q_1 + p_3q_2q_1 + \ldots \tag{19.3}$$

For training the risk LP-model of quality loss on the statistical data, it is necessary to introduce grades for the criteria Z_1–Z_9, which is quite obvious and not considered here.

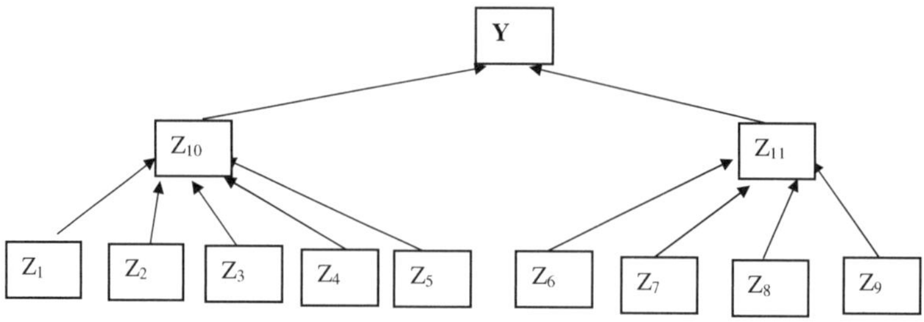

Fig. 19.2. Structural models of non-success risk:

As a result of self-training, probabilities of non-success of events Z_1–Z_9 are known. Then it is possible to calculate risks of all N companies, pretending to awards, and to plot their risks on the risk axis. Let us choose the admitted risk P_{ad} in such way that N_b objects from the general sampling N are bad and the number of good objects N_g from the general sampling N is equal to the number of awarded premiums. For each company i, one can calculate the relative number of companies a_i, having smaller risk:

$$a_i = N_i/N. \tag{19.4}$$

One can also calculate risks $P_{25}, P_{50}, \ldots$, which mean 25%, 50%, ... of risk of the best companies.

Example 1. Let us consider difference of estimations of quality of objects by traditional arithmetic and logic addition of estimations for the criteria $Z_1, Z_2, \ldots, Z_9$ (Fig. 19.2). According to statement of a general problem of quality, we suppose that the points $Z_1, Z_2, \ldots, Z_9$ are given in relative values. Now we consider them as probabilities. Let five objects be considered. They are sorted in the degree of deviation of probabilities from the average value, which for all these objects are equal to $P_m = 0.5$.

The objects 4 and 5 differ from the objects 2 and 3: in them the place of the points by the criteria Z_6 and Z_7 are transposed (Z_7 has the maximum number of the points). Logic estimations of objects 4 and 5 do not change, but arithmetic estimations do. The arithmetic estimations of quality of objects as the average of distribution of points are calculated. The logic estimations of quality of objects as probabilities of success are computed.

From the results given in Table 19.1, it is easy to see that the arithmetic and logic addition of initiating events lead to different ranking objects. From the common sense, the stability of quality estimation demands that the first place should be given to object 1 as does logic addition of events. It should be noted that the estimation is similar to that by the methods of maximum likelihood and information entropy.

Table 19.1. Quality of objects with arithmetic and logic addition of estimations

Ob-jects	x_1	x_2	x_3	x_4	x_5	x_6	x_7	x_8	x_9	Arith-metical	RA	Logi-cal	RL
1	0.5	0.5	0.5	0.5	0.5	0.5	0.5	0.5	0.5	500	3	0.00195	1
2	0.4	0.6	0.4	0.6	0.4	0.6	0.4	0.5	0.6	494	4	0.00166	2
3	0.3	0.7	0.3	0.7	0.3	0.7	0.3	0.5	0.7	484	5	0.00099	3
4	0.4	0.6	0.4	0.6	0.4	0.4	0.6	0.5	0.6	510	2	0.00166	2
5	0.3	0.7	0.3	0.7	0.3	0.3	0.7	0.5	0.7	520	1	0.00097	3

19.2 Particular problems of quality loss risk

As an example, let us consider quality for object of a type of "bridge" (Fig. 11.3 b). This is a complex object; the risk L-model has connections AND, OR, and repeated elements. Risk objects have signs 1, 2, 3, 4, 5, which correspond with the logic variables Z_1, Z_2, Z_3, Z_4, Z_5. The success risk L-model of the "bridge" is made on the basis of all shortest paths of successful functioning (11.18). After orthogonalization of (11.18), we obtain the risk P-model for the "bridge" (11.19).

The events Z_1, Z_2, Z_3, Z_4, Z_5 correspond with units of a product of "bridge" type. The units are made by brigades A_1, A_2, A_3, A_4, A_5 respectively with the number of workers (having personal marks), equal to the number of grades in a sign. The assembled products are tested; results of the test can be successful or non-successful. The protocol of testing form is Table 11.1. It is natural that instead of workers we can consider various process equipment, technological processes, and firms deliveringcomponent units.

For training the risk P-model of "bridge" type (Fig. 11.3b), the table with 1000 objects (700 good and 300 bad ones) is used. Signs have from 4 to 11 grades — 40 grades in aggregate. Results of training this model are in detail stated in Section 17.1 for the problem of bribes.

19.3 Modeling risk in problems of efficiency

The statistical data are given in the form of Table 19.2, containing in rows conditions of influencing parameters $Z_1, \ldots, Z_j, \ldots, Z_n$, and the efficiency parameter Y. The parameters have different nature and dimension. The number of conditions (or objects) in the table is equal to N.

The main principles of the LP-theory of efficiency coincide with the main principles of the risk LP-theory with GIE, stated in Chapter 10. Here we use another statement of optimization problems. We shall consider two different statements of problems of efficiency:
(1) Classification of object conditions into several classes;
(2) Determination of parameter weights, influencing the efficiency parameter.

Table 19.2. States, initiating parameters, and efficiency parameter

States, i	Parameter, Z_1	...	Parameter, Z_j	...	Parameter, Z_n	Efficiency, Y
1						
...	...	...	...	...	...	...
			Z_{ij}			
...	...	...	...	...	...	...
N						

General principles. In mechanics for solution of problems of analysis of motion and stability, we pass from the timing of continuous representation of functions to the discrete frequency one. In the risk LP-theory for solution of problems of efficiency, we also pass from the continuous distribution of random parameters to the discrete frequency distribution.

Quantization. The parameters $Z_1, \ldots, Z_j, \ldots, Z_n$ influencing the efficiency parameter are the system of random variables, which defines the efficiency parameter Y. The efficiency parameter Y, being a function of random values, has many-dimensional distribution.

In order to pass to discrete distributions, let us break ranges of parameter change $Z_1, \ldots, Z_j, \ldots, Z_n$ into intervals $N_1, \ldots, N_j$ and N_y of identical (or different) width. The number of intervals for different parameters are different. The numbered values of intervals for each parameter are considered as grades of the influencing parameter $r = 1, 2, \ldots, N_j$ and the efficiency parameter $r = 1, 2, \ldots, N_y$.

Any interval r of the parameter j we denote by logical variable (the random event) Z_{jr}. That is, the actual values of the parameters $Z_1, \ldots, Z_j, \ldots, Z_n$ on intervals we replace by the numbers of intervals Z_{jr} themselves. We shall consider the numbers of intervals as random events, which correspond with random grade-events forming GIE. Thus, we obtain Table 19.3.

Logic. The maximal number of combinations (different states of the efficiency parameter) is determined by (11.1). The logical function for possible conditions of the efficiency parameter is given in perfect disjunctive normal

Table 19.3. States and parameter grades

States i	Parameter grades Z_1	...	Parameter grades Z_j	...	Parameter grades Z_n	Parameter grades of efficiency Y
1						
...	...	...	...	...	...	...
			Z_{jr}			
...	...	...	...	...	...	...
N						

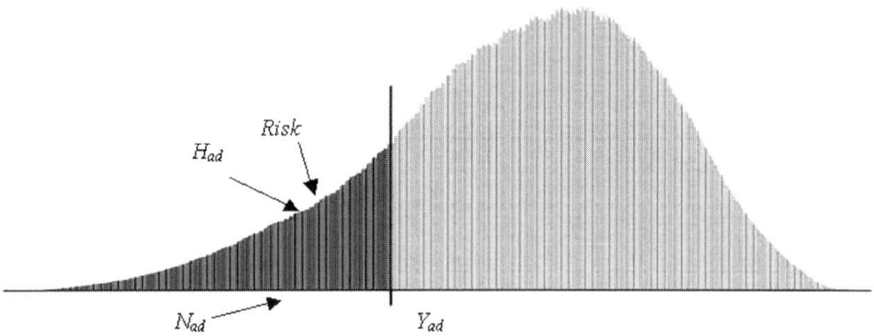

Fig. 19.3. Classification of objects to classes $1, 2, \ldots, k$

form (PDNF). Each condition of the efficiency parameter from the possible conditions N is defined by the logical function (11.15).

Classification of object conditions to several classes. By analogy with the problem of classification to two classes, by using the formulas (11.15, 11.17), the problem is stated as follows. It is required to determine probabilities of grade-events P_{jr}, $j = 1, 2, \ldots, n$; $r = 1, 2, \ldots, N_j$ and admitted risks $P_{1ad}, P_{2ad}, \ldots, P_{kad}$ (Fig. 19.3) dividing objects or states of parameter of efficiency into classes $1, 2, \ldots, k$ with the target function:

$$F = N_{1c} + N_{2c} + \ldots + N_{kc} \Rightarrow \max_{P_{jr}}, \tag{19.5}$$

where $N_{1c}, N_{2c}, \ldots, N_{kc}$ are numbers of correctly recognized objects in classes $1, 2, \ldots, k$.

The objects classification errors are as follows:

$$E_1 = (N_1 - N_{1c})/N_1; \quad E_2 = (N_2 - N_{2c})/N_2; \ldots; E_m = (N - F)/N; \tag{19.6}$$

where $E_1, E_2, \ldots, E_k$ are errors in classification of objects or states to classes $1, 2, \ldots, k$; E_m is the mean error in classification of objects or states; $N_1, \ldots, N_k$ are numbers of objects or states in classes $1, 2, \ldots, k$ in the table "Objects and parameters" or "Influential parameters and efficiency parameter" (Table 19.2).

The contributions of grade-events into risk:

$$\begin{aligned} E_{1jr} = (N_{1jr} - N_{1cjr})/N_{1jr}; \quad E_{2jr} = (N_{2jr} - N_{2cjr})/N_{2jr}, \ldots, \\ E_k = (N_k - N_{kc})/N_k; \quad E_{mjr} = (N_{jr} - N_{jrc})/N_{jr}, \end{aligned} \tag{19.7}$$

where $N_{1jr}, N_{2jr}, \ldots, N_{kjr}$ are numbers of objects or states with grades Z_{jr} in classes $1, 2, \ldots, k$; $N_{1jrc}, N_{2jrc}, \ldots, N_{kjrc}$ are numbers of objects or states with the grade Z_{jr} in classes $1, 2, \ldots, k$ with correct classification of objects.

Finding weights of parameters influencing the parameter of efficiency. We formulate the management problem of efficiency as the problem

Table 19.4. Conditions and grade probabilities

Conditions i	Probabilities of parameter grades Z_1	...	Probabilities of parameter grades Z_j	...	Probabilities of parameter grades Z_n	Probabilities of grades of efficiency parameter Y
1						
...	...	...	...	...	...	...
			P_{jr}			
...	...	...	...	...	...	...
N						

of estimation of weights of parameters, influencing the efficiency parameter. In the traditional problem of the efficiency, for example in the theory of accuracy, laws of distribution of parameters, influencing the accuracy, and the accuracy parameter itself, are assumed to be normal, generalized normal, or analytical [146]. These assumptions result in errors in accuracy analysis and in non-effective management of accuracy.

Probabilities. As was said above, the actual values of parameters $Z_1, \ldots, Z_n$ in intervals are replaced by numbers of the intervals Z_{jr}. Numbers of intervals Z_{jr} are regarded as random grade-events, which form GIE for any parameter.

From the table "Conditions and parameter grades" (Table 19.3) we construct the new table "Conditions and grade probabilities" (Table 19.4).

Probabilities (frequencies) of the grade-events r for the influencing parameter-events j are equal

$$P_{jr}^{tab} = N_{jr}/N, \quad j = 1, 2, \ldots, n; \quad r = 1, 2, \ldots, N_j, \tag{19.8}$$

where N_{jr} is the condition number for the parameter j with the grade r.

Let us also construct the discrete distribution for the efficiency parameter Y by the statistical tabular data:

$$P_{yr}^{tab} = N_{yr}/N, \quad r = 1, 2, \ldots, N_y, \tag{19.9}$$

where N_{yr} is the conditions number of the efficiency parameter with grade r.

We calculate the model (effective) values of probabilities of conditions i for the efficiency parameter (lines in the table) from the expression:

$$P_i^{mod} = x_1 \cdot a_1 \cdot P_{1J_1} + \ldots + x_j \cdot a_j \cdot P_{jJ_j} + \ldots x_n \cdot a_n \cdot P_{nJ_n}. \tag{19.10}$$

Here:

$J_1, J_2, \ldots, J_n$ are the indexes belonging to the corresponding value areas: $J_1 \in \{1, N_1\}$; $J_2 \in \{1, N_2\}$; ...; $J_n \in \{1, N_n\}$. That is, for each parameter we substitute into (19.10) probabilities of its grade in the condition i.

$x_1, \ldots, x_j, \ldots, x_n$ are related weights of influencing parameters Z_1, $Z_2, \ldots$, Z_n, which are needed to be determined;

Table 19.5. Model and tabular values of probabilities of conditions of the parameter Y

Conditions i	Tabular probabilities of the efficiency parameter P_i^{tab}	Model probabilities of the efficiency parameter P_i^{mod}	Values of the efficiency parameter Y_i
1			
...	...	...	...
...	...	...	...
N	...	...	...

$a_1, \ldots, a_j, \ldots, a_n$ are the correction coefficients of probabilities of the parameters $Z_1, Z_2, \ldots, Z_n$, which we determine by the condition of equality of their mean values in the GIE to the mean value of the probability of the efficiency parameter in its GIE:

$$a_1 = P_{ym}/P_{1m}; \quad \ldots \quad a_j = P_{ym}/P_{jm}; \quad \ldots \quad a_n = P_{ym}/P_{nm}, \tag{19.11}$$

where $P_{ym}, P_{1m}, \ldots, P_{jm}, \ldots, P_{nm}$ are mean values of probabilities of grades in GIE, which are equal to:

$$P_{ym} = 1/N_y, \; P_{1m} = 1/N_1, \ldots, P_{jm} = 1/Nj, \ldots, P_{nm} = 1/N_n. \tag{19.12}$$

Thus, the correction coefficients $a_1, \ldots, a_j, \ldots, a_n$ are equal to:

$$a_1 = N_1/N_y, \quad \ldots, \quad a_j = N_j/N_y, \quad \ldots, \quad a_n = N_n/N_y. \tag{19.13}$$

The mean value of the efficiency parameter Y_i can be calculated by the following formulae:

$$Y_{m1} = \sum_{i=1}^{N} Y_i \;/\; N;$$

$$Y_{m2} = \sum_{i=1}^{N} Y_i \cdot P_i. \tag{19.14}$$

Let us make the table "Model and tabular value of probabilities of conditions of the efficiency parameter" (Table 19.5), in which the model values of P_i^{mod} are determined by (19.9), and the tabular values of probabilities P_i^{tab} are taken to be equal to the probability values of corresponding grade-events of the efficiency parameter in the interval, to which Y_i belongs.

Closeness of distributions of probabilities P_i^{tab} and P_i^{mod} depend on weights $x_1, \ldots, x_j, \ldots, x_n$ of parameters $Z_1, \ldots, Z_j, \ldots, Z_n$, influencing the efficiency parameter. We determine optimal estimations of weights x_j, $j = 1, 2, \ldots, n$ by algorithmic solution of the optimization problem with the criterion function by the least-squares method:

Table 19.6. Weights of parameters influencing efficiency

Variants	Number of grades N_y	Value of function F	x_1	x_2	x_3	x_4	x_5
1	20	1.051	0.436	0.093	0.176	0.130	0.162
2	10	3.845	0.451	0.101	0.182	0.141	0.121
3	5	14.792	0.503	0.087	0.153	0.111	0.143

$$F = \sum_{i=1}^{N} (P_i^{tab} - P_i^{mod})^2 \rightarrow \min . \tag{19.15}$$

For finding $x_1, \ldots x_n$, we use the random search method and the small increments method, which we used before. The weights $x_j, j = 1, \ldots, n$ are needed for resource distribution to management of the efficiency parameter Y.

Values of the efficiency parameter Y_i and their probabilities P_i^{mod} are presented in Table 19.5. Let us sort this table by the value of the efficiency parameter Y_i and make normalization of values P_i^{mod} (the sum of probabilities of the array from N values should be equal to 1). Then, at the given admitted value of the efficiency parameter Y_{ad}, we can calculate the probability (or *Risk*) of the condition:

$$Y < Y_{ad}. \tag{19.16}$$

For this purpose, we sum up probabilities of those conditions of the efficiency parameter for which the condition (19.16) holds.

Estimation of risk *Risk* of the current condition of the efficiency parameter is carried out by constructing and sorting Table 19.5 using the above described method (see Chapters 10 and 15).

The forecasting the efficiency parameter is done as follows. For the prediction values of the parameters $Z_1, \ldots, Z_j, \ldots, Z_n$, influencing the efficiency parameter, the probability P_i^{mod} is calculated from (19.10) and the efficiency parameter Y is calculated by constructing and sorting Table 19.5.

Example. Numerical investigation for estimation of weights of parameters influencing the efficiency parameter Y are fulfilled with the earlier used data of $N = 1000$ credits. Only the first five signs of the credits, having from 4 to 11 grades, are taken into account. As the efficiency parameter, we take the computed values of risks of credits, obtained on the trained LP-model of the credit risk with all 20 signs (see Chapters 11 and 15).

In research, the efficiency parameter Y is represented by 20, 10, and 5 discrete values of grades, for which probabilities are computed by (19.9).

However, in research, the optimization problem (19.15) is solved algorithmically with use of the above described method of random search. It allows us to establish stability of the solution x with respect to the number of influencing parameters. By results of calculation (Table 19.6), we can make the following conclusions:

1. The weights of influencing parameters differ more than five times;

2. The weights of influencing parameters change with change of number of grades in the efficiency parameter Y;

3. The number of the analyzed parameters, influencing the efficiency parameter, should not exceed $4 \div 7$ because of decreasing stability of the solution.

20

LP-Models of Company Management Non-success Risk

In the memory of Peter F. Drucker, an outstanding economist.

We describe the task of building company management non-success risk models and provide the following developed logic-probability risk models: management non-success in functions; company non-success in the directions of its activities; management of the company as a complex object, non-success in the achievement of an objective or a group of objectives; loss of quality in the company operation. The developed non-success risk LP-models enhance the strategic company management efficiency.

The future manager should be capable of accomplishing new tasks [146]: to manage proceeding from the set objectives, to undertake longer term risks, to calculate all the risks, to choose the most justified risk option, to take strategic decisions, to create a team, to provide information promptly and accurately, to perform several functions and envision the business on the whole, to know the company products and to coordinate it and the branch with the outside world.

The manager can cope with all this provided there is some summing up the managerial experience describing facts, rules, situations (cases), and risk assessment procedures for all the decision-making aspects. The results of analysis works on strategic management, however, have revealed that there are no mathematical methods or models of risk-based business management, that common sense is not transformed to the non-success risk logics and model, that no scenario-based business management is used, that management strategies are considered separately rather than on the whole, and there are no company non-success risk models.

20.1 Problem statement

Works by Russian and foreign authors describe many cases of managerial decision making. The textbook for universities and colleges [147] provides a

E.D. Solojentsev, *Scenario Logic and Probabilistic Management of Risk in Business and Engineering*, DOI 10.1007/978-0-387-77946-1_20,

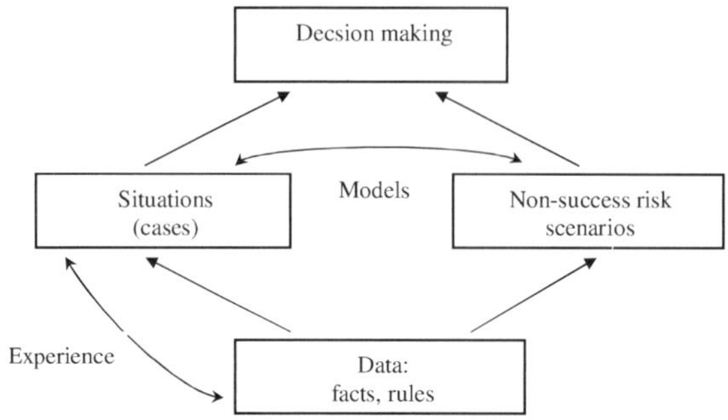

Fig. 20.1. Decision making, models, and data in management

characteristic example. Cases and precedents often become the contents of tasks, letters, topics of seminars, etc. Some of these materials are provided by the presidents of well-known companies. The most illustrious are the precedents provided by the following companies: *General Motors, U. Steel, IBM, Digital Equipment, McDonell Douglas, General Electric, Sears, Kodak, Toyota Motor, Austin, Ford, Chrysler, General Motors Cadillac, Intel, Adobe Systems Inc., Hewlett-Packard, etc.*

The cases describe successful or unsuccessful fragments of the company strategic management, these, however, have not been used for building mathematical models of management non-success risk, or for formalizing the company management process. Summarizing the management experience in the form of description of facts, rules of and cases is needed, primarily for the training, students and managers (Fig. 20.1). It is impossible to base a situational company management model or an expert system on the management situations of a single company because of the limited statistics. Indeed, each situation element may have several meanings, whereas the number of different possible situations is astronomic. This is why there are no expert systems or situational models in the management.

At the same time, we can build management non-success risk scenarios and appropriate LP-models (and this is what this chapter is devoted to). Probabilities of events initiating management non-success risk can be assessed for the LP-models by expertise on the basis of statistics on the company management situations and state of the world market and the company itself.

20.2 LP-models of management non-success risk

There cannot be strategic company management without the management risk quantitative assessment and analysis. The major achievements in the risk assessment and analysis are featured by the LP-theory [2]. The LP-theory's attractiveness is in its clearness and unambiguous quantitative risk assessment, extensive capabilities in analyzing the effect of any element, including the personnel, on the entire system security. The risk model may have *OR, AND, NOT* logical links between the system elements and cycles. The risk dynamics are taken into account by varying element failure probabilities in the course of time.

The LP-risk theory in the systems with incompatible groups of events [3, 30] enables risk to be modeled and analyzed in the systems whose elements have several states and LP-risk models to be constructed for economical and organizational systems.

The LP risk model with the groups of incompatible events (GIE) ensures transparency of the risk assessment and analysis results, and enables the risk to be controlled in keeping with the initiating events' contributions to this risk.

The notions of company non-success risk and management non-success risk are equivalent. Considered below are the following non-success risk models: management in functions; company in the directions of its activities; management of the company as a complex object; accomplishment of an objective or a group of objectives; assessment of the company functioning quality.

20.3 Model of management non-success risk in functions

Let us consider the scenario of the company management non-success risk in the functions: personnel management, strategic planning, marketing and sales, accounting, etc. Functions are designated by logical variables. The structural model of the management non-success risk is provided in Fig. 20.2. In words, the scenario of the management non-success risk in functions is formulated as follows: management non-success is caused by the non-success in any one, any two ... or all the functions. The non-success risk scenario in the functions is associative and takes into account all the possible management states.

Independent binary variables for the non-successfulness events in management functions assume the values of 1 (non-success) or 0 (success) with the following probabilities:

$$P\{Z_1 = 1\} = p_1, \ldots, P\{Z_n = 1\} = p_n;$$
$$P\{Z_1 = 0\} = 1 - p_1 = q_1, \ldots, P\{Z_n = 0\} = 1 - p_n = q_n. \quad (20.1)$$

Logical management non-success risk function

$$Y = Z_1 \vee Z_2 \vee \ldots \vee Z_j \vee \ldots \vee Z_n. \quad (20.2)$$

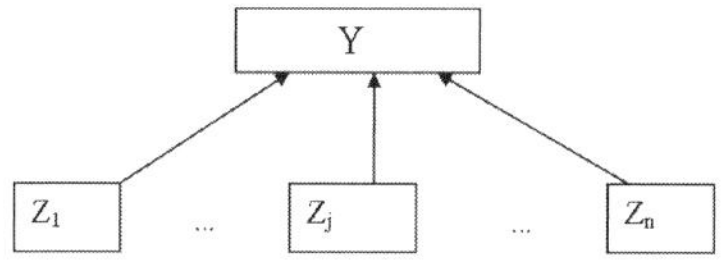

Fig. 20.2. Structural model of the management non-success risk in functions

Logical non-success risk function in the equivalent form after its orthogonalization

$$Y = Z_1 \vee Z_2\overline{Z_1} \vee Z_3\overline{Z_1}\ \overline{Z_2} \vee \ldots . \tag{20.3}$$

Orthogonalization means that the logical product of any two summands in (20.3) is equal to 0. For a simple logical function (20.2), orthogonalization procedure is easy and obvious. For complex logical functions with *NOT, OR, AND* variables links and cycles, the orthogonalization procedure cannot be performed without use of a computer and appropriate software facilities. After the orthogonalization of L-function (20.3), from the logical description of the non-success risk, we will pass to the arithmetic description. Probability model of the management non-success risk (P-model, P-polynomial)

$$P = p_1 + p_2 \cdot q_1 + p_3 \cdot q_1 \cdot q_2 + \ldots \tag{20.4}$$

"Arithmetics" in the B-model of the non-success risk is such that for the final event, the amount of the risk is within [0,1] with any values of the initiating event probabilities. Probabilities of the management non-success in functions (initiating events) can be determined by the expert assessment.

If the non-success probabilities of the initiating events are above 0.05, the company non-success risk becomes large and inadmissible. The saturation of probabilities (the final event risk approaching 1) occurs with the growth of the number of initiating events and their probabilities.

20.4 Model of management non-success risk in directions of activities

Let us consider the company non-success risk scenario in the directions of operation: coastal system, marine onboard equipment, integrated systems, avionics, aeronautical support, sea and flight simulators. The company activities directions will be denoted with $Z_1, Z_2, \ldots, Z_n$ variables, and the respective funds with $E_1, E_2, \ldots, E_n$. The structural model of the company non-success risk in the directions of activities is also shown in Fig. 20.2, and equations of (20.1–20.4) type are used in the non-success risk LP-model.

The company's possible losses resulting from the non-success are equal to

$$T = p_1 \cdot E_1 + p_2 \cdot E_2 + \cdots + p_n \cdot E_n, \tag{20.5}$$

where $P_1, P_2, \ldots, P_n$ probabilities of non-success in the company's directions of activities.

Let us calculate the company non-success risk for three P_3, four P_4, and five P_5 directions of activities with the risks added logically according to (20.4) formula and arithmetically. There are considerable differences between the results of the logical and arithmetic addition of the event risks. With the events added up arithmetically, which is often done in practice, weights are summed up, and an absurd result can be obtained: the company's non-success probability of more than 1. This example also demonstrates the necessity for the company disintegration with the growing number of the directions of its activities, as the summary non-success risk may become more than permitted.

The non-success risk LP-models above describe all the possible states and were the most complete and accurate ones. In a number of cases, however, it is unnecessary to take into account all the possible states of the system. For example, it is known from the statistic data that there was non-success when one and not more than two events occurred: $Z_1, Z_2, \ldots, Z_n$. Then, to simplify the model, you should use the risk model for a limited number of system states.

If we have a logical system risk model of four elements

$$Y = Z_1 \vee Z_2 \vee Z_3 \vee Z_4. \tag{20.6}$$

then for a limited number of states, when one or either of two events occur, the non-success risk model will be noted as

$$\begin{aligned} Y = {} & Z_1\overline{Z_2}\,\overline{Z_3}\,\overline{Z_4} \vee Z_2\overline{Z_1}\,\overline{Z_3}\,\overline{Z_4} \vee Z_3\overline{Z_1}\,\overline{Z_2}\,\overline{Z_4} \vee Z_4\overline{Z_1}\,\overline{Z_2}\,\overline{Z_3} \vee Z_1Z_2\overline{Z_3}\,\overline{Z_4} \vee \\ & \vee Z_1Z_3\overline{Z_2}\,\overline{Z_4} \vee Z_1Z_4\overline{Z_2}\,\overline{Z_3} \vee Z_2Z_3\overline{Z_1}\,\overline{Z_4} \vee Z_2Z_4\,\overline{Z_1}\,\overline{Z_3} \vee Z_3Z_4\overline{Z_1}\,\overline{Z_2}. \end{aligned} \tag{20.7}$$

In the risk L-model, all the logical summands are orthogonal in pairs, which allows the non-success risk P-model (P-polynomial) to be written directly:

$$\begin{aligned} P\{Y\} = {} & p_1q_2q_3q_4 + p_2q_1q_3q_4 + p_3q_1q_2q_4 + p_4q_1q_2q_3 + p_1p_2q_3q_4 + \\ & + p_1p_3q_2q_4 + p_1p_4q_2q_3 + p_2p_3q_1q_4 + p_2p_4q_1q_3 + p_3p_4q_1q_2. \end{aligned} \tag{20.8}$$

20.5 Management of company as complex object

The company development is regarded as the management of the company state (Fig. 20.3) in the direction from the initial state A to the set final state B along the chosen program track $A - B$ and with a correction of the company state in the case of deviation from this track. Such interpretation of the company management objective uses the following notions: $H(1, 2, \ldots, N)$ are development stages or years; $Y(Z_1, Z_2, Z_3, \ldots, Z_m)$ are monitored parameters or strategic objectives or their risks; $U(U_1, U_2, \ldots, U_n)$ are control actions or strategic decisions; $W(W_1, W_2, \ldots, W_n)$ are corrective actions to return the company state to the program track if it has deviated from it. The company management and manager training technology is shown in the form of the following logically closed series of cognition pattern procedures:

1. Forecasting the company non-success caused by Y parameters, i.e., exit of the parameters beyond the admissible value corridor.

2. Modeling or distribution of resources for monitoring Y parameters, control U and corrective W actions.

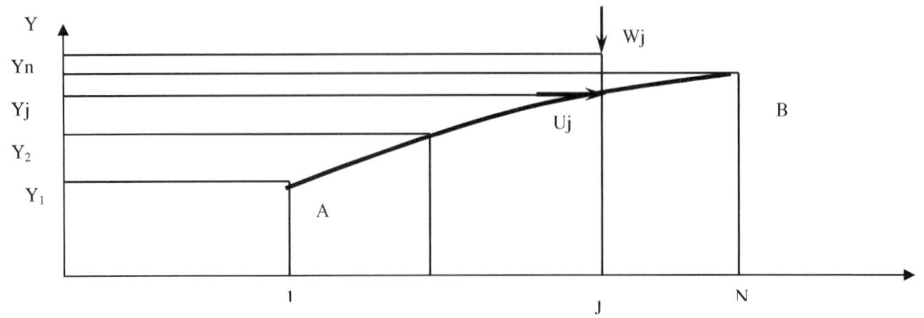

Fig. 20.3. The scheme of management of company as the complex object: Y are controlled parameters; U, W are manage and corrective actions

3. Elaboration of development program, values of monitored parameters Y, control U and corrective W action at H stages included.

4. Processing information in the course of functioning and making a decision on the selection of W correction.

5. Refinement of models to determine values of Y, U, W parameters, resources for the implementation of Y, U, W parameters, and losses in the case of failure to implement these parameters.

20.6 Models of non-success risk in accomplishing objective or group of objectives

Strategic objectives, e.g., of Transas in 2000–2005, were the development of business. Problems of strengthening market positions were solved through the creation of the Russian Federation's (RF) commercial image. Five groups of strategic objectives $Y_1, Y_2, \ldots, Y_5$ were singled out, each of them consisting of several targets:

Y_1 - to increase the RF defense potential: Z_{11} - to introduce new technologies in the defense market and market of dual-purpose products, Z_{12} - to improve performance characteristics of military equipment, Z_{13} - to expand sales market for the hi-tech products, Z_{14} - to enhance safety of military equipment operation.

Y_2 - RF integration in the world economic space: Z_{21} - to improve RF's business image; Z_{22} - to enhance investment attractiveness for the foreign capital, Z_{23} - to reduce the external debt, Z_{24} - to get into the market of the RF public debt and to include Transas in the state programs for the re-payment of the RF foreign debts.

Y_3 - to create conditions for including the RF in the World Trade Organization: Z_{31} to integrate in the world economic space, Z_{32} - to strengthen external economic links, Z_{33} - to improve transport infrastructure of the North-West area of Russia.

Y_4 - to contribute to the image of St. Petersburg as a cultural and science capital of Russia: Z_{41} - to attract investments to St. Petersburg, Z_{42} - to improve the appearance of the city's historical center, Z_{43} - to develop information technologies and create techno parks in St. Petersburg, Z_{44} - to obtain main means for centralizing administrative and maintenance personnel and Transas production facilities.

Y_5 - to centralize the Transas management staff: Z_{51} - to improve the company management efficiency, Z_{52} - to re-structure the company for the further successful development of business including the development in new directions.

Several groups of strategic objectives were considered at separate stages. In accordance with the diagram of managing the company as a complex object (Fig. 20.3), we will construct logic and probability models of the risk of the company non-success in accomplishing strategic objectives. We will consider $Y, Y_1, \ldots, Y_4, Y_5$ vectors and their components as random events that will be designated by the same identifiers as logical variables.

The scenario of risk of non-success in accomplishing one group of company objectives will be formulated on the basis of the common sense: the group of objectives will not be accomplished if any one objective in the group, or any two objectives ... or all the objectives in the group are not achieved.

The structural model of risk of non-success in accomplishing Y_i group of objectives is similar to that presented in Fig. 20.4. If objectives $Z_{i1}, Z_{i2}, \ldots, Z_{in}$ are included in Y_i group of objectives, then L model of non-success risk in accomplishing an objective will be

$$Y_i = Z_{i1} \vee Z_{i2} \vee \cdots \vee Z_{in}. \quad (20.9)$$

We will note down L-function of non-success risk in accomplishing an objective in an equivalent form after its orthogonalization

$$Y_i = Z_{i1} \vee Z_{i2}\overline{Z_{i1}} \vee Z_{i3}\overline{Z_{i1}}\ \overline{Z_{i2}} \vee \cdots . \quad (20.10)$$

P-model of non-success risk in accomplishing a target will be

$$P_i = P_{i1} + P_{i2} \cdot (1 - P_{i1}) + P_{i3} \cdot (1 - P_{i1}) \cdot (1 - P_{i2}) + \ldots \quad (20.11)$$

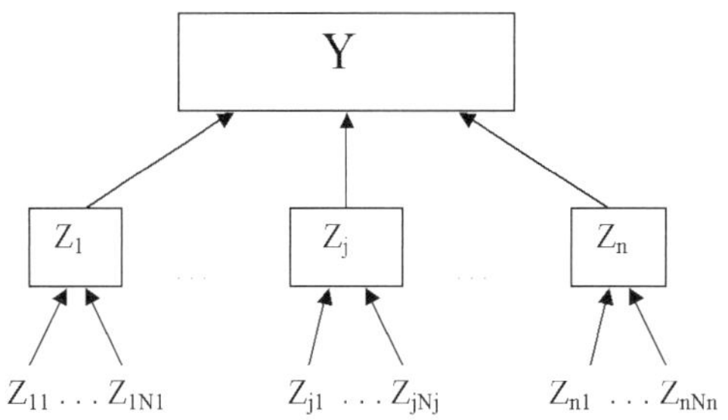

Fig. 20.4. Structural risk model in reaching several aims

Example 1. Let us consider the non-success risk in accomplishing the first group of objectives Y_1. Expert assessments of non-success probabilities for separate objectives are $P_{11} = 0.05; P_{12} = 0.04; P_{13} = 0.03; P_{14} = 0.06$. Then the non-success risk in accomplishing Y_1 group of objectives is equal to $P_1 = 0.05 + 0.04 \cdot 0.95 + 0.03 \cdot 0.96 \cdot 0.95 + 0.06 \cdot 0.95 \cdot 0.96 \cdot 0.97 = 0.16844$.

Let us consider the non-success risk in accomplishing all the groups of objectives without noting down expressions for achieving each of these objectives. The L-model of non-success risk in accomplishing a sum of company's objectives at a stage, will be

$$Y = Y_1 \vee Y_2 \vee \dots \vee Y_j \vee \dots \vee Y_n, \ j = 1, 2, \dots, n. \qquad (20.12)$$

We will now pass from the logical description of the non-success risk in accomplishing an objective, to the arithmetic description. P-model (P-polynomial) of non-success risk in accomplishing an objective is

$$P = P_1 + P_2 \cdot (1 - P_1) + P_3 \cdot (1 - P_1) \cdot (1 - P_2) + \dots \qquad (20.13)$$

Example 2. Considered here is non-success risk in accomplishing objectives Y_1 and Y_2. Probability of non-success in accomplishing Y_1 group of objectives was calculated in example 5 and was equal to $P_1 = 0.1684$; assessment of non-success probability in accomplishing objective Y_2 was equal to $P_2 = 0.075$. The non-success risk in accomplishing Y_1 and Y_2 groups of objectives is equal to $P = 1 - 0.8315 \cdot 0.925 = 0.2308$.

20.7 Model of quality loss risk in company operation

The model of quality management of the entire company operation including finances, resources, personnel, etc., was considered earlier in Section 19.1. The Russian quality assessment and management standard (close to the American, European, and Japanese standards) is used. The problem to be solved is the company's self-assessment of its own activities.

Essence of the task. The quality model criteria are divided into two categories: Possibilities and Results. Possibilities Category Z_{10} is determined from the criteria assessments: role of the management in organization of works Z_1, use of employees' potential Z_2, planning and strategy in the sphere of quality Z_3, use of resources Z_4, technological processes involved in manufacturing, advertising, service Z_5. Results Category Z_{11} is determined from the assessment of satisfaction criteria of the interested parties: employees Z_6, consumers Z_7, society on the whole Z_8, and financiers Z_9.

In many countries, quality estimation systems are known as national premiums and are used with the purpose of stimulation of improvement of quality and competitiveness of production (services) and encouragement of organizations, which use advanced and effective methods of quality management. For example in Russia, premiums by the Government of the Russian Federation in the field of the quality were founded in 1996.

Centers of quality control carry out consultation and estimation of the documentation applied for competition on the premium. The information, submitted by the companies, is strictly confidential and is not subject to disclosure. The estimation of quality of management systems of the companies is done by a group of independent, highly professional specialists who are specially prepared for this work. Usually not more than 12 premiums a year are given. Periodicity of self-estimation of quality of management systems is equal to one year.

The maximal numerical estimations of criteria in points for the Russian Federation Government Award model in the quality field are shown in Fig. 19.1. The sum of all maximal estimations is equal to 1000. The given distribution of estimates is recommended to be used for any enterprise and organization irrespectively of the kind of production and activity, of the sizes and forms of ownership.

The company's quality self-assessment is made by the company's own "aim group." It assesses each criterion $Z_1 \div Z_9$ in points and calculates this assessment's ratio to the maximum possible value from the standard. The system of participation in quality competitions allows the criteria level to be assessed in per cent of the maximum possible value. As this is done, the degree of the company's shortcoming in each criterion and each direction of operation can be objectively seen. The company's achievements are determined by comparing the criteria numeric values $Z_1 \div Z_9$ year after year.

Object of investigation. Object of investigation is the management of company TRANSAS (TRANsport SAfety Systems) - one of the leading manufacturers of hi-tech products well in demand worldwide. The company produces what it has itself invented, and has a high mission: to reduce the risk and cost of systems in the sea and air transport.

Transas customers are shipping and fishing companies, owners of boats and yachts, Navy and defense establishment, ports and coastal services, civil aviation and air force, schools and simulator centers, rescue services, and oil extraction companies.

Transas provides a complete range of commissioning services for the products it supplies. To ensure the high-quality customer support, a network of service representative offices is maintained and permanently expanded.

Transas, a Russian group of companies, is one of the leading manufacturers of hi-tech products well in demand in the world. The kernel of the group is Transas founded in 1990 in St. Petersburg. The staff of employees at the enterprise included in Transas group comes to about 700 people. The company distributor net is deployed in more than 100 countries of the world. Products made by the Russian group of companies Transas are known in 110 countries. The company's turnover reached 100 mln in 2005.

Directions of activities. Transas is engaged in the development of technologies and manufacture of hi-tech products in the following directions: coastal traffic safety systems, shipborne equipment and avionics, integrated navigational systems, sea and flight simulators.

Main company management functions: strategic planning, personnel management, marketing and sales, control of the electronic technologies and programs development, accounting and monitoring the company operation.

Transas is an electronic technology company. It does not have its own large scale manufacturing facilities, there is only small well fitted out pilot line production for small series and testing ready products. Transas cooperates with other companies — this is the practice accepted in the world.

Customers: state institutions (2); shipyards (9); shipping companies (14); sea ports (25); schools and simulator centers (15); research organizations (4); oil companies (2). The number of installed training center systems in the schools and academies comes to several hundred. The company occupies 10–15 sea simulator market. The number of competitors is from 6 to 10 companies.

At the end of the 1980s there were serious problems in all the branches of the national economy. The sea merchant fleet was no exception. Commercializing shipping companies generated chaos and lack of supervision. Many qualified specialists began to look for their own way-out from the then situation. Among them were three deep sea masters: N. Lebedev, N. Moujikov, E. Komrakov and the ship engineer V. Godounov who came together because of their interest in computers. Being experienced seafarers, they were very well aware of how much the ship captains needed navigational systems. They also knew what had to be changed and how to relieve the captains of their workload.

Rectangular components of the Transas company success are high quality of production, use of progressive computer technologies, flexible price politics, a wealth of practical experience, creating professionals group, training talented managers.

In not simple transitional period of business developing in Russia, the phenomenal success of the Transas company is needed in scientific reasoning, generalization and systematization of management experience. Developing production for ensuring safety on marine and aviation transports on ISO 9001 standard quality, Transas extends concepts and approach for ensuring safety (acceptable risk) and quality from the area of engineering to the area of management by developing the quality of the company itself.

Functioning Quality analysis of Transas

Assessment of the Transas company functioning quality was made by the company's own group of experts in the period between 2000 and 2004. The expert assessments for each criterion were averaged. The relative assessment of the company functioning quality is considered as the company quality probability in this criterion. These assessments (probabilities) have been used as the basis for calculating relative and arithmetic mean criteria of the company functioning quality.

On average yearly, the staff of the company increased by 55 persons. The assets volume increased many times over, which proves success of the company. At the same time the debtor indebtedness grew up. It is explained by big investment in building industrial houses. Because of this the profit of Transas is heavily oscillated and has even the minus value.

On the standard method, description in Chapter 19, the quality estimation of the functioning company Transas (criteria P_{score}) is made. Results are stated in Table 20.1 for period from 2000 to 2004. The expert estimations on every quality criteria are averaged; deviation of expert estimations from average values was not more than 15%. This testifies about equal qualification of experts and their understanding of conditions of company development.

They put the estimates in the ball N_i for every criteria. Every criteria has maximum possible estimate $N_{i\ max}$. The relative estimate $Q_i = N_i \ / \ N_{i\ max}$ is computed for every criteria. The relative estimate of functioning quality of the company's management can be viewed as an indicator of the company itself. On base of this, relative estimates (probabilities) are computed:

the relative criteria of functioning quality of company

$$P_{rel-score} = P_{score} \ / \ 1000. \tag{20.14}$$

the mean arithmetic criteria of functioning quality

$$P_{m-score} = (\sum_{i=1}^{9} Q_i) \ / \ 9. \tag{20.15}$$

In the research, values of the probability quality loss risk criterion and mean geometric probability have been calculated. Transas functioning quality criteria were growing from year to year, but this growth was different for different criteria. The logical quality criterion is more accurate and transparent [3, 4].

The logic and probabilistic quality model. Modernizing the company functioning quality model has been proposed. It consists in replacing the arithmetic addition of criterion events (in points) with logical addition of criteria probabilities (relative weights).

For the structural, logical, and probability model of the risk of losing the company functioning quality, random events are designated by the same logical variables $Z_1 - Z_9$. "Quality" property of the logical Y variable and derivative "Possibilities" and "Results" are Z_{10} and Z_{11}, respectively.

The structural risk model of quality loss is stated in Fig. 19.2. L-function of risk in a disjunctive normal form

$$Y = Z_1 \vee Z_2 \vee Z_3 \vee Z_4 \vee Z_5 \vee Z_6 \vee Z_7 \vee Z_8 \vee Z_9. \tag{20.16}$$

L-function of risk in an orthogonal disjunctive normal form

$$Y = Z_1 \vee Z_2\overline{Z}_1 \vee Z_3\overline{Z}_2\overline{Z}_1 \vee \ldots. \tag{20.17}$$

Table 20.1. Quality assessment of functioning Transas

Criteries	Name of criteria	2000 N_i / Q_i	2001 N_i / Q_i	2002 N_i / Q_i	2003 N_i / Q_i	2004 N_i / Q_i	Max
Z_1	Role of leaders in organization of works	0.7825	0.79	0.815	0.83	0.90	1.0
Z_2	Using potential of clerks	0.583	0.679	0.764	0.725	0.864	1.0
Z_3	Planning in the area of quality	0.885	0.88	0.912	0.9	0.942	1.0
Z_4	Rational using resource	0.77	0.773	0.77	0.77	0.757	1.0
Z_5	Management of technological processes and works in company	0.690	0.738	0.813	0.800	0.775	1.0
Z_6	Satisfaction of personnel	0.844	0.819	0.883	0.833	0.922	1.0
Z_7	Satisfaction of consumers by quality of production and service	0.575	0.585	0.621	0.639	0.605	1.0
Z_8	Influence of company on society	0.365	0.454	0.637	0.633	0.837	1.0
Z_9	Finance results of company work	0.663	0.712	0.762	0.792	0.879	1.0
P_{score}	Criteria of quality in balls	684.2	713.2	768.0	764.0	812.2	1000
$P_{rel-score}$	Criteria of quality in relative balls	0.684	0.713	0.768	0.764	0.812	1.0
$P_{m-score}$	Criteria of quality mean arithmetics	0.679	0.715	0.775	0.769	0.831	1.0
P_{m-prob}	Criteria of quality of mean probabilistic	0.639	0.703	0.765	0.761	0.825	1.0

P-function of risk (P-polynomial of risk)

$$P\{Y = 1\} = p_1 + p_2 q_1 + p_3 q_2 q_1 + \dots \tag{20.18}$$

Corresponding P-model of success functioning company on quality we describe as follows

$$P_{prob} = 1 - P\{Y = 1\}. \tag{20.19}$$

Earlier we yet determined the probabilities of event success $Z_1 - Z_9$. Now let us compute by formulas (20.19) the values of probabilistic risk criteria of quality losses P_{prob} and the mean geometrical probability P_{m-prob} as root of degree $n = 9$ from P_{prob} (the last line in Table. 20.1).

Thus, for estimation of the functioning quality of the company, we can use the following criteria (Table 20.1):

P_{score} – in balls,

$P_{rel-score}$ – in relative balls;
$P_{m-score}$ – mean arithmetics;
P_{prob} – probability;
P_{m-prob} – mean geometric probability.

Criteria of functioning quality of the company "Transas" increased by years, but increase was non-monotonic and different for criterion. Comparison of different criteria for estimation of "Transas" functioning quality shows that mean geometric (logical) criteria of the estimation of the company functioning quality $Pm - prob$ has more accuracy and transparency.

20.8 Conclusions

Summation of formulated and proved concepts in this work, the constructed LP-models of management of non-success risk and also practice results, are presented as decision of the actual scientific and economic problem of increasing efficiency of the strategic management.

The main results of the work are the following:

1. The scheme of taking decisions in management is developed; non-success risk scenarios and corresponding logic and probabilistic LP-models for taking decisions on the risk criteria are suggested.
2. The scheme of management of company development as complex object with motion on the program strategic trajectory and correction in the case deviation from it is suggested.
3. The procedures of company management and manager training, which are procedures of cognition scheme, are named.
4. The LP-model of functioning quality of Transas using real data is developed.
5. The following logic probability risk models have been developed:
    - management of non-success in functions;
    - company non-success in directions of activities;
    - non-success of management of company as a complex object;
    - non-success in accomplishing an objective or a group of objectives;
    - loss of the company functioning quality.
6. Different criterion for estimation of functioning quality of company Transas are compared and the bigger accuracy and transparency of the quality LP-model is shown.
7. LP-models of the management non-success risk can be used for the control of the company by risk criterion.
8. The developed LP non-success risk models improve efficiency of the strategic management.

21

LP-Models of Fraud and Interaction of Companies

Everyone is capable of fraud if under vital circumstances, values are badly taken into account, and it is possible to hide the fact of plunder for a while.
W. Albrecht, G. Wernz, T. Williams

21.1 Fraud of manager

Fraud is not a crime that is made a parade. There is no question about "corpus delicti" at a robbery of a bank that is witnessed by employees or clients. Fraud differs from other kinds of crime by difficulty of its revealing. However, frauds have a mass character, and there are many data on frauds both in judicial law-court and in the staff departments of large firms and banks.

For each type of fraud, it is possible to find signs [27, 65] that are characteristic. Each of such signs has at least 2 grades. The fraud risk P-model can be trained on statistical data by methods described in Chapter 12.

Existing methods [65] allow one to estimate the presence of fraud only qualitatively and do not give reliable recommendations, whether it is necessary to begin investigation. Such investigation can be carried out only in the case when there are serious reasons to believe that the fraud took place in the past or that the fraud is done now. The value of this "seriousness" is not quantitatively estimated, and the final decision is accepted by the company head, who can be biased.

Special signs testify to the fraud of the manager [27, 65]. They are divided into the following basic groups: (1) anomalies in activity of the enterprise; (2) personal qualities of the manager; (3) anomalies in organizational structure of the company; (4) special relationship with other partners. In their turn,

E.D. Solojentsev, *Scenario Logic and Probabilistic Management of Risk in Business and Engineering*, DOI 10.1007/978-0-387-77946-1_21,

these groups include other warning signs of "deeper level." Corresponding events have probabilities. In total, it is suggested there are about 40 signs of fraud, which are divided into some groups.

Signs of frauds that reveal as:

1. Special relationships with partners,
2. Special relationships with banks and financial organizations,
3. Special relationships with managers,
4. Special relationships with parent companies and tax services,
5. Special relationships with auditors,
6. Presence of private interests,
7. Financial needs,
8. Predilection for gambling and risky deals,
9. Vague or criminal past,
10. Dependence of financial well-being only from activity of the given firm,
11. Dishonest or unethical behavior at work.

Signs of frauds that reveal as unnecessary complicated structure of the company:

12. Absence of an effectively working division of internal company audit,
13. Belonging to the branch, related to high risk, and adherence to highly risky operations,
14. Use of out-of-date or worn-out production means,
15. Frequent changes among the top managers and directors,
16. Large transactions with related partners.

Signs of frauds that reveal through financial accounting and other documents:

17. Inexplicable changes in the balance report,
18. Work on the verge of bankruptcy,
19. Untrustworthy high profitableness in the report,
20. Unusually good bargains at the end of the accountable period,
21. Deterioration of the quality of the profit,
22. Insufficiency of a basic capital,
23. High debts or the big parts of overhead charges,
24. Difficulties in collecting debts and problems, related to movement of finances,
25. Faster growth of charges in comparison with incomes,
26. Dependence of manufacture on one or two products,
27. Participation of the company in legal processes.

Other signs of frauds are

28. Frequent change of auditors,
29. Refusals or delays in information given to auditors,
30. Refusals of auditors from pronouncement of judgment or disagreement with the data of submitted financial documents,
31. Withdrawals of licences,

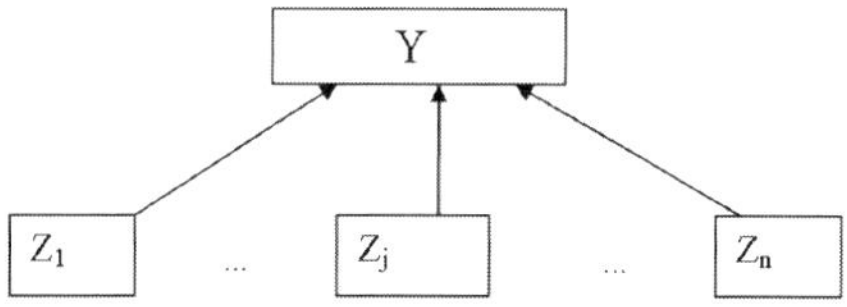

Fig. 21.1. Scenario model risk of type "knot"

32. Frequent checks by supervising companies,
33. Information that the given firm is engaged in risky operations,
34. Frequent or significant hiding assets from taxation,
35. Continual problems with taxation authority,
36. High personnel fluctuation among managers,
37. New top managers,
38. Breach of admissible debts,
39. Use of several banks,
40. Inability to provide financing by formation of loans or credits,
41. Continual pressure to merging, sale, or absorption of one's firms by others,
42. Reorganization of the structure of partner relations;
43. Frequent change of legal counsel,
44. Plenty of new customers or suppliers,
45. Pleading pressure by politicians, defenders of environment, and public opinion.

The structural fraud model is presented in Fig. 21.1. The scenario of the manager fraud is described as follows: the fraud can take place if any one sign-event or any two sign-events or ... all sign-events occur.

In disjunctive normal form, the logical risk function of fraud is described as follows:

$$Y = Z_1 \vee Z_2 \vee \ldots \vee Z_n. \tag{21.1}$$

The risk L-function of the fraud in the orthogonal disjunctive normal form:

$$Y = Z_1 \vee Z_2\overline{Z}_1 \vee Z_3\overline{Z}_2\overline{Z}_1 \vee \ldots \tag{21.2}$$

The risk P-function of the fraud

$$P = p_1 + p_2(1 - p_1) + p_3(1 - p_1)(1 - p_2) + \ldots \tag{21.3}$$

21.2 Fraud of worker

Suspicions in swindle of hired workers cause certain attributes. They are divided in the following basic groups: non-standard data in accounting documents; weaknesses inside the firm's control; deviations from average (normal) values sizes of those or other parameters; irregular living; unusual behavior; presence of complaints.

Attributes of swindle are non-standard data in accounting documents:

1. Unusual deviations from norm of analytical parameters.
2. Changes in the financial reporting.
3. Manipulations with registration registers.
4. Non-standard data in accounting documents.
5. Strange changes of financial parameters.

Attributes of swindle are weaknesses of the internal control:

6. Disappearance of separate documents.
7. The postings delayed concerning an extract under the bank account.
8. Plenty of annulments or discounts.
9. Concurrence of names and addresses of payers or customers.
10. Increase in number of the delayed accounts.
11. Increase in number of the corrected documents.
12. Substitution of documents.
13. Double payments.
14. Transfer inscriptions on checks to address of the second persons.
15. Sequences of documents not making sense.
16. Doubtful inscriptions by hand on documents.
17. Representations not originals, and copies of documents.

Attributes of swindle are deviations from average (normal) values of sizes of parameters:

18. Records in registration registers without documentary confirmation.
19. Inexplicable additions to the accounts received or subject to payment, incomes or expenses.
20. Absence of balance on records in registers of book keeping.
21. Realization of records by the person who does not usually make them.
22. The records made shortly before the end of the accounting period.
23. Discrepancy of a result under any register of balance.
24. Discrepancy of analytical accounting to synthetic.
25. Discrepancy of increase on cash department to expectative.
26. Inexplicable increase/reduction of receipts of means.
27. Inexplicable increase/reduction of material stocks.
28. Increase in payments at reduction of stocks.
29. Sale/purchase of actives.
30. Occurrence of means for payment of dividends.

Attributes of swindle are unusual behavior:

31. Irregular living.
32. Unusual behavior.
33. Presence of complaints.
34. Biographic attributes.

Attributes of swindle are bad organization of work:

35. Absence of division of duties.
36. Absence of physical protection.
37. Absence of independent checks.
38. Absence of corresponding powers.
39. Absence of corresponding documents and records.
40. Neglect existing rules.
41. Inadequate system of document circulation.

Other attributes of swindle:

42. Unexpected shortage or surpluses.
43. Deviations from specifications.
44. Increase in breaks.
45. Superfluous purchases.
46. Plenty of accounts of notes payable or creditor reminders.
47. Substantial growth/reduction of the sums by the current accounts.
48. Not clear divergences in physical characteristics.
49. Shortage or excess on cash department.
50. Payments with delay.
51. Unreasonable expenses and payments.
52. Increase in incomes with reduction of material stocks.
53. Increase in incomes at reduction of amount of transferred money and a revolution on cash department.
54. Increase in incomes at increase in stocks.
55. Increase in incomes at reduction of creditor debts.
56. Increase in incomes at reduction of number of sales.
57. Increase in incomes at increase in the price per unit of output.

Some attributes of swindle on the part of hired workers correspond with attributes of swindle on the part of the manager. In logic function of swindle of the hired worker, all initiating events are connected with each other the attitude relation *OR*. Thus, we use the structural model in Fig. 21.1 and the risk LP-model (21.1–21.3).

21.3 Fraud with investigation

Usually the financial reporting is falsified to show overestimated incomes of the enterprise. These overestimated incomes are used to increase the market rate of shares of the given company and incomes by these shares. Doing such frauds, swindlers frequently give false promises and deliberately distort the facts. All this is done to attract additional capital investments. In frauds with investments, the deceit, as a rule, concerns the verbal or written promises of high dividends.

Signs of frauds are

1. High annual interest rates;
2. Investments that do not have business interest;
3. Pressure on investors so that they should invest means as soon as possible;
4. Use of all ways for evasion or reduction of taxes;
5. The area of business that is new for the given city or district; and it is with held where these managers appear from and where they were engaged earlier;
6. Actions, having been accompanied with bankruptcies or scandals;
7. Exposing figures and financial documents that have not gone through the public checking;
8. Projects, based on reception of "unofficial" reward, involving schemes of marketing, searching special approaches to people, having money, and verbal agreements, about which it is "impossible" to speak, because of local laws;
9. Financial accounts or advertising applications, not confirmed by auditor check;
10. Investments, which mean continuation of inflation or preliminary fixation of attractive interest rates on the laid-down capital, which in the given moment cannot be realistic;
11. Promises of success of investments, based on someone's "unique" abilities or opportunities to achieve the financial successes (type of prediction of the future prices for the goods);
12. Exposing special appeal of these investments, based on emotions;
13. Necessity of reception of large loans for achievement of success;
14. Search of investors for payment of existing debts;
15. Luxurious style of life of top officials, despite the fact that their business began rather recently;
16. Pressure to invest all money to one enterprise;
17. Impossibility to leave business or to bring back the investments;
18. Complaining to cause the investor to sympathize with the company and to involve additional means for overcoming allegedly temporary problems.

The structural fraud model has the kind of "node" and it is analogous to that presented in Fig. 21.1. The scenario of the fraud with investments is described as follows: the fraud can take place if any one sign-event takes place or any two sign-events take place or ... all sign-events take place.

In disjunctive normal form, the risk function of the fraud is described:

$$Y = Z_1 \vee Z_2 \vee \ldots \vee Z_n. \tag{21.4}$$

The risk L-function of the fraud in the orthogonal disjunctive normal form:

$$Y = Z_1 \vee Z_2\overline{Z}_1 \vee Z_3\overline{Z}_2\overline{Z}_1 \vee \ldots \tag{21.5}$$

The risk P-function of the fraud

$$P = p_1 + p_2(1 - p_1) + p_3(1 - p_1)(1 - p_2) + \ldots \tag{21.6}$$

21.4 Struggle of building firms for profitable contract

Let us consider the example using the risk LP-model for estimation of risk and efficiency at struggle of two building firms for profitable contract with counteraction of the third building firm.

The building firms A and B would like to get the profitable contract. The building firm C can prevent them. The firm C (event Z_3 with probability p_3) will enter the struggle for the contract (Fig. 21.2). Counteractions of the firm C can make the firm A (event Z_4 with probability p_4) and the firm B (event Z_5 with probability p_5) to give up their intentions. If the firm B (event Z_1 with probability p_1) and the firm A (event Z_2 with probability p_2) could get the building contract, then the profit of the firm A (B) would be $E = 6$ million ($E = 2$ million).

In the example, the initial probabilities $p_1, p_2, p_3, \ldots$ are fixed by the method of expert estimation, taking into account the external factors and the capital of firms A, B, and C. There are the inverse connections from the nodes Z_5 and Z_4 in this risk model. They define conditions of realization of the functions Y_1 and Y_2 by the elements Z_1 and Z_2. These functions consist in preservation of intentions under counteraction. We write down the logic equations separately for the success:

$$\begin{gathered} Y_1 = Z_1 \wedge \overline{Y}_5; \quad Y_2 = Z_2 \wedge \overline{Y}_4; \\ Y_3 = Z_3; \quad Y_4 = Z_4 \wedge Y_3; \quad Y_5 = Z_5 \wedge Y_3 \end{gathered} \tag{21.7}$$

and for the non-success:

$$\begin{gathered} \overline{Y}_1 = \overline{Z}_1 \vee Y_5; \overline{Y}_2 = \overline{Z}_2 \vee Y_4; \overline{Y}_3 = \overline{Z}_3; \\ \overline{Y}_4 = \overline{Z}_4 \vee \overline{Y}_3; \overline{Y}_5 = \overline{Z}_5 \vee \overline{Y}_3. \end{gathered} \tag{21.8}$$

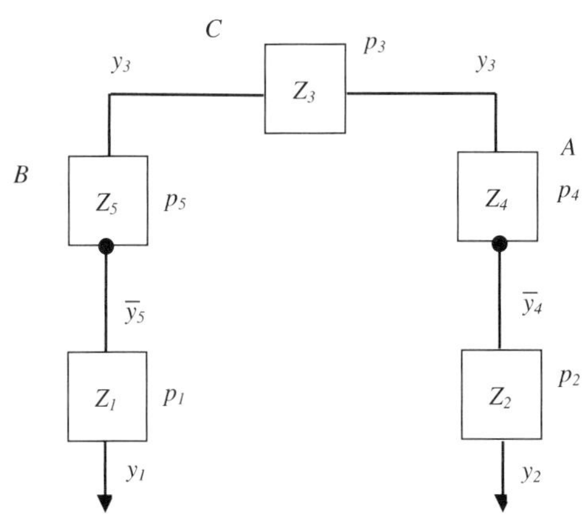

Fig. 21.2. Struggle of three civil engineering firms for two expedient orders

Let us assume that probabilities of events are as follows: $p_1 = 0.85$; $p_2 = 0.95$; $p_3 = 0.7$; $p_4 = 0.4$; $p_5 = 0.5$. Then we obtain the LP-model of achievement of the purpose,

$$Y = Y_1 \vee Y_2 = Z_1 \wedge \overline{Z}_3 \vee Z_1 \wedge \overline{Z}_5 \vee Z_2 \wedge \overline{Z}_3 \vee Z_2 \wedge \overline{Z}_4. \tag{21.9}$$

Polynomial of the probabilistic function of purpose achievement is

$$P_c = p_1 p_3 q_4 + p_2 p_3 q_4 + p_1 q_3 + q_1 p_2 q_3 - p_1 p_2 p_3 q_4 q_5 = 0.825. \tag{21.10}$$

Let us introduce the efficiency parameters of achieving three purposes:

1. $E_1 = 6$, if only the element Z_1 achieves the purpose $(Y_1 \wedge \overline{Y}_2)$,
2. $E_2 = 2$, if only the element Z_2 achieves the purpose $(\overline{Y}_1 \wedge Y_2)$,
3. $E_3 = 8$, if both elements Z_1 and Z_2 achieve the purposes $(Y_1 \wedge Y_2)$.

Using the above mentioned scheme of construction of the logic and probability functions, probabilities of achievement of these three purposes are determined: $P_1 = 0.141; P_2 = 0.272; P_3 = 0.412$. Total efficiency of achievement of the purposes is equal:

$$W = E_1 * P_1 + E_2 * P_2 + E_3 * P_3 = 4.271 \text{ million.}$$

21.5 Financing building projects with reservation

Let us consider the risk LP-model for estimation of risk and efficiency of financing building projects with reservation. In this case, the danger model or the scheme of functional integrity contains fictitious tops.

Consider the following situation. The bank A should finance the building project 1 (event Z_1 with success probability p_1). The bank B should finance

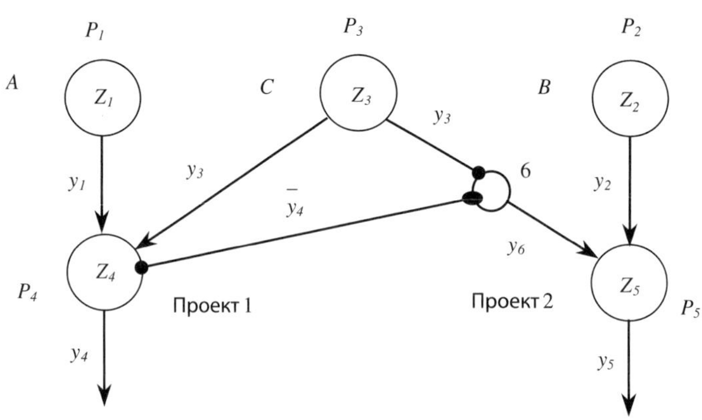

Fig. 21.3. Financing building projects with reserving

the project 2 (event Z_2 with success probability p_2) (Fig. 21.3). The bank C is in a reserve. The bank C can finance both the project 1 and the project 2, but only one project (event Z_3 with success probability p_3). The unconditional priority for bank C is the project 1 (event Z_4 with success probability p_4). Only if the project 1 fails, the bank C can finance the project 2 (event Z_5 with probability p_5). Write down the logic equations separately for the success:

$$Y_1 = Z_1; Y_2 = Z_2; Y_3 = Z_3; Y_4 = Z_4 \wedge (Y_1 \vee Y_3);$$
$$Y_5 = Z_5 \wedge (Y_2 \vee Y_6), Y_6 = Y_3 \wedge \overline{Y}_4 \tag{21.11}$$

and for the non-success:

$$\overline{Y}_1 = \overline{Z}_1; \overline{Y}_2 = \overline{Z}_2; \overline{Y}_3 = \overline{Z}_3; \overline{Y}_4 = \overline{Z}_4 \vee \overline{Y}_1 \wedge \overline{Y}_3;$$
$$\overline{Y}_5 = \overline{Z}_5 \vee \overline{Y}_2 \wedge \overline{Y}_6; \overline{Y}_6 = \overline{Y}_3 \vee Y_4. \tag{21.12}$$

On this scheme, the node 6 is fictitious. It is used for the graphic description of the complex logic condition for the reserve mode of maintenance of the building project 2 by the bank C. The arches, ending with points, represent the conjunctive connections. The arches, ending with arrows, represent the disjunctive logic conditions of maintenance of functioning elements. We use the direct and inverse outputs from the node Z_4, designating conditions of realization and non-realization of functions by the corresponding element of the system. Let us consider 5 variants of the logic functions:

1. Realization of at least one building project $(Y_4 \vee Y_5)$,
2. Realization of neither building project $(\overline{Y}_4 \wedge \overline{Y}_5)$,
3. Realization of both building projects $(Y_4 \wedge Y_5)$,
4. Realization of only the building project 1 $(Y_4 \wedge \overline{Y}_5)$,
5. Realization of only the building project 2 $(\overline{Y}_4 \wedge Y_5)$.

Each of the obtained logic functions precisely defines the set of conditions in which the given criterion is realized. These functions define corresponding complex random events; probabilistic functions are the sought mathematical models.

The considered examples show that under complex logic conditions and purposes, the numerical estimation of risk of success or non-success of organizational systems can also be done by using the risk LP-models.

22

The Formal LP-Theory of Non-success Risk with GIE

Any constructed algorithm is the mirror of a certain general theory, from which it normally follows as special case.
A. V. Yaroshenko

Now, after consideration of theoretical and applied aspects of the non-success risk LP-theory with GIE in Chapters 11–21, we will explain the formal logic and probabilistic theory of non-success risk with GIE on the basis of doctrines on the formal theory by academician A. I. Maltsev [84] as well as its development in works of A. V. Yaroshenko [43].

The futures of the explained formal risk LP-theory with GIE — the database (DB), knowledgebase (KB), and sets are considered as the tabular forms of the different presentation of the same information for the decision of various tasks of the manager.

22.1 Connection of database, knowledgebase, and sets

Tabular database (Table 11.2) contains statistics on congeneric objects (credits) or states of one object at different points of time (securities portfolio). The values of parameters (characteristics) in DB can be integral or non-integral and are considered as statistics on objects or states of one object. Table cells show values (quantitative or qualitative) characterizing the object or its state. Table 11.2 shows characteristics values z_{ij} and efficiency characteristic value y_i in the last column.

We changed the initial representation of statistics data replacing values of characteristics by their grades (numbered intervals). In scenarios and LP risk models of classification, investment, efficiency, management, corruption, and bribery problems, there exist a great variety of objects N (up to 1000 and more), parameters-events n (up to 20 and more), and grades-events within each parameter-event (from 2 to 40).

E.D. Solojentsev, *Scenario Logic and Probabilistic Management of Risk in Business and Engineering*, DOI 10.1007/978-0-387-77946-1_22,

Thus, the modified tabular DB (Table 11.2) now has denumerable sets with finite number of elements (grades) for each characteristic. Unlike Table 11.2 where characteristics could have an indefinite number of different values, now each range for each parameter has a denumerable number of elements equal to the number of grades it is broken down into.

In the modified DB (Table 11.2) parameters describing the object shall be called events-parameters or logical variables. Cells of Table 11.2 hold events-grades $Z_{jr}, j = 1, \ldots, n; r = 1, \ldots, N_j$ for parameters $Z_1, \ldots, Z_j, \ldots, Z_n$. The last column holds events-grades $Y_r, r = 1, \ldots, N_y$ for efficiency parameter Y.

As a result, we get a set of N logic equations with the left-hand side of the equation and the right-hand side of the equation. Let's correlate the probability of validity or invalidity of each logic variable of the right- and the left-hand side with it. This logic equations set shall be called a knowledge base (KB). This system shall be also considered a set of logical propositions and shall be used to acquire new knowledge.

In scenarios and LP risk model, events-parameters are linked with the help of logic operations AND, OR, NOT and can be arranged in cycles. Events-descriptors (triggering events) correspond with logical variables that can be dependent though not initially but just because they are within some definite logical formula determining connection between them. Events-grades for each group of incompatible events are dependent. They form a group of incompatible events for each characteristic [3, 30]. We have obtained the connection of objects set, objects, parameters and grades, presented at Fig. 11.1.

Below we explain the formal non-success risk LP-theory with GIE for the construction of the LP-models for classification of objects or its states (credit risk of physical and juridical persons, bribes, swindle, etc.). The formal non-success risk LP-theory is diffused to another different risk area, by changing of the signature.

For example, we give the credit risk LP-models that are specific for each bank are analyzed as an illustration. Because each bank has its own clients, provides service to a certain branch of industry, is located in a certain area, a different number of various parameters and number of grades is used to describe the credit (loan).

22.2 Sets

The non-success risk LP-theory includes the following problems:

Problem-1. Construction of the risk LP-model: formulation of the risk scenario, record of the logic risk function and its orthogonalization, record of the probabilistic risk function. Probabilities of grade-events are unknown.

Problem-2. Identification (training) of the risk LP-model based on statistical data: determination of probabilities of grade-events and accepted risk.

Problem-3. Risk forecasting: estimation of new objects' risk.

Problem-4. Risk analysis and management: determination of contributions to the risk from sign-events and grade-events, management of the risk by alteration of the scenario, grades' number, and signs. *Definition.* The model is a set with relations on it. In set M, which consists of a finite number of elements, we will define relations w, signatures W, axioms A for construction of the risk LP-model, as well as the attributes of the set elements. Thus the top indexes of M, w, and W will indicate their dimension.

Designations of sets' elements of:

i is an index of the object, $i = 1, 2, \ldots, N$;

j is an index of the parameter, $j = 1, 2, \ldots, n$;

r is an index of the grade, $r = 1, 2, \ldots, N_j$;

N is a number of objects;

n is the number of parameters;

N_j is a number of parameter grades

In the description of sets and their element attributes, the basis for the parameter determination will be specified in brackets as follows:

statistics: it is determined from statistics;

problem-1, problem-2, problem-3, problem-4: it is determined by problems solution 1, 2, 3, or 4;

axiom-1, axiom-2, axiom-3, ...: it is determined by means of axioms $A_1, A_2, A_3, \ldots$.

22.2.1. Set of objects M^N

Set M^N is a set of objects in statistical data (statistics).

Attributes. Each object of set M^N (Fig. 11.1 and Table 11.2) has attributes:

i is the number of the credit *(statistics)*;

$Z_{ij}, j = 1, 2, \ldots, n$ are grades of object parameters (columns in Table 11.2);

$Z_{iy\ st}$ are grades of the object efficiency parameter based on statistics (the sign good or bad object, 1 or 0) *(statistics)*;

$Z_{iy\ mod}$ are grades of the object efficiency parameter based on the model (the sign good or bad object, 1 or 0) *(problem-2)*;

P_i is the object risk *(problem-2)*;

S_i is the object price *(problem-2)*;

T_i is risk price *(problem-3)*.

Relation 1.

w_1 is a set of object attributes;

$w_1(w_{11}, w_{12}, w_{13}, \ldots)$ is attributes' vector.

Relation 2.

w_2^2 is a ratio between the object risk P_i and the accepted risk P_{ad};

$w_2^2(P_{ad} > P_i)$.

The relation is used for division of model-based objects into two classes (subsets): good objects and bad objects.

Relation 3.

w_3^2 is a ratio between mean risks of objects described by statistics P_{av} and by model P_m;

$w_3^2(P_{av} = P_m)$.

The relation is used for training (identification) of the risk model based on statistics and provides real frequency dimension to the probabilities $P_{jr}, j = 1, 2, \ldots, n; r = 1, 2, \ldots, N_j$.

Attributes of objects and the relation w_2^2 allow us to set out the following subsets from credits (loans) set M^N:

M^{Nb} is bad objects based on statistics *(statistics)*;

M^{Ng} is good objects based on statistics *(statistics)*;

M^{Nbc} is bad objects based on the model *(problem-2)*;

M^{Ngc} is good objects based on the model *(problem-2)*;

M^{Nbb} is correctly recognized bad objects (based both on statistics and on model) *(problem-2)*;

M^{Ngg} is correctly recognized good objects (based both on statistics and on model) *(problem-2)*;

M^{Ngb} is incorrectly recognized good objects (good based on statistics and bad based on model) *(problem-2)*;

M^{Nbg} is incorrectly recognized bad objects (bad based on statistics and good based on the model) *(problem-2)*,

Where: N_g is the number of good objects in statistics *(statistics)*,

N_b is the number of bad objects in statistics *(statistics)*;

N_{gc} is the number of good objects on the model *(problem-2)*;

N_{bc} is the number of bad objects on the model *(problem-2)*;

N_{gg} is the number of correctly recognized good objects *(problem-2)*;

N_{bb} is the number of correctly recognized bad objects *(problem-2)*;

N_{gb} is the number of incorrectly recognized good objects *(problem-2)*;

N_{bg} is the number of incorrectly recognized bad objects *(problem-2)*.

The integrated parameters calculated for set M^N:

CF is the criterion function of identification: $CF = N_{gg} + N_{bb}$ *(problem-2)*;

P_m is mean risk of object based on model *(problem-2)*;

P_{av} is mean risk of object based on statistics *(statistics)*;

P_{ad} is accepted risk *(problem-2)*;

N_{ad} is the number of dangerous objects *(problem-2)*;

S_m is population mean of financial losses *(problem-2)*;

S_{ad} is allowable financial losses *(problem-2)*;

H_{ad} is probabilities entropy of dangerous objects *(problem-2)*;

E_m is a mean error in recognition of objects *(problem-4)*;

E_g is an error in recognition of good objects *(problem-4)*;

E_b is an error in recognition of bad objects *(problem-4)*;

E_{gb} is asymmetry in recognition of good and bad objects *(problem-2)*.

22.2.2. Set of parameters and logic variables Z^{n+1}

Let's regard parameters of object as random variables (parameter-events) and designate them as logic variables. Thus, set Z^{n+1} will consist of logic variables $Z(Z_1, \ldots, Z_n, Y)$ describing the object.

Relation 4.

w_4^n is a set of logic variables $Z(Z_1, \ldots, Z_n)$ describing the object. The relation is used for the logic description of the non-success risk scenario.

Relation 5.

$w_5^2(Y = Z)$ is the relation between logic variables Z describing the object, and the logic variable Y describing the efficiency parameter of the object.

w_5^2 is the logic risk model (risk L-model) in a general form:

$$Y = f(Z_1, \ldots, Z_j, \ldots, Z_n). \qquad (22.1)$$

Probabilistic risk model (P-model) in general form:

$$P = f(P_1, \ldots, P_j, \ldots, P_n), \qquad (22.2)$$

where: $P_1, \ldots, P_n, P$ are probabilities for logic variable (events).

Attributes of element (variable j) of sets:

j is the number of a random logic variable;

P_j is probability of the object's non-success based on initiating parameter-event j;

P_e is probability (risk) of the object's non-success;

P_{jm} is mean value of probability P_j on objects set,

$\triangle\, P_{jm}$ is contribution of parameter-event Z_j to the object's risk;

$\triangle\, F_{jm}$ is contribution of parameter-event Z_j to accuracy of the risk model (to the error of recognition or in value of criterion function of training).

22.2.3. Sets for grades of the object parameters Z_j^{Nj}

Let's set out the following sets for the parameters' grade-events:

Z_1^{N1} is the grade set of the parameter Z_1; N_1 is the number of grades;

...

Z_j^{Nj} is the grade set of the parameter Z_j; N_j is the number of grades;

...

Z_n^{Nn} is the grade set of the parameter Z_n; N_n is the number of grades;

Y^{Ny} is the grade set of the efficiency parameter Y; N_y is the number of grades.

Relation 6.

w_{6j}^{Nj} is the of grade-events set $Z_{j1}, \ldots, Z_{jr}, \ldots, Z_{jNj}$ for parameter-events $Z_j, j = 1, 2, \ldots, n$;

$w_{6j}^{Nj} = (Z_j, Z_{jr}), r = 1, 2, \ldots, N_j$.

Grade-events of sets $Z_1, \ldots, Z_n, Y$ included in GIE.

Relation 7.

w_{7j}^5 is the relation between probabilities in GIE by modified of Bayes' formula;

$w_{7j}^5 = (P2_{jr}, P1_{jr}, P_{jr}, P2_{jm}, P1_{jm}, P_{jm}), r = 1, 2, \ldots, N_j$.

Relation 8.

$w_{8y}^{N_y}$ is the set of grade-events $Y_1, \ldots, Y_r, \ldots, Y_{Ny}$ for the parameter-event of efficiency Y.

$w_{8y}^{N_y} = (Y, Y_1, \ldots, Y_{Ny})$.

Attributes of the sets' element $Z_1^{N1}, \ldots, Z_n^{Nn}$ included in GIE:

r is the number of the grade-event in the parameter-event j;

$P2_{jr}$ is the frequency (probability) of the grade-event in credits based on statistics (the sum of these probabilities for every GIE it is equal to 1) *(statistics)*;

$P1_{jr}$ is the probability of the grade-event in GIE (the sum of these probabilities for every GIE it is equal to 1) *(problem-2)*;

P_{jr} is the probability of the grade-event substituted in the formula of probability instead of probability P_j *(problem-2)*;

E_{jm} is the mean error of grade-event in recognition of objects *(problem-4)*;

E_{gjr} is the error of grade-event in recognition of good objects *(problem-4)*;

E_{bjr} is the error of grade-event in recognition of bad objects *(problem-4)*.

Integrated characteristics of sets $Z_1^{N1}, Z_2^{N2}, \ldots, Z_n^{Nn}$:

$P1_{jm}$ is the mean probability in GIE for probabilities $P1_{jr}$ *(problem-2)*;

P_{jm} is the mean probability in GIE for probabilities P_{jr} *(problem-2)*.

22.3 Relations

Definition. Relation w on set is defined as the ordered collections or trains of k the set's elements. The area of relation is specified in the top index, for example w_k.

Let's express these relations in the general form, keeping in mind that the formal LP-theory of non-success risk should be invariant for recognition (classification) of objects (conditions of object) in various applications: credit risk of physical persons and legal persons, risk of bribes, risk of the manager's swindle, risk of investments fraud, etc.

While describing the relations, we will use the general term "object" and keep in mind that the definitions of "relations" should not depend on the sets' capacity, the number of parameters and grades in the parameters describing the object.

The relation 1: w_1 is a set of attributes for each object.

The relation 2: w_{22} is a relation between the object risk P_i and allowable risk P_{ad}; $w_{22}(P_{ad} > P_i)$. The relation is used for division of objects based on the model into two classes (subset): good objects and bad objects (Fig. 11.5).

The relation 3: w_3^2 is a relation between mean risks of objects based on statistics P_{av} and based on the model P_m; $w_3^2(P_{av} = P_m)$. The relation is used for training (identification) of the risk model based on statistics and also provides real frequency meaning to the probabilities $P_{jr}, j = 1, 2, \ldots, n; r = 1, 2, \ldots, N_j$.

The relation 4: w_4^n is a set of logic variables $Z(Z_1, \ldots, Z_n)$, describing the object's behavior. The relation is used for the logic description of the non-success scenario.

The relation 5: $w_5^2(Y = Z)$ is a relation between a logic variable of the efficiency parameter Y and logic variables of influencing parameters Z; $w_5^2 : Y = f(Z_1, \ldots, Z_j, \ldots, Z_n)$.

Relations 6: w_{6j}^{Nj} are the grade set for initiating parameters $Z_{j1}, \ldots, Z_{jr}, \ldots, Z_{jNj}$, $j = 1, 2, \ldots, n$; $w_{6j}^{Nj} = (Z_j, Z_{jr}), r = 1, 2, \ldots, N_j$.

Relation 7: w_{7j}^5 is the relation between probabilities in GIE under modified Bayes' formula; $w_{7j}^5 = (P2_{jr}, P1_{jr}, P_{jr}, P2_{jm}, P1_{jm}, P_{jm}), r = 1, 2, \ldots, N_j$.

Relation 8: w_{8y}^{Ny} is the grade set for efficiency parameter $Y_1, \ldots, Y_r, \ldots, Y_{Ny}$; $w_{8y}^{Ny} = (Y, Y_r)$.

22.4 The signature

In the given definition of 1 model there is no compulsory defining word — the model of what? In order to answer this reasonable question, the concept "signature" is introduced. This concept expresses a certain quality inherent in the model of a particular object. For this purpose, it is agreed to distinguish the relation designated w from its name, which is designated as W. Thus the relation is compared with the name of the relation. In a symbolic form it is

$$\alpha(W) = w, \tag{22.3}$$

where α designates the procedure of this comparison.

The set of relations names in this model is referred to as the mode's signature L. In a symbolic form it is written as:

$$\{W_1^{i1}, W_2^{i2}, \ldots W_n^{in}\} \subseteq L. \tag{22.4}$$

With the introduction of the "signature" concept, it is possible to give another definition of the model.

The the train is referred to as the model $\overline{M}$ in signature L:

$$\overline{M} = (M, \alpha) \tag{22.5}$$

The signatures for the relations description in the credit risk LP-model are the following:

W_1^2 is a set of attributes (characteristics) of each credit;

W_2^2 is a ratio between the risk of credit P_i and accepted risk P_{ad};

W_3^2 is a ratio between mean risks of credits based on statistics P_{av} and model-based risks P_m;

W_4^n is a set of logic variables $Z(Z_1, Z_2, \dots, Z_n)$ for parameter-event;

W_5^2 is a ratio between a logic variable of the efficiency parameter Y and logic variables of influencing parameters Z;

W_{6j}^{Nj} is a set of grade-events for initiating parameter-events $Z_{j1}, \dots, Z_{jr}, \dots, Z_{jNj}$, $j = 1, 2, \dots, n$;

W_{7y}^{Ny} is a set of grade-events for the efficiency parameter-event $Y_1, \dots, Y_r, \dots, Z_{Ny}$.

22.5 Axioms of the formal risk theory

The concept of the model and modeling has proved to be one of the most effective, diverse, and universal methods of scientific research. Modeling requires identification abstraction allowing one to identify the real object under investigation with its artificially created model. Mathematical theory never investigates properties of natural objects as such. The objects of its research are only abstract objects, which in the most general case are regarded as sets with relations on them.

Science never studies the object on the basis of observations alone. There should be some theoretical concept of this object, and the object itself is treated as a representative of the class of objects for which this concept holds. In modeling, the object is compared not only with its mathematically identical object — i.e., its model — but also with other objects that fit into this concept.

In other words, mathematical logic considers mathematical models of various objects of the real world as practical applications or embodiments of some theories. For example, if the researcher already has some mathematical model, it can be implied, that somewhere (though at present it may be not clear where exactly) there is a theory from which the given model follows as its special case. The task of the researcher is to find this still unknown theory. In order to solve this problem, the researcher should know the structure or the doctrine of the formal theory.

The structure of the formal theory. Construction of mathematical models is not the ultimate purpose. They are essential for application in various algorithms designed for the solution of all types of applied problems. The essence of these algorithms is the following. Applying a set of formulae F

$$\{f_1, f_2, \dots, f_m\} \in F \tag{22.6}$$

in a certain sequence on set of elements $x_i, i = 1, \dots, k$ of the models $\overline{M}$

$$\{x_1, x_2, \dots, x_k\} \in \overline{M}, \tag{22.7}$$

we get required results yield. In a general form each formula

$$f_1(x_1, x_2, \dots, x_k) \tag{22.8}$$

contains k various free variables, which are the elements x_i of model $\overline{M}$. The subset F^A is selected from the common set of formulae F. On each of formulae $f_i^A, i = 1, 2, \ldots, n$ included in this subset $\{f_1^A, f_2^A, \ldots, f_w^A\} \in F^A$ consistently imposing quantifiers $\forall$ (generality) and $\exists$ (existence) and thus reducing the number of free variables $x_j, j = 1, 2, \ldots, k$

$$\begin{gathered} f_i^A(x_1, x_2, \ldots, x_k); \\ \forall x_1 f_i^{A(1)}(x_2, x_3, \ldots, x_k); \\ (\forall x_1)(\exists x_2) f_i^{A(2)}(x_3, x_4, \ldots, x_k); \\ \ldots \\ (\forall x_1)(\exists x_2) \ldots (\forall x_{k-1}) f_i^{A(k-1)}(x_k) \end{gathered} \tag{22.9}$$

Then in a limit we will yield the formula without free variables

$$(\forall x_1)(\exists x_2) \ldots (\forall x_{k-1})(\exists x_k). \tag{22.10}$$

This formula expresses the statement concerning not a particular set of free element variables $x_1, x_2, \ldots, x_k$ set M, but some unconditional general properties of the model $\overline{M}$ for which this formula was constructed. Such formulae are referred to as axioms. *The axiom* is some absolute unconditional statement on which the model $\overline{M}$ is based.

On the basis of set of axioms $(A_1, \ldots, A_t) \in A$, derived from formulae F^A by means of set of derivation rules Y

$$\{Y_1, Y_2, \ldots, Y_p\} \in Y, \tag{22.11}$$

it is possible to construct all other formulae F

$$\overline{F} = F/F^A; \overline{F} = A \cap Y. \tag{22.12}$$

It is important that formulae have an identical form for any model $\overline{M}$ with the same signature L. Therefore it is possible to apply such formulae irrespective of a specific model. In particular, by means of derivation rules, it is possible to deduce other formulae from the initial ones, regardless of the model for which they were originally designed. Now there are all essential elements for the definition of the formal theory.

Formal theory. T_F is set of set of axioms A in signature L together with a set of derivation rules Y.

The symbolic record of the formal theory is the train-like

$$T_F = (A, L, Y). \tag{22.13}$$

The model $\overline{M}$ is referred to as the model of the formal theory T_F, if:
(1) model signature L_M coincides with the signature of theory L_T

$$L_M = L_T; \tag{22.14}$$

(2) interpretation of each name of the relation in the theory as same relation in the model allows one to regard each axiom $A_i, i = 1, 2, \ldots, m$ theories as the true statement for the given model.

The main feature of the formal theory concept is that there is no base set M. There are some axioms and based "on nothing," i.e., axioms "in the pure form." *The formal theory* T_F describes some properties of objects, but does not specify these objects.

Axioms appear to fix an investigated class of models in some way, hence the theory can have many various models. The fact that the model M_1 is the model of theory T_F implies that any model isomorphic to the model M_2 is also a model of the theory T_F.

Axioms. After the originally unknown parameters were excluded, the following axioms yield for LP-model of credit risk:

A_1. Statistical data (credits in the bank statistics) should be not less than in the regress-based recognition (classification) techniques.

A_2. Continuous distribution of random parameter values on statistical data can be replaced by discrete distribution;

A_3. The risk scenario is assumed to be known, for example, the credit non-success takes place if there is any one or any two or ... all initiating events.

A_4. Values of the parameters describing the credit (loan), efficiency parameter of and gradations of these parameters are random events.

A_5. Asymmetry of recognition of good and bad credits is assumed as predetermined.

A_6. Mean credit risks according to statistics and the model should be equal.

A_7. Structural risk model (Fig. 11.1) and risk L-model can be constructed on the basis of the risk scenario;

A_8. The risk logic model can always be written in the orthogonal form and, further, in the form of probabilistic polynomial (risk P-model)

22.6 The mathematical apparatus of derivation

To pass over from the concept of the formal theory to the required risk LP-theory, it is necessary:

(1) to define the base set of elements of the mathematical model M;

(2) to define the relations and the signature of relations L on base set of elements of M;

(3) to define the set of axioms A;

(4) to define the set of derivation rules Y.

The formal non-success risk LP-theory with GIE applies derivation rules that have been already stated partially in the previous sections of this chapter and will be considered in detail in the subsequent sections of the book. First of all, we have to denote algorithms and methods of identification and analysis of risk LP-models on statistical data, stated in Chapters 12–21.

We suggested the algorithmic iterative methods, which guarantee the possibility of the solving of risk problems regardless of:

- The number of objects N (or states of the object) in statistical data,
- The number of parameters n describing of an object (a state),
- The number of grades N_j in every parameter,
- The structures and complexity of logical functions of risks $Y = f(Z)$.

23

Training Course "Modeling, Estimation, and Analysis of Risks in Economics"

You can leave the science, but not before you write a textbook.
The precept to Professors

The classifier of scientific area — YDK 519.862.6, Econometrics.

The problem - Modeling and analysis of risk in complex systems.

The essence of techniques — Introduction in statistical database (DB) the groups of incompatible events or finite sets that allows one to receive the LP-equation system (knowledgebase (KB)), to use the LP-calculus, and to solve new tasks: risk, efficiency, management.

23.1 Features and advantages of the risk LP-theory

Features:

- Using logic addition of events;
- Adequate logic formulation of risk scenarios;
- Application of the KB in the form of the logic equation system;
- Construction of logic and probabilistic risk models;
- Calculation of probabilities of events taking into account GIE and Bayes' formulas;
- Correct formulation of the criterion function for identification of the risk model according to statistical data;
- Using special logic software.

Advantages:

- Almost a twofold accuracy in recognition of good and bad objects;
- A seven times higher robustness (stability of object classification);

E.D. Solojentsev, *Scenario Logic and Probabilistic Management of Risk in Business and Engineering*, DOI 10.1007/978-0-387-77946-1_23,

- An absolute transparency in estimation and analysis of the risk of objects, sets of objects, and the risk model;
- Opportunity to manage the risk, changing asymmetry of recognition of good and bad objects, number of parameters and grades, describing object.

New problems:

1. Determination of risk attributes of the credit on the trained risk LP-model: estimation of the risk of the credit; classification of the credit as good or bad; determination of a price for risk; analysis of the risk of the credit.
2. Determination of risk attributes of credit set of the bank on the trained risk LP-model: contributions of signs, describing the credit, the average credit risk of the bank; contributions of grades of signs of credits to the average credit risk of the bank; contributions of the signs, describing the credit, to accuracy of recognition of good and bad credits and entropy of risks of bad credits.
3. Identification and estimation of the quality of the risk LP-model on statistical data of the bank: determination of probabilities of grade-events; determination of the admitted risk; determination of the average risk; estimation of accuracy and robustness of the risk LP-model.
4. Management of credit risk of bank: determination of the optimum number of the signs, describing the credit and the client, and the optimum number of grades for each sign; determination of the asymmetry optimum factor of recognition of good and bad credits and optimum width of intervals at allocation of grades for such signs, as a credit sum, a term, a client age, determination of the optimum number of classes for classification of credits; use of the non-success credit risk model with limited set of credit non-success events.

Types of mathematical problems:

1. *Direct tasks of risk* — construction of the risk LP-model and calculation of the risk of accident or non-success at the given probabilities of events;
2. *Return tasks of risk* — estimation of probabilities of initiating events on statistical data at known structure of the risk model and the risk analysis.

23.2 Application of the risk LP-modeling and analysis

Accidents and failures in engineering. Scenarios and LP-models for estimation and analysis of risk of accidents in different technical objects at all stages of their life cycle are considered. Concepts, principles, experience, and examples of management by risk on design stages, debugging, and operational tests, running systems on the basis of monitoring and on dangerous areas of manufacture are systematized.

Crises and defaults in economy. We consider scenarios and risk LP-models for estimation, analysis, and management of: credit risk of physical and juridical persons, security portfolio risk, non-success risk of company management, risk in management of a condition and development of the company, risk of loss of quality of functioning, risk in problems of efficiency, etc.

Non-contact wars. The concept of creation of an ecological catastrophe is put in the basis of non-contact wars in territory of opponents. For effective application of cruise missiles (war of the sixth generation) or acts of sabotage (war of the seventh generation), it is necessary to have the adequate data on aims: communication lines, pipeline, automobile and railway transport, plants, chemical plants, metallurgical combines, power stations, etc. On the constructed risk LP-model of the engineering infrastructure of the opponent country, we find functionally and topological weak and dangerous places and personnel actions by corresponding software.

Bribes and corruption. The phenomena of bribes and corruption is close with phenomena of failure and accident in engineering and the phenomena of default, bankruptcy, damage in economy. Their consequences are financial, material and moral losses for society and state. The concept risk of bribes coincides with the concept of reliability and safety in engineering and risk in economy and business. More often, bribes take place at reception of licenses (formation, tourism, medicine, building), sanctions (GAI, customs), education (certificates, examinations), registration (administration, local authorities).

23.3 Purpose and problems of the training course

The training course is reading on the lectures of the economic department of the state university of aero-space instrument making. The task of the training course — formation of knowledge of students on the basic concepts, terminology, and methods of estimation, modeling, analysis, and management of the risk in business and engineering, and also knowledge on corresponding software.

The task of the training course is to study: basic concepts of the risk theory of non-success and accidents; principles and methods of management of the risk on stages of design, test, and operation of complex systems; bases of the logic and probabilistic calculus; bases of the risk LP-theory with groups of incompatible events; scenario, structural, logic, and probabilistic risk models; methods of risk analysis, risk management; identification of risk LP-models on statistical data; models of credit risks of physical and juridical persons; models of security portfolio risk; risk models of loss of quality and efficiency; risk models of bribes and fraud; non-success risk models of company management; the formal risk LP-theory.

23.4 Themes of lectures

1. Management and risk.
2. Person and risk.
3. Principles of management of risk in designing.
4. Management of risk during tests.
5. Management of risk in running on the basis of monitoring.
6. Management of risk on dangerous manufactures.
7. Transparency of techniques for estimation of the risk.
8. Bases of logic-and-probability calculus.
9. LP-modeling and analysis of risk in engineering.
10. The automated structure logic modeling.
11. The logic-and-probabilistic risk theory with GIE.
12. Identification of the risk LP-model with GIE.
13. The risk LP-analysis in systems with GIE.
14. Software for LP-estimation, analysis, and management of risk.
15. LP-models of credit risk of natural persons.
16. LP-models of credit risk of juridical persons.
17. Scenario risk LP-models of bribes and corruption.
18. Risk LP-models of security portfolio.
19. Risk LP-models of loss of quality and efficiency.
20. Risk LP-models of company management non-success.
21. LP-models and scenarios of the fraud risk.
22. The formal risk LP-theory with GIE.

23.5 Laboratory works

Section 1. The LP-theory of the credit risk: estimation, identification, analysis, and management [156]:

1. Identification of the risk P-model on statistical data;
2. Estimation and analysis of the credit risk;
3. Choice of recognition asymmetry of bad and good credits.
4. Analysis of credit risk LP-models;
5. Analysis of credit activity of the bank.

Section 2. The risk LP-theory of security portfolio: choice, analysis, and management [157]:

1. LP-choice of security portfolio structure;
2. LP-analysis of security portfolio, its yield and parameters;
3. LP-optimization of security portfolio;
4. Estimation of efficiency of LP-management of security portfolio.

Section 3. The automated structure logic modeling of risks [158]. Independently develop the scenario and risk LP-model; to execute researches on one of the themes:

1. Risk of non-success of the presidential activity;
2. Risk of non-success of the presidential election;
3. Risk of falling EURO;
4. Risk of non-success of a manager of a company;
5. Risk of fraud of hired clerks (no more than 4 incoming edges);
6. Risk of fraud of managers (no more than 4 incoming edges);
7. Risk of fraud with investments (no more than 4 incoming edges);
8. Risk of bribes (the scheme with bridges);
9. Credit risk of natural persons (no more than 4 incoming edges);
10. Credit risk of juridical persons (no more than 4 incoming edges);
11. Risk of non-success of a company management (separately and totally on functions, directions of activity and purposes);
12. Risk of loss of quality in functioning company;
13. Risk of non-success of financing by three banks to two projects;
14. Risk of non-success of civil engineering firms in struggle for two favorable orders;
15. Risk of non-success in development of a company (totally on stages);
16. Risk of an apartment robbery;
17. Risk of a collector machine robbery;
18. Risk of a bank robbery;
19. Risk of a crash of two cars on crossroads;
20. Risk of being late on lectures;
21. Non-success risk of debugging tests of a complex engineering system;
22. Risk of flooding a vessel;
23. Risk of explosion and fire on gas-oil pumping stations;
24. Risk of an accident on the railway.

Each student develops and investigates the risk LP-model including some scenarios of accidents. The scenario of a "big" accident is created on the basis of association of scenarios on several themes. For example, the new scenario established at association by 1 and 3, 2 and 16, 9 and 10. We write mathematically: $Y1 \wedge Y3; Y2 \wedge Y16; Y9 \wedge Y10$.

Generally, if we have two outputs Y1 and Y2 from two different scenarios, so logic criterions of functioning (LCF) can be written down those:

1. $(Y1 \vee Y2)$ is is a realization of the criterion $Y1$ or the criterion $Y2$;
2. $Y1 \wedge Y2$ is a realization of the criterion $Y1$ and the criterion $Y2$;
3. $Y1 \wedge \overline{Y2}$ is a realization of the criterion $Y1$ and not realization of the criterion $Y2$;
4. $\overline{Y1} \wedge Y2$ is not a realization of criterion $Y1$ and realization of the criterion $Y2$;
5. $\overline{Y1} \wedge \overline{Y2}$ is not a realization of the criterion $Y1$ and the criterion $Y2$.

23.6 The list of indexes

The following alphabetic index is used in the training course "Modeling, estimation, and analysis of risks in economics" as the basis of concepts, definitions, and terminologies of discipline on modeling and risk analysis of accidents and bribes.

1. Algebra,
2. Analysis of risk,
3. Attributes of grade risk,
4. Attributes of object risk,
5. Attribute of object set risk,
6. Attributes of risk,
7. Attributes of sign risk,
8. Bribes,
9. Construction of risk LP-model,
10. Contributions of grade-events,
11. Contributions of sign-events,
12. Criteria of optimization,
13. Criteria of the risk LP-model quality,
14. Efficiency,
15. Errors,
16. Events,
17. Formal risk theory,
18. Logic functions,
19. Groups of incompatible events,
20. Identification of risk models,
21. Information and risk,
22. Knowledge of risk,
23. LP-calculus,
24. Management of risk,
25. Methodology of coordinate switching,
26. Operations of risk LP-technology,
27. Risk,
28. Risk LP-model in business,
29. Risk LP-model in engineering,
30. Risk LP-model with GIE,
31. Risk LP-theory with GIE,
32. Risk LP-theory,
33. Risk of credits,
34. Risk of non-success management,
35. Risk of security portfolio,
36. Scenario management of risk,
37. Scoring's methods,
38. Software for risk,

39. Structural risk models,
40. Structurally complex systems,
41. Structure logic modeling,
42. Tabular presentation of data,
43. Technology of risk management,
44. The human and risk,
45. Transparency in risk problems.

23.7 Software for identification of risk LP-models with GIE

A demo version of software is developed and presented on www.inorisklab.com for training and demonstration purposes [156] as an addition to the commercial version. The demo version uses the following fixed data: Western package from 1000 credits of natural persons, among which 700 are good and 300 are bad credits:

- The credit is described by 20 signs (parameters).
- Each sign has grades. The general number of grades in signs equal to 96.

Estimation of risk. We solve problems of determination of risk attributes of the credit on trained risk LP-model on statistical data.

Analysis of the risk of the credit. After estimation of the risk of the credit, we analyze the risk: determine what grades are selected and what its probabilities. For calculation of the risk of the credit, these probabilities logically sum according to the risk L-model. Contributions of grades to the risk of the credit are proportional to these probabilities.

Identification of LP-models of credit risk is fulfilled on the Western statistical data consisting of 1000 credits among which 300 were bad. There is a screen form for identification of the risk LP-model.

The analysis of the bank credit activity. Contributions of attributes to criterion function are calculated by pressing button Analysis on the screen form. We consistently automatically exclude one sign only and the model is retrained. The number of optimizations for retraining should be set less than at initial training the risk LP-model. The results of researches we allocate the most significant signs.

Contributions of signs to average risk are determined for set of all credits. Contributions of signs to the average risk differ more than two times. The bank must pay special attention to signs bringing the greatest contributions to the average risk. Probabilities of grade-events and its contributions to accuracy of the LP-model are contained in the file $FmaxLast.txt$ for all 20 signs.

23.8 Software for LP-modeling of security portfolio risk

The software carries out following functions [157]:

- Automated input of action quotes of Western and Russian companies from open resources of Internet in database.
- Construction of graphs of prices and yields under actions of chosen companies or market indexes. We give the time period and interval of observation.
- Formation of the security portfolio. We choose needed actions and give initial parts of capital invested in them.
- Support simultaneously several portfolios.
- Automatic recalculation of the portfolio cost according to last data on prices.
- Display of diagrams illustrating distribution of money resources on activities.
- Calculation of parameters of statistics for actions included in the portfolio.
- Modeling of the discrete distribution of the portfolio yield for the given time horizon by several methods: with the full account of dependence between actives, without taking into account dependence; taking into account dependence on the factor; with taking into account a correlation matrix.
- Graphic display of the distribution of the portfolio yield and calculation of its parameters: the average yield, the standard deviation, the minimal admitted yield for the given risk, the risk for the minimal admitted yield, minimally admitted cost of the portfolio for the given risk, VaR in money and in percentage.
- Verification of modeling of yield distribution according to historical data.
- Optimization of the portfolio structure on one of the criteria: maximization of admitted yield for the given risk; minimization of risk at the given admitted yield.
- Calculation of characteristics of the management efficiency by security portfolio (Sharp's factors, the dispersion of the portfolio, etc.) and comparison of them with market standard and among themselves.
- Calculation of contributions of action grades in risk and yield of security portfolio.

23.9 Software for automated structurally logic modeling of risks

Developed in SPIK SZMA, software for the automated structure logic modeling risk (decision of direct problems) has the name ARBITER [108]. It passed examination in the Council of Software Certification of the federal service on ecological, technological, and nuclear supervision of the Russian Federation. Software has the passport of certification No. 222 from February 21st, 2007, for application on objects of Russian technical supervision. The theoretical basis of ARBITER is the general logic-and-probabilistic method (GLPM) of analysis of structurally complex objects of various kinds, classes, and purposes. Uniqueness of the development consists in the following [158]:

1. GLPM has functionally full basis of L-operations AND, OR, NOT and all opportunities for modeling of logic algebra. It allows one to build automatically all kinds both monotonic and non-monotonic models of reliability and safety of structurally complex objects for various purposes.
2. Software ARBITER uses new graphic means — schemes of functional integrity (SFG) for the structural description of properties of reliability and risk of systems. By the graphic means, we can represent both monotonic (block diagrams, graphs of connectivity, trees of refusals and events) and non-monotonic structural models of reliability and safety of systems.
3. Purposes of ARBITER are the following:
   - Automated modeling and calculation of parameters of reliability of systems, including objects of atomic energy and other dangerous industrial objects;
   - Automated modeling and calculation of probabilities of occurrence of emergencies and failures of dangerous industrial objects.
4. Practical application of ARBITER is based on technology of the automated structurally logic modeling, which includes following stages for investigated systems: formalized statement of the risk of the analyzed problem; calculations of probabilistic parameters of reliability or safety. At these stages, software provides:
   - automatic construction of L-functions representing MWF, MSF, or their non-monotonic combinations;
   - automatic construction of P-functions, parameters of reliability providing exact calculation of safety and risk;
   - calculation of realization probability of given LCF of non-failure operation, refusal and technical risk of functioning of the system and/or its separate subsystems;
   - calculation significances, positive and negative contributions of all elements of system to probability of realization given LCF;
   - approached calculation of probabilistic parameters (without construction of P-function) with cutting or non-cutting of little significant ways and sections;
   - calculation of realization probability of separate MWF or MSF system;
   - calculation of the importance and the total importance of refusal sections;
   - calculation of the importance factors of reduction and increase in risk of elements;
   - structural and automatic account of refusals of groups of elements;
   - account of various kinds of dependencies of elements represented by GIE;

- account of two-level decomposition of the block diagram, disjunctive and conjunctive repetition factor of complex elements (subsystems);
- account of unlimited number of cyclic connections between elements of system;
- account of combinatory attitudes between groups of elements and subsystems.

Literature

1. Ryabinin I. A. Reliability and safety of structurally complex systems. Second edition. SPb.: Polytechnics, 2007, 276 p.

2. Mozhaev A. S. Certification of the program "Arbiter" of the automated calculation of safety and technical risk of systems. // Proc. of the Seventh Int. Scien. School / by Editors I. Ryabinin and E. Solojentsev. 2007, September, St. Petersburg: PUASE.

3. Yaroshenko A. V. Contactless wars of the seventh generation. SPb. The sea newspaper. The former war will not be. Special issue, No 11, December, 2006.

4. Solojentsev E. D. Scenario Logic and Probabilistic Management of Risk in Business and Engineering. Springer: 2004.

5. Solojentsev E. D. Scenario logic and probabilistic models of bribes risk. Finance and Business, No 1, 2007; p.p. 125-138.

6. Solojentsev E., Stepanova N., Karasev V. Transparency of methods for assessment of credit risks and ratings. SPb.: St. Petersburg University Press, 2005. 197 p.

7. Methodical instruction for laboratory works "Logic-and-probabilistic theory of credit risks" (N. Medvedeva, E. Solojentsev, D. Strokov). St. Petersburg, SUAI, 2007, 25 p.

8. Methodical instruction for laboratory works "Logic-and-probabilistic theory of security portfolio risk" (V.Alexeev, E. Solojentsev, V. Shokolov). St. Petersburg, SUAI, 2007, 48 p.

9. Methodical instruction for laboratory works "Automated structurally logic modeling of risks" (A. Babenkov, E. Solojentsev). St. Petersburg, SUAI, 2007, 32 p.

Conclusion

In studying sciences, examples are not less instructive than rules.

Isaac Newton

We have stated methodological and theoretical principles of the scenario LP-management of risk and efficiency, LP risk theory with GIE and its applications, and let's formulate briefly the obtained results. Let's describe applications, advantages, differences, models, and peculiar properties of the LP risk theory with GIE in a brief tabular form.

Classifier of scientific area — management and econometrics;

Problem — risk management in complex systems;

Idea — introduction of finite sets in database (DB) permits to form a set of L-equations or knowledgebase (KB), to use LP-calculus and solve problems of risk, efficiency, and management.

Applications. LP risk theory with GIE is applied in problems of classification, investment, efficiency, quality, management, bribes, development management (Table 23.1).

Table 23.1. Applications of the LP risk theory with GIE

N	Application	Contents
1	Classification	Credit risks, ratings, monitoring
2	Investment	Security portfolio
3	Efficiency	Social processes management
4	Management	Failure risk management according to functions, subject areas, achievements of goals, and functioning quality
5	Bribes and corruption	Identification of bribes with use of statistics of parameters of organization's functioning, behavior of officials, service
6	Development and tests management	Organizational and economical systems; machines technologies and systems

Table 23.2. Advantages of LP risk theory with GIE

N	Advantages
1	Twice more accuracy in recognition of bad and good objects
2	Seven times more robustness (stability) in recognition of objects
3	Absolute transparency in estimation and analysis of risk and LP risk model
4	Decisions of new problems of risk analysis for objects and risk model
5	Possibility to manage risk and efficiency

Advantages. In above-mentioned applications, LP risk models with use of "standard" statistical data have shown advantages in: accuracy, robustness, transparency of results, decisions of new problems, possibility to manage risk and efficiency (Table 23.2).

LP risk models. For above-mentioned applications, we are offering next types of LP risk models with GIE: with total number of events, with limited number of events, dynamic, and integrated (Table 23.3).

Differences of the LP risk theory with GIE. LP risk theory differs by: type of information used, type of connections between variables, distribution functions of variables, methods of inverse tasks solution, complexity of logical functions and descriptions of objects, use of statistical and dynamic risk models, estimation methods, use of integrated risk models, methods of management and scientific fundamentals (Table 23.4).

Features of the LP risk theory with GIE. Let's formulate features of LP risk theory with GIE (Table 23.5). These features define advantages of LP risk theory with GIE and areas of its applications.

The list of applications of the LP risk theory. The area of applications of the scenario logical and probabilistic management of risk failure have not been completely defined. We have only single examples. Let's consider LP risk models presented on First, Second, Third, Fourth, Fifth, Sixth and Seventh International Scientific School "Modeling and Analysis of Safety and Risk in Complex Systems" (2001–2007 years, Saint-Petersburg, IPME RAS).

Table 23.3. LP risk model with GIE

N	Types of LP risk models	Contents
1	With total number of events	L-function of risk in perfect disjunctive normal form (PDNF)
2	With limited number of events	L-function of risk in a form of: shortest routes of success, minimal sections of failure and number of events, limited by scenario
3	Dynamic	With changing probabilities of events and time-parameter in a technical analysis
4	Integrated	With logical integration of separate scenarios by connections OR, AND, NOT and risk L-functions

Table 23.4. Differences of the LP risk model with GIE

N	LP risk theory is used	Other theories are used
1	Several values for parameters	Two values for parameters
2	Knowledgebase (KB)	Database (DB)
3	Logical dependence between variables	Functional and correlative dependences between variables
4	Discrete tabular distributions	Normal and analytical distributions
5	Algorithmic iterative methods of solving inverse and and optimization tasks	Analytical methods of solving inverse and optimization tasks
6	L-functions of any complexity with any number of objects in statistics, parameters and grades of parameters	Functions of limited complexity with small number of objects, parameters and grades of parameters
7	Statistical and dynamic risk models	Statistical risk models
8	Estimations on statistical data	Expert estimations
9	Integrated L-models of risk	Separate risk models
10	Management with use of contributions of initiating events in risk and efficiency	Management on values of risk and efficiency
11	Logic and discrete mathematics	Theory of statistics and continuous mathematics

Applications of the LP risk theory in economics, business, and banks:

- Credit risk of natural persons.
- Credit risk of juridical persons.
- Analysis of credit activity of bank.
- Choice of optimal investment portfolio.
- Management of activities and development of a company according to risk and efficiency criteria.
- Risk of manager's fraud.
- Shady transaction with investments.
- Failure risk in a competition among construction firms to get profitable order.
- Failure risk in financing of projects with reservation.
- Risk management in a problem of efficiency of social processes.
- Bank staff management.
- Risk of lease business.
- Risk of loss of quality and market.
- Insurance risk.
- Risk analysis at money and goods markets.
- Failure risk in company's management according to functions and directions of business.

Table 23.5. Features of LP risk theory with GIE

N	Features	Contents
1	Objects of research	Structural complex, multicomponent and multilevel systems: banking, economical, and organizational systems
2	LP risk models with GIE	With total number of events, with limited number of events, integrated risk models, dynamic risk models
3	Areas of application	Classification problems, investment, efficiency, management, bribes and corruption, development management
4	Decided tasks	Quantitative estimation and analysis of risk of object and set of objects, risk management, risk and portfolio yield management, modeling of management failure risk, identification of bribes
5	Methodical bases of LP risk theory with GIE	Logic, sets theory, and probability theory, combinatorial calculus, discrete mathematics, LP-calculus with logical connections OR, AND, NOT between elements and cycles
6	Transition from database to knowledgebase	GIE or finite sets are entered, and statistical DB is transformed into KB as a set of L-equations, it allows one to use LP-calculus and decide tasks of risk, efficiency, and management
7	Risk determination	Admissible risk is entered, dividing objects into bad and good. Risk of every level is determined by its attributes.
8	Quality criteria of risk L-model	Accuracy (classification mistakes), robustness, transparency of results and risk analysis, transparency of risk model
9	Distributions of variables	Discrete tabular distributions of casual variables
10	Connection and dependence of variables	L-variables are dependent, not primordially, but only because they are in certain L-formula, defining dependence among them grade-events for every parameter are dependent and considered as GIE
11	Type of decided calculations	Algorithmic interactive methods. Inverse optimization task is not depending on number of objects in DB, parameters in object, grades in every parameter, logical complexity of risk model.
12	Construction of risk model	Scenario, graph-model of risk, L-function of risk, orthogonal L-function, P-function of risk, identification of P-function of risk
13	LP risk model identification	Identification of P-model of risk with use of statistical data is performed by solving of optimization task by algorithmic interactive methods, use of Bayes' formula
14	Logical differences	On every step of optimization it is necessary to define an increment for every probability. At the same time, it is required to normalize probabilities in GIE—their sum has to be equal to 1.
15	Risk analysis	We are calculating contributions in object's risk, in risk of set of objects, and criterion function of parameter-events and grade-events. These contributions are risk attributes and allow one to manage risk and efficiency.
16	Risk management	Active risk management with use of contributions of parameter-events, grade-events in risk and efficiency
17	Software	Special software for identification of P-model of risk, choice of security portfolio, and structural logical modeling

- Failure risk in company's management as a risk in complex object's control.
- Failure risk in company's management in achievement of one or several goals and functioning quality.
- Risk in problems of identification of bribes and corruption.
- Risk management in a restaurant business.
- Risk of explosion inside a submarine.

- Risk of automobile collision accidents at a crossing without a traffic light.
- Risk management in a testing of systems, machines, and technologies.
- Safety of a railway section.
- Risk of ordnance depot explosion.
- Risk of catastrophe in a highly explosive lodgment.
- Risk of water penetration inside a submarine.
- Analysis of safety of ship energy installation.
- Explosion-proof and fire safety in gas- and oil-transfer plants.
- Modeling of risk of three-channels safety system of nuclear power plant.
- Economical safety of military plant.
- Safety of exploitation of ground-based space structure.
- Risk analysis in man-caused safety.
- Estimation of safety of ships divergence trajectory in a sea.
- Monitoring of functioning risk at starting ground-based equipment.
- Modeling and estimation of risk of nuclear dangerous objects under adverse external influences.
- Modeling and risk analysis at power systems.
- Safety control for nuclear reactor.
- Risk of prolongation of depreciated equipment resource.

Future trends of the LP risk theory with GIE. Consideration and analysis of applications, advantages, differences, features, and models of LP risk theory with GIE allow one to declare a new scientific direction in theory and practice of risk management. This direction is scenario logical and probabilistic management of risk of failure in complex systems with groups of incompatible events.

Based on above-mentioned materials, the special course "Scenario Logical and Probabilistic Management of Risk and Efficiency in Economics" is lectured at Faculty of Economics in Saint-Petersburg State University of Aerospace Instrumentation.

Research of risk problems in economics and finance, engineering and ecology, state and national safety is impossible without construction of scenarios, graphs, logical and probabilistic risk models, quantitative modeling and analysis of risk and efficiency in order to provide the possibility to manage.

At the same time, there are unfounded solutions and unsolved problems. We suppose, future trends of research in the area of LP risk theory for goals of management have to be the following:

1. Development of LP risk models with GIE for other applications;
2. Development of training course "LP risk theory with GIE";
3. Improvement of techniques and calculating algorithms;
4. Improvement of software for LP modeling and risk management;
5. Expertise and certification of LP risk models and software.

References

1. Ryabinin I. A. *Reliability of Engineering Systems. Principles and Analysis.* Moscow: Mir, 1976.
2. Ryabinin I.A. *Reliability and safety of structure–complex systems.* Second edition. SPb.: Polytechnics, 2007, 276 p.
3. Solojentsev E. D. *Scenario Logic and Probabilistic Management of Risk in Business and Engineering.* Springer: 2004.
4. Solojentsev E. D., Stepanova N. V., Karasev V. V. *Transparency of methods for assessment of credit risks and ratings.* Saint Petersburg.: St. Petersburg University Press, 2005.
5. Mojaev A. S., Gromov V. N. *Theoretic basis of common logic and probabilistic methods of automated modelling systems.* Saint Petersburg: VITU, 2000.
6. Mojaev A. S. Modern state and some directions of developing the logic and probabilistic methods of analysis of systems / Theory and information technology of modeling safety of complex systems. Issues 1–5 /Edited by I. Ryabinin and E. Solojentsev. Saint Petersburg, IPME RAS, 1994–95.
7. Vasiliev V. D., Solojentsev E. D. *Cybernetic methods at development of piston machines.* Moscow: Mashinostroenie, 1978.
8. Winer N. *Cybernetics, or control and connection in animal and machine.* Trans. from Eng. M.: Sovjet Radio, 1968.
9. John von Neumann. *Theory of self-reproducing automata.* M.: Mir, 1971.
10. John von Neumann. Probabilistic logics and systems of reliability organisms from non-reliability components. In the book of papers by edition K. Shannon and John Makkarty "Automata". Trans. from Engl. Edited by A. A. Lyapunov. Moscow, 1956.
11. Shanon Clod. *Works in theory information and cybernetics.* Moscow, 1962.
12. Keynes J. M. *A treatise on probability.* London, 1921. v. 7.
13. Morgenstern O., Neumann John. *Game theory and economic behaviour.* M.: Nauka, 1970.
14. Kolmogorov A. N. Sulla determinazione empirica di uns legge di distribuzione.– Giornaledell'Istituto Italiano degli Attuari, 1933, t.4.
15. Kolmogorov A. N. *Success of mathematics sciences*, issue 5, M.–L., 1938.
16. Kolmogorov A. N., Draganov A. G. *Introduction in mathematical logics.* M.: MGU, 1982.

17. Markowitz H. Portfolio selection. Journal of Finances, 1952; 7:77–91.
18. Sharp W., Alexander G., Boily D. *Investigation.* Moscow: Infra, 2001.
19. Glushkov V. M., etc. *Encyclopaedia of cybernetics.* Kiev: 1975, v.v. 1–2.
20. Skurihin V. I., Morosov A. A. Integrated automation systems of management. Features of construction and directions of development. Controlling systems and machines, 1976; 2:5–11.
21. Pospelov D. A., *Logic and linguistic models in system control.* Moscow: Energoizdat, 1976.
22. Ivanishev V. V., Marley V. E., *Introduction in the theory of algorithmic networks.* Saint Petersburg: STU, 2000.
23. Ryabinin I. A., Cherkesov G. N. *Logic and probabilistic investigation methods of structure-complex systems reliability.* Moscow: Radio and communication, 1981.
24. Ryabinin I. A., Parfenov Yu. M. Determinition "weight" and "importance" of individual elements at estimation of reliability of complex system. Power and Transport, 1978; 6: 22–32.
25. Ryabinin I. A. Leningrad Scientific School of Logical and Probabilistic Methods of Investigation of Reliability and Safety of Structure-Complex Systems. In *Science of Saint-Petersburg and Marine Power of Russia*, v.2. Saint-Petersburg: Nauka, 2003.
26. *Nilsson N. J. Probabilstic Logic.* Artifical Intelleqence, vol. 28 (1986), Elsevier Science Publ., North Holland, pp. 31–56.
27. Solojentsev E. D., Karassev V. V., Solojentsev V. E., *Logic and probabilistic models of risk in banks, business and quality.* Saint Petersburg: Nauka, 1999.
28. Solojentsev E. D., Karassev V. V. Identification of logic and probabilistic models of risk of structure–complex systems with groups of incompatible events. Automation and telemechanics, 2002; 3:97–113.
29. Solojentsev E. D., Karassev V. V. Logic and probabilistic risk models in business with groups of incompatible events. Economics and mathematics methods, 2003; 1: 90–105.
30. Solojentsev E. D. *Scenario Logic and Probabilistic Management of Risk in Business and Engineering.* Saint Petersburg: Business-Press, 2006.
31. Vitlinsky V.V., Velikoivanenko G. I.*Rizkology in economics and business* Kiev: Khey, 2004.
32. Yusupov R. M., Zabalotsky V. P.*Scientific and methodological bases of informatisation.* Saint Petersburg: Nauka, 2002.
33. Alexandrov V. V. *Developing systems in sciency, engineering, society and culture.* Saint Petersburg: St. Petersburg University Press, 2000.
34. Alexandrov V. V., Arsentieva A. V. *Information and developing structures.* Saint Petersburg: LIAAN, 1984.
35. Kalyagina L. V. Taxanomia of economical objects on the bases of measure of dispersion. // In book of KRAU. Krassnoyarsk. 2005. No 9. pp. 361–364.
36. Pechenin N. K. Concept of managing risk on historical examples. Alternative energetic and ecology, 2000; 1:116–129.
37. Krasnov O. V., *Safety operation of complex engineering systems.* Saint Petersburg: VISA named A.F. Mojaisky, 2002.
38. Nikolai V. Hovanov, James W. Kolari, Mikhail V. Sokolov. Syntetic Money. In book: Multidisciplinary Economics (Edited by Peter de Gusel). Springer, 2005, p.p. 293–305.

39. *Risk management: Risk, Stability development, Synergistic. (Series Cybernetics: unlimited possibilities and possible limitations).* Moscow: Nauka, 2000.
40. Yaili E. A., Musalevsky A. A. : , , // By editor Karlin L. N.–Saint Petersburg: RGGMU, VVM, 2005. 234 p.
41. *Accidents and catastrophes. Preventing and estimation of consequences.* Four volumes, Edited K.E.Kochetkova, V.A.Kotlyarovskogo. Moscow: Association of building Institutes,1995–1997.
42. Frolov K. V., Mahutov N. A. "Problems of safety of complex engineering systems". In book: *Problems of machine-building and reliability of machines.* Moscow: Nauka,1992.
43. Yaroshenko A. V.*Methodology of co-ordinate switching.* Saint Petersburg: Naval academy, 2006.
44. Yaroshenko A. V. *Contactlee wars of the seventh generation.* Saint Petersburg. The sea news paper. The last war will not be.- Special release, 11, December, 2006.
45. Mahutov N. A. The problem of risk decrease of beginnings of extraordinary conditions of technogenic nature. Problems of safety at extraordinary conditions, 2001; 3:29–41.
46. Victorova V. C., Kuntsher H., Petruhin B. P., Stepanyanz A.C. Relex – program of analysis of relaibility, safety, risks. // M.: Relaibility.– 2003.– No 4 (7).– P. 42–64.
47. Lukjyanov V. D., Melnikov V. A. On approximate formulas for optimal distribution resources on elements of engineering systems for maitance of its maximum relaibility.–News of RAS.– Theory and systems of controls – 2005. No 2. P. 105–111.
48. Matrosov V.M., Izmodenova – Matrosova K.V.. *Doctrine on nono-sphere, global modeling and stability development: Course of lectures.* / M.: Academia, 2005.– 368 p.
49. *Problems of destruction, resource and safety of engineering systems.* Krasnoyarsk: Siberia, 1997.
50. *Russia safety. Legal, social and economic, research – engineering aspects. Dictionary of terms and definitions*", v.2. Moscow: Znanie, 1999.
51. Belov P. G., *Methodology aspects of national Russia safety.* Moscow: FCNTP Safety, 2002.
52. Grazhdankin A. I., Lisanov M. V., Pechorkin A. S. Using of the probabilistic estimations during analysis of safety of dangerous factories. Safety of labor in factories. 2001; 5: 33–36.
53. Ryabinin I. A. On using apparatus of logic algebra for researches of questions of relaibility. // Ship's electric technics and svyaz.– No 28,– 1965. P. 30–35 .
54. Control of Oscillations and Chaos // Proceedings of Second Int. Conf., July 5–7, 2000 / Edited by F. L. Chernousko, A. L. Fradkov. St. Petersburg: IPMash, 2000.
55. Guding D., Lennoks Dg. *Weltanschauung. For what do we live in the world and what is our place in the world ?*, Transl. from Engl., Edited T. V. Barchunoboy. Yaroslavl': TF Nord, 2001.
56. Ivchenko B. P., Martishenko L. A. *Information ecology*, v. 1 and 2. Saint Petersburg: Nordmed, 1998,2000.
57. Han G., Shapiro S., *Statistical models in engineering tasks.* Transl. from English. Moscow: Mir, 1969.

58. Tihonov V. I., Himenko V. I., *Kicks of trajectories of random processes.* Moscow: Nauka, 1987.
59. Kendall M. J., Stjuart A. *Statistical conclusions and connections.* M.: Nauka, 1973.
60. Nelsen R. B. *Introduction to Copulas*, New York, Springer-Verlag, 1999.
61. Jouini M. N., Clemen R. T. Copula Models for Aggregating Expert Opinions, Operations Research, Vol. 44, No. 3, 1996, pp. 444–457.
62. *Falconer K. J.* Fractal Geometry: Mathematical Foundations and Applications. John Wiley & Sons, 2003.
63. Grehem R., Knut D., Patashnic O. *Konkretnaya mathemstics. Bases of informatics* M.: Mir, 1998.
64. Susumu Sato, Hiromitsu Kumamoto, *Reengineering the Environment.* Ney–York: Vantage Press, 1995.
65. Albrecht W., Wernz G., Williams T. *Fraud. Bringing Light to the Dark Side of Business.* Transl. from English. Saint Petersburg: Piter, 1995, 396 p.
66. *Andrey Nicolaevich Tupolev. Verge of venture creation.* Moscow: Nauka, 1988.
67. Solojentsev E. D. *Introduce in intellectual AWS and expert systems in machine-building; textbook.* Saint Petersburg: LIAP 1991.
68. Ivanchenko N. N. Researches in the area of work process, gaseous exchange and blast of diesels. In *Dieselestroenie*, Saint Petersburg: Mashinostroenie, 1974.
69. Encyclopaedia of finance risk-managment. By eritors A. Lobanov and A. Chugunov. M.: Alpina Publisher, 2003.
70. Solojentsev E. D. *Bases of system development for automated debugging of complex objects of machine-buildings.* The thesis of Dr. Sc. Kiev: Institute of Cybernetic, 1982.
71. Solnitsev R. I. *Computer–aided design of systems of automation control.* Moscow: Height School, 1991.
72. *Analysis statistical methods of safety of complex technical systems*, textbook, edited V. P. Sokolov. Moscow: Logos, 2001.
73. *Stevens R.* Operational test and evaluation: A Systems Engineering Process. New York: John Wiley and Sons, 1978.
74. Yusupov R. M. and etc. *Elements of the test theory and checking of quality of engineering systems.* Saint Petersburg: Energetica, 1978.
75. Solojentsev E. D. Problems and models of the debugging theory of piston machines. Dieselestroenie, 1980; 4:11–14.
76. Solojentsev E. D. The system of automated debugging of complex objects — volumetric energetic machines. Control systems and machines, 1981; 1: 118–123.
77. Solojentsev E. D., Korobizin I. A. Information technology of ensuring of reliability of complex machines on step of debugging test. Problems of machine–building and reliability of machines, 1990; 6:92–96.
78. Solojentsev E. D. Scenario management of the debugging non-success risk of complex objects. Problems of machine–building and reliability of machines, 2003; No. 6.
79. Prohorovich V. E. *Condition forecasting of complex technical systems.* Saint Petersburg: Nauka, 1999.
80. Prohorovich V. E. *Forecasting of condition of complex engineering installations.* Saint Petersburg: Nauka, 1999.

81. Annikeichik N. D., Devyatkin A. M., Krasnov O. V. Analysis of fire models at accident of rocket launching pad. Journal of VISA named A. F. Mojaisky, 2000; 7:54–59.
82. Annikeichik N. D., Sokolov E. I. Development of method of monitoring of safety of operation of technical systems of rocket launching pad. Proceedings of the First International Scientific School: Modelling and Analysis of Safety and Risk in Complex Systems; 2001 June 18-22; Saint Petersburg: Omega, 2001.
83. Krasnov O. V. *Methods and models of investigation of safety of operation of rocket launching pad.* Saint Petersburg: VISA named A. F. Mojaisky, 2000.
84. Malzhev A. I.*Algebraical systems.* M.: Nauka, 1970.
85. Andreeva G. V. Scoring as the method of estimation of credit risks. Banking technology. 2000. No 6. $http://www.bizcom.ru/bank-business/2000-06/02.html.$
86. Stroev A. A. Application of the credit scoring systems in a bank $http://www.bdc.ru/raschet/2004/ras6-1.shtml.$
87. *Henley W. E.* Statistical aspects of credit scoring. Ph.D. thesis. Open University. 1995.
88. *Boyle M., Crook J. N., Hamilton R., Thomas L. C.* Methods for credit scoring applied to slow payers in Credit Scoring and Credit Control. Oxford University Press. 1992.
89. Pomazanov M. Quantitative analysis of credit risks. Banking technology. 2004, No 2.–P. 22–28.
90. Pomazanov M. Credit-risk management and modeling the new active on the portfolio. Finance and credit, No 6 (144); 2004, P. 12–18.
91. Solnitsev R. I., Slusarenko A. S., Bondarev A. G., Vasiliev Yu. P., Tereshenko S. N. Software with open codes for interests of maintance of country safety. Prociding of Fourth Intern. conf. "Instrument-making in ecology and safety human being". Saint Petersburg: SPbGUAP ITMO, 2004. P. 262–269.
92. Solojentsev E. D., Stepanova N. V. Credit Risk: Tips for Identification and Assessment. Global Association of Risk Professionals. July/August 05 Issue 25, 2005. P.p.42–43.
93. Solojentsev E. D., Stepanova N. V. Requirements for quality of methods of estimation of credit risks. Quality, Innovations, Education. 2005. No 4.
94. *Stepanova N. V., Solojentsev E. D.* Requirements for methods of credit risk assessment. Modeling and Analysis of Safety and Risk in Complex Systems. Proc. of the Fifth Int. Scien. School by Editors I. Ryabinin and E. Solojentsev. 2005 June 28 – July 1. Saint Petersburg: SPbSUASI, 2005.
95. International Convergence of Capital Measurement and Capital Standards. A Revised Framework. Basel Committee on Banking Supervision, Bank for International Settlements Press & Communications. Basel, 2004.
96. Solojentsev E. D., Stepanova N. V., Rybakov A. V. Management of finance risks (Publishing house by Grebennikova). No 4. 2005. P. 30–43.
97. *Seitz J., Stickel E.* Consumer Loan Analysis Using Neural Network. In Proc. of the Bankai Workshop: Adaptive Intelligent Systems. Brussels, 14–19 October 1996.
98. Ryabinin I. A. Concept of the logic and probabilistic safety. Instrumentation and Systems of Management, 1993; 10:6–9.

99. Volik B. G., Ryabinin I. A. Effectiveness, reliability and survivability of control systems. Automation and Telemechanics, 1984; 12:
100. Volik B. G. On concepts of technogenic safety. Automation and Telemechanics, 1988; No 2:
101. Henly E. I., Kumamoto H. *Reliability engineering and risk assessment.* New York: Prentice–Hall, 1985.
102. Kulik B. A. System of logic programming based on algebra of corteges. Technical Cybernetic, 1993; 3: 226–239.
103. Kulik B. A. 1.Presentation of logical systems in probabilistic space on basis of algebra corteges: Basis of algebra of corteges; 2. Measuring of logical and probabilistic systems. Automation and telemechanics,1997; 1:126–136; 2: 169–179.
104. Mozhaev A., Demidov Ya. Algorithmic bases of technology of structure - logical modelling in problems of systemic analysis of reliability, safety and risk. Proceedings of the Second International Scientific School: Modelling and Analysis of Safety and Risk in Complex Systems; 2002 July 2-5; Saint Petersburg: Business Press, 2002.
105. Solojentsev E. D., Karasev V. V., Rybakov A. V. *Logic and probabilistic estimation and analysis of credit risks.* Certificate on official registration Software No. 2006610004. 10 January 2006.
106. Mozhaev A. S. Program complex of automated structure logical modelling of systems (PC ACM 2001). Proceedings of the First International Scientific School: Modelling and Analysis of Safety and Risk in Complex Systems; 2001 June 18-22; Saint Petersburg: Omega, 2001.
107. Mozhaev A. S., Alekseev A. O., Gromov V. N., *Automated logic and probabilistic modelling of engineering systems, Handbook of User of PC ACM, Version 5.0.* Saint Petersburg: VITU, 1999.
108. Mozhaev A. S. Certification of the program "Arbiter" of the automated calculation of safety and technical risk of systems. Proc. of the Seventh Int. Scien. School, by editors I. Ryabinin and E. Solojentsev. 2007, September, St. Petersburg: PUASE.
109. *Aven T., Jensen U.* Stochastic models in reliability. New York: Springer–Verlag, 1999.
110. *Lisnianski A., Levitin G.* Multi-state system reliability. Assessment, Optimization and Applications. New Jersey, London, Singapore, Hong Kong: Word Scientific, 2003.
111. Guk Yu. B., Losev E. A., Myasnikov A. V. *Estimation of relaibility of electric plants.* Moscow: Energy. 1974. P. 200.
112. Erlih A., *Hand-book in Technical Analysis for commodity and financial markets.* Moscow: Infra, 1996.
113. Solojentsev E. D., Karassev V. V. Risk logic and probabilistic models in business and identification of risk models. Informatica, 2001, No 25. P. 49–55.
114. Wasserman Philips D. *Neural Computing Theory and Practice.* New York: ANSA Research, 1990.
115. Taxa X., *Introduce in operation research, v. 1,2.* Moscow: MIR, 1985.
116. Aivasyn S. I., Mhitaryan V. S., *Applied statistics and basis of econometrics.* Moscow: Uniti, 1998.
117. Heckman James J., Leamer Edward. *Handbook of Econometrics*, v.5, 2002.

118. Solojentsev E., Maksimov A. Statistical, combinatoric and logical and probabilistic risk analysis in complex systems with GIE. Proc. of the Second Intern. Scaint. School: Modelling and Analysis of Safety and Risk in Complex Systems; 2002 July 2–5; Saint Petersburg: Business Press, 2002.
119. Ryabinin I. A. A suggestion of a new measure of system components importance by means of a Boolean difference. Microelectron. Reliability, v. 34, 1994, No. 4. P. 603–613.
120. Uryasev S., Rockafellar R. T. Conditional value-at-risk for general loss distributions. Journal of Banking & Finance, 2002. N 26. P. 1443–1471.
121. Melnikov A. V. Finance market: stoxastic analysis and calculation of derivative securities. M.: TVP, 1997.
122. Kibsun A. I., Kuznetsov E. A. Optimal management by the security portfolio. Automation and Telemechanics, 2001; 9:101–113.
123. Barone-Adesi G., Giannopoulos K., Vosper L. Backtesting Derivative Portfolios with Filtered Histirical Simulation. European Financial Management, 2002; 8:31–58.
124. Giannopoulos K. Var modeling on long run horizons. Proc. of Int. Scien. School "Modeling and Analysis of Safety and Risk in Complex Systems", 2002. July 2–5. St. Petersburg: Business Press, 2002.
125. Solojentsev E. D. Logic and probabilistic theory of security portfolio risk. Finance India, vol. XVIII No. 3, September 2004, p.p. 1269–1293.
126. Solojentsev E. D., Alekseev V.B. Logic and probabilistic theory of security portfolio risk. Proc. of the Int. Scien. School: Modeling and Analysis of Safety and Risk in Complex Systems, 2003. August 20–23. St. Petersburg: SPUASE, 2003.
127. Zakrevsky A. D. Matrix apparatus of the logic deduction in final predicates. Proc. of seminar in logic: Philosophic basis the non-classical logic. Moscow: Institute of Philosophy RAS, 1990.
128. *Introduction to management of credit risks.* Price Waterhouse: 1994.
129. Internal method for determination of category of finance state of borrower-issuer. Decision of board of directors OAO "PSB" from 4.08.2004.
130. Solojentsev E., Rybakov A. Optimization in problems of identification of logical and probabilistic risk models. Automation and Telemechanics, 2003; No 7.P. 51–63.
131. Shiryaev A. I., *Probability.* Moscow: Nauka, 1980.
132. Pervozvanskiy A. A., Pervozvanskaya T. N., *Finance market: calculation and risk.* Moscow: Infra, 1994.
133. Marshal G. F., Bansal V. K., *Finance engineering. The total textbook for the finance innovation.* Moscow: Infra, 1998.
134. Vasyutovich A., Sotnikova Yu. Market risk: measuring and management. Banking Technology, 1998; 1: 4–10.
135. Voloshin I. VaR–approach to the search of the optimal portfolio of actives. http://bankclub.ru/library.htm.
136. Sobol I. M., *Quantitative methods by Monte-Carlo.* Moscow: Nauka, 1973.
137. Pugachev V. S. *Introduction in the theory of probabilities.* Moscow: Nauka, 1968.
138. Alexeev V. V., Chokolov V. V. Logical and probabilistic risk management of security portfolio. Modeling and Analysis of Safety and Risk in Complex

Systems. Proc. of the Fourth Int. Scien. School, by editors I. Ryabinin and E. Solojentsev. 2005. June 28–July 25. Saint Petersburg: SPUASE.
139. Krymsky V. G., Akhmedjanov F. M., Markert F. A Probabilistic Model for Closing Gas Transmission Pipelines on Destruction or Leakages. Proc. of Intern. Scient. School on Modeling and Analysis of Safety and Risk in Complex Systems (MASR'04), Saint Petersburg, IPME RAS Publ., 2004, pp. 624–629.
140. Cooke R., Jager E. A Probabilistic Model for the Failure Frequency of Underground Gas Pipelines. Risk Analysis, vol. 18, No. 4, 1998, 511–527.
141. MacKenzie G. Approximately Maximum-Entropy Multivariate Distributions with Specified Marginals and Pairwise Correlations. Ph.D. Dissertation, University of Oregon, the USA, 1994.
142. Kendall M. J., Stjuart A.*Multivariate statistical analysis.* M.: Nauka, 1976.
143. Solojentsev E. D. Logical and Probabilistic models of Quality and Market Loss Risk. Proceedings Volume. The 6-th World Congress for Total Quality Management "Business Excellence — What is to be done". 20–21 June 2001. Stockholm School of Economics in Saint Petersburg. Saint Petersburg, 2001. P. 604–612.
144. Solojentsev E. D. To logic and probabilistic theory. In book: Problems of machine-building. SPb.: Nauka, 2005. P. 90–102.
145. Lebedev N. Yu., Solojentsev E. D. Logic and probabilistic models of nonsuccess risks of managment of company. Management of finance risks. 2005. No 4. P. 30–43.
146. Peter F. Drucker *Practice of Managment:* Trans. from Engl.: "Williams" Publishing House, 2003.
147. Robbins, Stephen, P., Coulter, Mary. *Management*, 6th Edition, Translated from English: M: "Williams" Publishing House, 2002. 880 p.
148. Lebedev N. Yu. Experience and models of non-success risks of managment (of company Transas). The dissertation of candidate econ. sciences. Saint Petersburg: NW NIIK, 2005.
149. Satarov G. A. *Anticorruption politics.* The manual, M.: RA "SPAS", 2004, 368 p.
150. Eliseeva I. I. *General statistics theory.* The text book. M.: Finance and statistics, 2004. 656 p.
151. J. D. Agarwal, Aman Agarwal. (2006) "Money Laundering: New Forms of Crime Victimisation (Current Trends and Modus Operandi)". Finance Indian, Vol. XX No. 4, December 2006.
152. Solojentsev E. D., Mahutov N. A. Logic and probabilistic risk models in multi component systems with groups of incompartible events for the problems of classification, investment, effiency and management. Problems of safety at emergency situations. No. 2, 2006.
153. Solojentsev E. D. Scenario logic and probabilistic models of bribes risk. Finance and Business, No. 1, 2007; pp. 125–138.
154. *Safety of Russia. Risk Analysis and safety problems* (in four parts). MGF "Knowledge", 2006–2007.
155. *Modeling and Analysis of Safety, Risk and Quality in Complex Systems.* Proceedings of the International Scientific School / by editors I. Ryabinin and E. Solojentsev. 2001–2007. St. Petersburg: SPUASE.

156. *Methodical instruction for laboratory works "Logic-and-probabilistic theory of credit risks"* (N. Medvedeva, E. Solojentsev, D. Strokov). St. Petersburg: SUAI, 2007, 25 p.
157. *Methodical instruction for laboratory works "Logic-and-probabilistic theory of security portfolio risk"* (V. Alexeev, E. Solojentsev, V. Shokolov). St. Petersburg: SUAI, 2007, 48 p.
158. *Methodical instruction for laboratory works "Automated structurally logic modeling of risks"* (A. Babenkov, E. Solojentsev). St. Petersburg: SUAI, 2007, 32 p.

Subject Index

Printed in the United States of America